Springer Proceedings in Mathematics & Statistics

Volume 177

Springer Proceedings in Mathematics & Statistics

This book series features volumes composed of selected contributions from workshops and conferences in all areas of current research in mathematics and statistics, including operation research and optimization. In addition to an overall evaluation of the interest, scientific quality, and timeliness of each proposal at the hands of the publisher, individual contributions are all refereed to the high quality standards of leading journals in the field. Thus, this series provides the research community with well-edited, authoritative reports on developments in the most exciting areas of mathematical and statistical research today.

More information about this series at http://www.springer.com/series/10533

Tao Qian · Luigi G. Rodino
Editors

Mathematical Analysis, Probability and Applications – Plenary Lectures

ISAAC 2015, Macau, China

Springer

Editors
Tao Qian
Faculty of Science and Technology
University of Macao
Taipa
Macao

Luigi G. Rodino
Department of Mathematics
University of Turin
Turin
Italy

ISSN 2194-1009 ISSN 2194-1017 (electronic)
Springer Proceedings in Mathematics & Statistics
ISBN 978-3-319-41943-5 ISBN 978-3-319-41945-9 (eBook)
DOI 10.1007/978-3-319-41945-9

Library of Congress Control Number: 2016945107

Mathematics Subject Classification (2010): 35QXX, 46EXX, 60GXX

This work is published under the auspices of the International Society of Analysis, its Applications and Computation (ISAAC).

Printed on acid-free paper

This Springer imprint is published by Springer Nature
The registered company is Springer International Publishing AG Switzerland

Preface

The present volume is a collection of papers devoted to current research topics in mathematical analysis, probability and applications, including the topics in mathematical physics and numerical analysis. It originates from plenary lectures given at the 10th International ISAAC Congress, held during 3–8 August 2015 at the University of Macau, China.

The papers, authored by eminent specialists, aim at presenting to a large audience some of the attractive and challenging themes of modern analysis:

- Partial differential equations of mathematical physics, including study of the equations of incompressible viscous flows, and of the Tricomi, Klein–Gordon and Einstein–de Sitter equations. Governing equations of fluid membranes are also considered in this volume.
- Fourier analysis and applications, in particular construction of Fourier and Mellin-type transform pairs for given planar domains, multiplication and composition operators for modulation spaces, harmonic analysis of first-order systems on Lipschitz domains.
- Reviews of results on probability, concerning in particular the bi-free extension of the free probability and a survey of Brownian motion based on the Langevin equation with white noise.
- Numerical analysis, in particular sparse approximation by greedy algorithms, and theory of reproducing kernels, with applications to analysis and numerical analysis.

The volume also includes a contribution on visual exploration of complex functions: the technique of domain colouring allows to represent complex functions as images, and it draws surprisingly mathematics near the modern arts.

Besides plenary talks, about 300 scientific communications were delivered during the Macau ISAAC Congress. Their texts are published in an independent volume. On the whole, the congress demonstrated, in particular, the increasing and major role of Asian countries in several research areas of mathematical analysis.

Taipa, Macao Tao Qian
Turin, Italy Luigi G. Rodino
March 2016

Contents

A Review of Brownian Motion Based Solely on the Langevin Equation with White Noise

L. Cohen

Abstract We give a historical and mathematical review of Brownian motion based solely on the Langevin equation. We derive the main statistical properties without bringing in external and subsidiary issues, such as temperature, Focker-Planck equations, the Maxwell–Boltzmann distribution, spectral analysis, the fluctuation-dissipation theorem, among many other topics that are typically introduced in discussions of the Langevin equation. The method we use is the formal solution approach, which was the standard method devised by the founders of the field. In addition, we give some relevant historical comments.

Keywords Brownian motion · Langevin equation · History · Einstein · White noise

1 Introduction

The two seemingly simple equations (as originally written)

$$\frac{\partial f(x,t)}{\partial t} = D\frac{\partial^2 f(x,t)}{\partial x^2} \tag{1}$$

and

$$m\frac{d^2x}{dt^2} = -6\pi\mu a\frac{dx}{dt} + X \tag{2}$$

revolutionized our understanding of the of the universe and ushered an incredible number of physical and mathematical ideas [11]. The first equation is due to Einstein [13], whose aim was to show that atoms exist; the second is due to Langevin [26], who brought forth a new perspective regarding both the physics and mathematics of Einstein's idea.

L. Cohen (✉)
City University of New York, 695 Park Ave., New York, NY 10065, USA
e-mail: leon.cohen@hunter.cuny.edu

T. Qian and L.G. Rodino (eds.), *Mathematical Analysis, Probability and Applications – Plenary Lectures*, Springer Proceedings in Mathematics & Statistics 177, DOI 10.1007/978-3-319-41945-9_1

It is often said that Brown [3] discovered Brownian motion, Einstein explained it, Langevin simplified it, and Perrin [31] proved it; this listing misses totally the motivations and wonderful history of the subject [11]. Brown did not discover Brownian motion, but did study it extensively. Einstein was not aware of Brownian motion; he *predicted* Brownian motion to obtain a macroscopic manifestation of atoms that could be measured. Equation (1) is the equation for the probability density for the Brownian particle at position x and time t. He derived the standard deviation of the visible Brownian particle that could be experimentally verified if indeed atoms exist. He solved explicitly for the standard deviation of position,

$$\lambda_x = \sqrt{\overline{x^2}} = \sqrt{2Dt} \tag{3}$$

and connected the parameters with the temperature of the medium, and the yet unnamed Avogadro number, a number that few believed in, and had never been measured or estimated at that time. Of course, Eq. (1) was known for 100 years before Einstein; it is the famous heat equation first derived by Fourier. However that is not relevant. What is important is that Einstein derived the probability density for position of the Brownian particle. Perrin had already been working on the issue of the existence of atoms, and his motivation was certainly heightened by Einstein's results. He experimentally verified Eq. (3), and hence verified the Einstein idea that the random "invisible" microscopic atoms can manifest a macroscopic effect which can be measured [31].

Equation (2) was the start of the field of random differential equation and is now called the Langevin equation. The way it stands, it is Newton's equation of motion where the left hand is mass times the acceleration, the first term on the right is the force of "friction" which is proportional to the velocity, and the second term, X, is an additional force. In Langevin's words: "X is indifferently positive and negative". X is what we now call the random force. Langevin's insight was to realize that to obtain the main result of Einstein, Eq. (3), one does not have to solve and derive the probability density, but one can obtain the second moment simply from Newton's equation and moreover that one can obtain it in a relatively simple manner. The Langevin equation has been applied to numerous fields and to a wide variety of physical situations. Random differential equations have become standard in many branches of science and has produced rich mathematics [7, 21–23, 27, 28, 35, 36, 39, 45].

1.1 The Aim of This Article

The author's involvement with Brownian motion [1, 2, 8] started with his attempt to understand and apply the theory of Chandrasekhar and von Neumann [4–6] regarding the random motion of stars, a subject that is fundamental in stellar dynamics, because it is the random motion that is important in the evolution of a collection of stars, such as globular clusters. In reading the review articles of that time I found considerable difficulty in that the articles mixed in a plethora of ideas which depended both on the interests of the author writing the article and relevant field of the article. Typically, historical and recent review articles and books mix together the Langevin equation with Fokker–Planck equations, temperature, mobility, master equations, the Maxwell–Boltzmann distribution, spectral analysis, Wiener processes, random walks, the fluctuation-dissipation theorem, white noise, Gaussian with noise, among many other topics. Of course, these are important to specific fields, but in my opinion, often detract from understanding the Langevin equation and its consequences as it stands. The aim of this article is to review and derive the relevant results of the Langevin equation without the encumbrance of other ideas. We give derivations of the main results based *solely* on the Langevin equation where the random force is taken to be white noise (not Gaussian white noise). Of course, most of the results we derive are known, but we hope that the presentation and derivations are of interest.

1.2 Notation

Expectation values of quantities that depend on time will be denoted in two equivalent ways

$$\langle x(t)\rangle = \langle x\rangle_t \tag{4}$$

Which notation is used is motivated by aiming at clarity of the equation and the historical usage.

We use the delta function, $\delta(t)$, routinely. The basic property is,

$$\int_a^b f(t)\delta(t-s)dt = f(s) \qquad a < s < b \tag{5}$$

If s is one of the end points then we will take half the value,

$$\int_a^b f(t)\delta(t-a)dt = \frac{1}{2}f(a) \qquad a < b \tag{6}$$

1.3 Deterministic and Random Initial Conditions

There is considerable variation in articles on Brownian motion regarding the initial conditions for the velocity $v(t)$, and position, $x(t)$. The two general approaches is to take them to be deterministic or random. When they are taken to be random, which is important in some fields, one very often averages over the initial conditions; this produces results which are seemingly different than if one takes them to be deterministic. We shall take them to be symbolically random, but we will not do any averaging over them. The deterministic case can be obtained from the random case in a manner that we now discuss. Our notation for the random initial conditions shall be, for example $\langle v(0)\rangle$ or $\langle v\rangle_0$. With this notation, to go to the deterministic case one just lets $\langle v(0)\rangle \to v_0$, where v_0 is the deterministic initial condition, etc. For $\left\langle v^2(0)\right\rangle \to v_0^2$ and for standard deviation of velocity $\sigma_v^2(0) \to 0$. Similarly for position.

2 The Langevin Equation with a White Noise Driving Force

In modern notation, the Langevin equation is generally written as

$$\frac{dv(t)}{dt} = -\beta v(t) + F(t) \tag{7}$$

It is a random differential equation for the velocity $v(t)$, but depending on the field, it could be any random variable that satisfies Eq. (7). The term $F(t)$ is called random force, and since it is random, the unknown $v(t)$ will also be random. The main issue is: given the statistical properties of $F(t)$, what are the statistical properties of $v(t)$. The standard statistical properties of F are taken to be

$$\langle F(t)\rangle = 0 \tag{8}$$

and

$$\left\langle F(t')F(t'')\right\rangle = 2D\delta(t'-t'') \tag{9}$$

The first indicates that the average at any one time of the random force is zero, and the second is that the force at two different times are uncorrelated except for equal times. Random processes that satisfy Eq. (9) are called white noise. Very often it is assumed that the statistical properties of the force are what is called Gaussian white noise. We will not assume so, and limit ourselves to results that follow only from Eqs. (7)–(9). We point out that a general view is that Eq. (9) implies a stationary process for the random force. That is not so. One can construct random process that satisfy Eq. (7) but are non-stationary.

How can one solve for $v(t)$? By solve we mean to obtain the statistical properties of $v(t)$. Historically, the first method was to solve the Langevin equation as if it were an ordinary differential equation, and then take appropriate expectation values. This method was implied by Langevin and developed by others, in particular, Ornstein, Uhlenbeck, Wang, and Chandrasekhar, among others. This is the procedure we will follow, and we discuss it further in Sect. 3.11. However, historically certain difficulties were pointed out and the first to do so was Doob [12].

Position. One also wants to study the statistical properties of position, $x(t)$, which is related to the velocity by

$$\frac{dx(t)}{dt} = v(t). \tag{10}$$

There are two approaches one can take. One is to first obtain the statistical properties of $v(t)$, and then consider Eq. (10) as a random differential equation for $x(t)$. Alternatively, one can combine Eqs. (7) and (10) to obtain a single differential equation for $x(t)$, namely

$$\frac{d^2x(t)}{dt^2} = -\beta\frac{dx(t)}{dt} + F(t) \tag{11}$$

and consider it a random differential equation with a driving force $F(t)$. Both approaches are interesting and are used.

It is important to appreciate that historically the calculation of $\langle x^2(t)\rangle$ was the focus, because it was the only measurable quantity of the Brownian particle.

3 Comments and Historical Notes

In this section we discuss some historical issues, motivations, and contributions of the many authors that developed the field of Brownian motion that was initiated by Einstein. This section may be skipped, as none of the results and discussions here are explicitly used in the subsequent derivations.

3.1 *The Classic Review Articles*

There are three classic historical review articles which are still the best review articles. In order of appearance, the first is by Uhlenbeck and Ornstein [38], titled "On the theory of Brownian motion" [38]; the second is the monumental article by Chandrasekhar [6], "Stochastic problems in Physics and Astronomy" [6]; and the third is that of Wang and Uhlenbeck [41], "On the theory of Brownian motion II" [41]. All these articles are much more than review articles, because they addressed new approaches and obtained new results.

Ornstein was among the first to solve the Langevin equation, and the paper by Uhlenbeck and him extended and simplified some of the results [38]. We discuss the Chandrasekhar paper below. The paper by Wang and Uhlenbeck, while having the same title as the paper by Uhlenbeck and Ornstein, is much more than a review of Brownian motion. It is a formulation of stochastic processes in general and a careful discussion of the Focker-Planck formulation.

The above three papers and three other important papers are collected in *Selected Papers on Noise and Stochastic Processes*, edited by Nelson Wax [42]. The three other papers are "Mathematical Analysis of Random Noise" by Rice [33]; "Random Walk and the Theory of Brownian Motion", by Kac [24]; and "The Brownian Movement and Stochastic Equations" by Doob [12]. At one time, ownership of this book was mandatory for anyone interested in stochastic process. The book is still in print.

We also mention that Einstein's five papers on Brownian motion were collected in a very short book, in 1926, edited by Furth [16]. The book is short because all of the Einstein articles are very short. The book was translated and published in English in 1956 and continues to be available. We also point out that Einstein published many papers on what we now call time series and stochastic processes. In fact, what is commonly called the Wiener-Khinchine theorem was first given by Einstein in 1914, in two papers entitled "A Method for the Statistical Use of Observations of Apparently Irregular, Quasiperiodic Process" and "Method for the Determination of Statistical Values of Observations Regarding Quantities Subject to Irregular Observations" [14].

3.2 Chandrasekhar

Chandrasekhar was one of the greatest scientists and astronomers of the last century, and received the Nobel Prize for the remarkable discovery of electron degeneracy in stars. He made monumental contributions to almost all fields of astronomy. He was one of the clearest scientific writers ever and while his famous article is often considered a review article, it is much more than that. It is perhaps one of the most remarkable articles written on the subject of stochastic process. The range is remarkable, ranging from the random walk to the recurrence theorem of Poincare.

Chandrasekhar wrote many articles on Brownian motion, but what is particularly important is that to the best of my knowledge, he was the first to *derive* the statistics of the random forces for the appropriate physical situation; in his case, the random force on a star. Subsequently, he and von Newman *derived* additional statistical properties for the random force, such as the two-time autocorrelation function.

While in his 1943 article, he derives the main results Brownian motion, most of the article concerns the issue of fluctuations, probability, and stochastic processes in general. Unlike previous works, he considers the three-dimensional case. He derives a number of new results regarding the transition from the Langevin equation to the problem of obtaining the probability densities. He obtains, in a very simple and elegant way the probability densities of position and velocity, and the equations of motion they satisfy. Moreover, he derives and discusses the joint position-velocity distribution, derives the partial differential equation that satisfies it, and gives a number of ways to solve it. Another part is a comprehensive discussion of the Langevin equation with an additional external force. Further, he makes connections between the Boltzmann equation, stochastic processes, Poincare cycles, and the fundamental deterministic equation of dynamics, the Liouville equation.

3.3 Smoluchowski

Smoluchowski developed the theory of Brownian motion and in fact he did considerably more than Einstein [37]. However he published his results in 1906, a year later than Einstein. He was the first to consider Brownian motion when there are external forces and in particular he considered Brownian motion under the influence of gravity. This was very important for the experimental procedures used by Perrin. He developed many of the mathematical issues. In the words of Chandrasekhar "The theory of density fluctuations as developed by Smoluchowski represents one of the most outstanding achievements..." [6].

3.4 White Noise, Gaussian White Noise, and Non-stationary White Noise

The power spectrum measures the intensity as a function of frequency. If the power spectrum is uniform, then it is called white noise. The power spectrum corresponding to Eq. (9) is indeed independent of frequency, and hence uniform [32]. White noise is called white because at one time it was thought that for the white light we perceive, the intensity as a function of frequency is more or less uniform. Of course that is not strictly the case, but the phrase has stuck.

If the statistics of the random force are Gaussian, then one says that we have Gaussian white noise. It is generally assumed that white noise is stationary. That is not necessarily the case. A process is stationary (more precisely, a second order stationary process) if the autocorrelation function depends on the difference of the two times,

$$\langle X(t_1)X(t_2)\rangle = R(t_2 - t_1) \qquad \text{(a function of } t_2 - t_1\text{)} \tag{12}$$

However white noise as defined by Eq. (12) is not necessarily a stationary process. Explicit methods for constructing non-stationary processes that are nonetheless white noise are given in [20].

3.5 The O-U Process

What has come to be called the Ornstein-Uhlenbeck process is the random process governed by the Langevin equation. The reason for the name is that Ornstein and later Ornstein and Uhlenbeck are the ones that derived the main results. However, we point out that depending on the field, the statistics of the standard driving force may be white noise, Gaussian white noise, or indeed, an arbitrary correlation function.

3.6 Brownian Motion: Stationary or Non-stationary?

The Langevin equation considered as a deterministic equations clearly produces a time dependent solution for $v(t)$ and $x(t)$, which of course depends on the time-dependence of the driving force. Considering it as a random differential equation also produces random quantities that evolve in time. In fact, Brownian motion is perhaps the most important and simplest example of a non-stationary random process. Nonetheless theorems which only apply to a stationary process, such as the Weiner Khinchine theorem, are often applied. The justification, often unstated, is that for large times, the autocorrelation function for the Brownian motion process does go to an autocorrelation that implies a stationary process. This will be further discussed in subsequent sections.

3.7 Spectral Properties

One of the major advances of noise theory is the work of Rice [33], who emphasized the spectral properties of a stochastic process. Wang and Uhlenbeck understood the importance of the spectral point of view, and calculated the power spectrum for velocity as given by the Langevin equation. Their often quoted result is that the power spectrum of velocity, $S(\omega)$, goes as [38]

$$S(\omega) \sim \frac{1}{\beta^2 + \omega^2} \tag{13}$$

Since $S(\omega)$ does not depend on time, the implication is that the process is stationary, but as we have discussed above Brownian motion is not stationary. Equation (13) is achieved by waiting an infinite amount of time, and this was achieved historically be starting the motion at minus infinity in time; hence for any finite time an infinite amount of time, will have already passed. Alternatively, the system is started at a finite time, and then one lets time go to infinity. This will be discussed further in Sect. 8, where we obtain the time-dependent power spectrum as defined by the Wigner distribution.

3.8 Brownian Motion in a Force Field

If in addition to the frictional force and the random force we have an external deterministic force which may be space and time dependent then the Langevin equation becomes [25, 37],

$$\frac{dv(t)}{dt} = -\beta v(t) + F(t) + K(x(t), t) \tag{14}$$

$$\frac{dx(t)}{dt} = v(t) \tag{15}$$

The equations are now coupled. We can no longer consider the velocity process by itself.

3.9 Focker-Planck Equations and Random Differential Equations

Focker-Planck equations are partial differential equations for evolution for the probability density [17, 34]. Einstein's equation, Eq. (1) is a Focker-Planck equation for position. The relation of the probability density to the Langevin equation is fundamental. A probability distribution is determined by its moments (except for some unusual circumstances), and hence if the Langevin equation can give us all the moments, say for the velocity, then indeed we could obtain the probability density of velocity. What moments can one obtain from the Langevin equation depends on the statistics of the random force. If one assumes white noise only, that is Eq. (7), then the probability distribution cannot be obtained, because only a few moments of velocity may be determined from the Langevin equation with white noise. However if one assumes that the random force is Gaussian white noise, then all the moments may be obtained, and hence so may the probability density.

3.10 Weiner and the Weiner Process

Weiner was not only a great mathematician, but also made major contributions to physics and engineering, and indeed is one the founders of noise theory and of modern electrical engineering. Weiner was a child prodigy, and as a young man aimed at making a contribution commensurate with his child prodigy status. He followed the major developments of his time both in mathematics and physics and was particularly interested in the then exciting development of Brownian motion and in so called pathological functions; functions that are continuous everywhere but differentiable nowhere. These functions were considered totally irrelevant to the real world. However, based on a hint by Perrin, he realized that the path of a Brownian particle may be a pathological function! So, he defined a mathematical idealization of Brownian motion based on measure theory [43]. To quote Wiener: "There were fundamental papers by Einstein and Smoluchowski that covered it, but whereas these papers concerned what was happening to any given particle at a specific time, or the long-time statistics of many particles, they did not concern themselves with the mathematical properties of the curve followed by a single particle. Here the literature was very scant, but it did include a telling comment by the French physicist Perrin in his book Les Atomes, where he said in effect that the very irregular curves followed by particles in the Brownian motion led one to think of the supposed continuous non-differentiable curves of the mathematicians. He called the motion continuous because the particles never jump over a gap, and non-differentiable because at no time do they seem to have a well-defined direction of movement."

What is currently called the "Wiener process", $W(t)$, is defined in various ways. Most commonly it is defined as a process governed by

$$\frac{dW(t)}{dt} = F(t) \tag{16}$$

where $F(t)$ is white noise, or sometimes Gaussian white noise. Hence we see that it is the Langevin equation without friction. One can also define it as process where the mean is zero and where the variance of $W(t) - W(t')$ is proportional to $t - t'$, and further that for $t_1 < t_2 \ldots < t_n$, then $W(t_2) - W(t_1), \ldots, W(t_n) - W(t_{n-1})$, are mutually independent.

3.11 The Wiener–Khinchine Theorem

The so called the Wiener–Khinchine theorem, originally derived by Einstein in 1914, relates the autocorrelation to the power spectrum. The theorem applies only to a stationary processes. In particular, if $X(t)$ is a stationary random process where the autocorrelation function depends only on the difference in times

$$\langle X(t_1)X(t_2)\rangle = R(t_2 - t_1) \tag{17}$$

then the power spectrum is given by

$$S(\omega) = \frac{1}{2\pi}\int R(\tau)\, e^{-i\omega\tau}\, d\tau \tag{18}$$

Often this theorem is applied to the Langevin equation, but that is not strictly proper because in the case of Brownian motion we do not have a stationary process. However, for large times, it does become a stationary process. See Sect. 8.

3.12 The Formal Solution Approach

The fundamental method of solution, that is, finding the statistical properties of $v(t)$ from the Langevin equation is to pretend it is an ordinary deterministic equation, and after solving it as such, one takes expectation values of the appropriate quantities.

This method was implied in Langevin's paper, but was developed by Ornstein, and Wang and Ornstein, Chandrasekhar, and others. This will be the general approach that will be taken in subsequent sections.

3.13 Brownian Motion Applied to Extended Bodies and Random Partial differential Equations

Vibrations of structures due to random forces is of great practical and theoretical interest. For example the vibrations of structures due to wind, earthquakes, etc., are often modeled as the response to random forces. Historically, most of the work on Brownian motion was on the random motion of a single particle, or more generally, one degree of freedom. van Lear and Uhlenbeck extended the results of standard Brownian motion to the case of a string, where the random force acts on each point of the string [40]. One can view this as the beginning of random *partial* different equations. For example, for the string one has

$$\frac{\partial^2 s(x,t)}{\partial x^2} - \frac{\partial^2 s(x,t)}{\partial t^2} = F(x,t) \tag{19}$$

where the driving random force $F(x, t)$ is given statistically.

3.14 Generalized Langevin Equation

In the standard Langevin equation, the friction term $\beta v(t)$ is directly proportional to the velocity at time t only, that is, it has no memory. One way to generalize it, is to make the friction coefficient β a function of time $\beta(t)$, so that the past is taken into account,

$$\frac{dv(t)}{dt} = -\int_{-\infty}^{t} \beta(t-\tau)v(\tau)d\tau + F(t) \tag{20}$$

3.15 Derivations of the Langevin Equation

Of fundamental importance is the derivation of the Langevin equation from first principles. There have been many approaches, and perhaps the first was that of Ford et al. [15]. The general idea is to start with the most fundamental equations of motion for N coupled particles, and focus on one of them. The rest are considered as the "heat bath". One then averages over the particles in the heat bath and aims to obtain the equation of motion for the particle we are focused on [30]. This continues to be an active area of research for both the classical and quantum case.

4 Statistical Properties of Velocity

Considering

$$\frac{dv(t)}{dt} = -\beta v(t) + F(t) \tag{21}$$

as an ordinary differential equation, the formal solution is (See Appendix 1)

$$v(t) = e^{-\beta t} v(0) + e^{-\beta t} \int_0^t e^{\beta t'} F(t')dt' \tag{22}$$

where $v(0)$ is the initial velocity. We obtain the statistical properties of $v(t)$ by manipulating Eq. (22) and then taking averages.

4.1 Average Velocity

We take the mean of both sides of Eq. (22) and assume that the averaging operator can be brought inside the integral; then we have

$$\langle v \rangle_t = \langle v \rangle_0 \, e^{-\beta t} + e^{-\beta t} \int_0^t e^{\beta t'} \left\langle F(t') \right\rangle dt' \tag{23}$$

Using Eq. (8), we have that

$$\langle v \rangle_t = \langle v \rangle_0 \, e^{-\beta t} \tag{24}$$

The limits of $\langle v \rangle_t$ at zero and infinity are,

$$\langle v \rangle_{t \to \infty} \to 0 \tag{25}$$

$$\langle v \rangle_{t \to 0} \sim \langle v \rangle_0 \, (1 - \beta t) \tag{26}$$

4.2 Second Moment

To obtain the second moment, we square the deterministic solution, Eq. (22), to obtain

$$v^2(t) = e^{-2\beta t} v^2(0) + e^{-2\beta t} \int_0^t e^{\beta t'} F(t')dt' \int_0^t e^{\beta t''} F(t'')dt'' + 2e^{-2\beta t} v(0) \int_0^t e^{\beta s} F(t')dt' \tag{27}$$

and again take expectation values of both sides, which yields

$$\left\langle v^2 \right\rangle_t = e^{-2\beta t} v^2(0) + e^{-2\beta t} \int_0^t \int_0^t e^{\beta(t'+t'')} \left\langle F(t')F(t'') \right\rangle dt'dt'' + 2e^{-2\beta t} \int_0^t e^{\beta s} \left\langle v(0)F(t') \right\rangle dt' \tag{28}$$

Assuming that

$$\langle v(0)F(t')\rangle = 0 \tag{29}$$

then

$$\langle v^2\rangle_t = e^{-2\beta t}\langle v^2\rangle_0 + e^{-2\beta t}\int_0^t\int_0^t e^{\beta(t'+t'')}\langle F(t')F(t'')\rangle\,dt'dt'' \tag{30}$$

Evaluation of the second term is done in Appendix 2,

$$e^{-2\beta t}\int_0^t\int_0^t e^{\beta(t'+t'')}\langle F(t')F(t'')\rangle\,dt'dt'' = \frac{D}{\beta}(1-e^{-2\beta t}) \tag{31}$$

Substituting this result in Eq. (30) we have

$$\langle v^2\rangle_t = e^{-2\beta t}\langle v^2\rangle_0 + \frac{D}{\beta}(1-e^{-2\beta t}) \tag{32}$$

which can also be written as

$$\langle v^2\rangle_t = \frac{D}{\beta} + \left(\langle v^2\rangle_0 - \frac{D}{\beta}\right)e^{-2\beta t} \tag{33}$$

Limits. The limiting value for time going to infinity is

$$\langle v^2\rangle_{t\to\infty} \longrightarrow \frac{D}{\beta} \tag{34}$$

For small times we have

$$\langle v^2\rangle_{t\to 0} \to \frac{D}{\beta} + \left(\langle v^2\rangle_0 - \frac{D}{\beta}\right)(1-2\beta t) = v_0^2 - 2\beta t\left(\langle v^2\rangle_0 - \frac{D}{\beta}\right) \tag{35}$$

giving

$$\langle v^2\rangle_{t\to 0} \to \langle v^2\rangle_0 (1-2\beta t) + 2Dt \tag{36}$$

Derivative of $\langle v^2\rangle_t$. We will see in Sect. 4.6 that the derivative of $\langle v^2\rangle_t$ plays an important role. We calculate it here. Differentiation of Eq. (33) gives

$$\frac{d}{dt}\langle v^2\rangle_t = -2\beta\left(\langle v^2\rangle_0 - \frac{D}{\beta}\right)e^{-2\beta t} \tag{37}$$

and we note that

$$\frac{d}{dt}\langle v^2\rangle_t \to 0 \qquad \text{as } t\to\infty \tag{38}$$

4.3 Standard Deviation

The standard deviation of velocity at time t is defined by

$$\sigma_v^2(t) = \left\langle (v(t) - \langle v \rangle_t)^2 \right\rangle = \left\langle v^2 \right\rangle_t - \langle v \rangle_t^2 \tag{39}$$

Using the values given by Eqs. (33) and (24), we have

$$\sigma_v^2(t) = \frac{D}{\beta} + \left(v_0^2 - \frac{D}{\beta} \right) e^{-2\beta t} - \langle v \rangle_0^2 \, e^{-2\beta t} \tag{40}$$

or

$$\sigma_v^2(t) = \frac{D}{\beta} \left(1 - e^{-2\beta t} \right) \tag{41}$$

Limits. The limit for infinite time is

$$\sigma_v^2(t) \to \frac{D}{\beta} \qquad t \to \infty \tag{42}$$

It is important to appreciate that indeed $\sigma_v^2(t)$ goes to a constant for infinite time. This is important because the standard deviation of velocity is proportional to temperature. The fact that $\sigma_v^2(t)$ goes to constant value is consistent with what is called the equipartition theorem. In this case, it implies that the Brownian particles achieve the same value as that of the atoms that are causing the movement of the Brownian particle.

For the small time limit we have

$$\sigma_v^2(t) \sim 2Dt \qquad t \to 0 \tag{43}$$

Deviation from initial velocity. It is also of interest to define the deviation from the initial velocity. We define it by way of

$$\lambda_{v_0}^2 = \left\langle (v(t) - \langle v \rangle_0)^2 \right\rangle = \left\langle v^2 \right\rangle_t - 2 \langle v \rangle_t \langle v \rangle_0 + \langle v \rangle_0^2 \tag{44}$$

Using Eqs. (33) and (24) we obtain

$$\lambda_{v_0}^2 = e^{-2\beta t} \left\langle v^2 \right\rangle_0 + \frac{D}{\beta}(1 - e^{-2\beta t}) - 2 \langle v \rangle_0 e^{-\beta t} + \langle v \rangle_0^2 \tag{45}$$

$$= e^{-2\beta t} \left\langle v^2 \right\rangle_0 + \langle v \rangle_0^2 \left(1 - 2e^{-\beta t} \right) + \frac{D}{\beta}(1 - e^{-2\beta t}) \tag{46}$$

and therefore

$$\lambda_{v_0}^2 = e^{-2\beta t} \left\langle v^2 \right\rangle_0 - \langle v \rangle_0^2 \left(2e^{-\beta t} - 1 \right) + \frac{D}{\beta}(1 - e^{-2\beta t}) \tag{47}$$

4.4 Correlation and Covariance of Velocity

If two random variables are independent, then the joint probability distribution is the product of the individual distributions for each of the variables. A cruder but more accessible measure is the expected value of the product of the two random variables, $\langle XY \rangle$, where X and Y are the random variables. If the probability distribution is a product of the two distributions, one says that the two variables are independent, in which case, $\langle XY \rangle = \langle X \rangle \langle Y \rangle$. Therefore, a measure of dependence is the excess of $\langle XY \rangle$ over $\langle X \rangle \langle Y \rangle$. This is called the covariance

$$\mathrm{Cov}(X, Y) = \langle XY \rangle - \langle X \rangle \langle Y \rangle \tag{48}$$

We point out that zero covariance does not necessarily imply that the two variables are independent, but the covariance does give a measure of the dependence of two variables.

For our case we take the two variables as the velocities at two different times, namely at time t and time s. We write

$$\mathrm{Cov}(v(t), v(s)) = \langle v(t)v(s) \rangle - \langle v \rangle_t \langle v \rangle_s \tag{49}$$

When we have a stochastic process such as the one we are considered here, quantities such as $\langle v(t)v(s) \rangle$ are called two-time autocorrelation functions.

Writing Eq. (22) for times t and s, and multiplying the two expressions we have

$$v(t)v(s) = \left(e^{-\beta t} v(0) + e^{-\beta t} \int_0^t e^{\beta t'} F(t') dt' \right) \left(e^{-\beta s} v(0) + e^{-\beta s} \int_0^s e^{\beta t''} F(t'') dt'' \right) \tag{50}$$

Taking expectation values gives

$$\langle v(t)v(s) \rangle = \left\langle v^2 \right\rangle_0 e^{-\beta(t+s)} + e^{-\beta(t+s)} \int_0^t \int_0^s e^{\beta t'} e^{\beta t''} \left\langle F(t')F(t'') \right\rangle dt' dt'' \tag{51}$$

In Appendix 3 we evaluate the second term of Eq. (51) to give

$$\int_0^t \int_0^s e^{\beta t'} e^{\beta t''} \left\langle F(t')F(t'') \right\rangle dt' dt'' = \frac{D}{\beta} \begin{cases} \left(e^{2\beta s} - 1 \right) & t > s \\ \left(e^{2\beta t} - 1 \right) & t < s \end{cases} \tag{52}$$

Therefore

$$\langle v(t)v(s) \rangle = e^{-\beta(t+s)} \left\langle v^2 \right\rangle_0 + \frac{D}{\beta} e^{-\beta(t+s)} \begin{cases} \left(e^{2\beta s} - 1 \right) & t > s \\ \left(e^{2\beta t} - 1 \right) & t < s \end{cases} \tag{53}$$

$$= e^{-\beta(t+s)} \left\langle v^2 \right\rangle_0 + \frac{D}{\beta} \begin{cases} \left(e^{-\beta(t-s)} - e^{-\beta(t+s)} \right) & t > s \\ \left(e^{-\beta(s-t)} - e^{-\beta(t+s)} \right) & t < s \end{cases} \tag{54}$$

This can be written as

$$\langle v(t)v(s)\rangle = e^{-\beta(t+s)}\left\langle v^2\right\rangle_0 + \frac{D}{\beta}\left(e^{-\beta(|t-s|} - e^{-\beta(t+s)}\right) \tag{55}$$

For t positive, we have

$$\langle v(t)v(0)\rangle = e^{-\beta t}\left\langle v^2\right\rangle_0 \tag{56}$$

It is sometimes useful to consider $\langle v(t)v(t+\tau)\rangle$ which can be obtained form Eq. (55) by setting

$$s = t + \tau \tag{57}$$

in which case we have

$$\langle v(t)v(t+\tau)\rangle = e^{-\beta(2t+\tau)}\left\langle v^2\right\rangle_0 + \frac{D}{\beta}\left(e^{-\beta|\tau|} - e^{-\beta(2t+\tau)}\right) \tag{58}$$

which may also be written as

$$\langle v(t)v(t+\tau)\rangle = e^{-\beta(2t+\tau)}\left\langle v^2\right\rangle_0 + \frac{D}{\beta}\begin{cases}\left(e^{\beta\tau} - e^{-\beta(2t+\tau)}\right) & \tau < 0\\ \left(e^{-\beta\tau} - e^{-\beta(2t+\tau)}\right) & \tau > 0\end{cases} \tag{59}$$

Limits. Consider the large time limit. Taking $t \to \infty$ in Eq. (59) we obtain

$$\langle v(t)v(t+\tau)\rangle_{t\to\infty} = \frac{D}{\beta}\begin{cases} e^{\beta\tau} & \tau < 0\\ e^{-\beta\tau} & \tau > 0\end{cases} \tag{60}$$

which shows that for large times, the autocorrelation function becomes independent of time.

The covariance. Using Eqs. (54) and (24) we have that

$$\mathrm{Cov}(v(t), v(s)) = \langle v(t)v(s)\rangle - \langle v\rangle_t \langle v\rangle_s \tag{61}$$

$$= e^{-\beta(t+s)}\left\langle v^2\right\rangle_0 + \frac{D}{\beta}\left(e^{-\beta(|t-s|} - e^{-\beta(t+s)}\right) - \langle v\rangle_0^2 e^{-\beta(t+s)} \tag{62}$$

giving

$$\mathrm{Cov}(v(t), v(s)) = e^{-\beta(t+s)}\sigma_v^2(0) + \frac{D}{\beta}\left(e^{-\beta(|t-s|} - e^{-\beta(t+s)}\right) \tag{63}$$

where we have defined the standard deviation of velocity at time zero as

$$\sigma_v^2(0) = \left(\left\langle v^2\right\rangle_0 - \langle v\rangle_0^2\right) \tag{64}$$

We also have

$$\mathrm{Cov}(v(t), v(t+\tau)) = e^{-\beta(2t+\tau)}\sigma_v^2(0) + \frac{D}{\beta}\begin{cases} \left(e^{\beta\tau} - e^{-\beta(2t+\tau)}\right) & \tau < 0 \\ \left(e^{-\beta\tau} - e^{-\beta(2t+\tau)}\right) & \tau > 0 \end{cases} \tag{65}$$

For large times

$$\mathrm{Cov}(v(t), v(t+\tau))_{t\to\infty} = \frac{D}{\beta}\begin{cases} e^{\beta\tau} & \tau < 0 \\ e^{-\beta\tau} & \tau > 0 \end{cases} \tag{66}$$

4.5 Correlation of Velocity and Force

It is particularly interesting to evaluate the two-time cross correlation of force and velocity. Langevin implied that it is zero, but Manoliu and Kittel showed it is not [29]. Multiplying the velocity equation, Eq. (22), at time t, by the force at time t', we have

$$v(t)F(t') = e^{-\beta t}v(0)F(t') + e^{-\beta t}\int_0^t e^{\beta' t''}F(t'')F(t')dt'' \tag{67}$$

Taking expectation values we have

$$\left\langle v(t)F(t')\right\rangle = e^{-\beta t}\left\langle v(0)F(t')\right\rangle + e^{-\beta t}\int_0^t e^{\beta' t''}\left\langle F(t'')F(t')\right\rangle dt'' \tag{68}$$

Assuming that $\left\langle v(0)F(t')\right\rangle = 0$, we have

$$\left\langle v(t)F(t')\right\rangle = e^{-\beta t}\int_0^t e^{\beta' t''}\left\langle F(t'')F(t')\right\rangle dt'' \tag{69}$$

$$= 2De^{-\beta t}\int_0^t e^{\beta' t''}\delta(t'-t'')dt'' \tag{70}$$

Clearly

$$\int_0^t e^{\beta' t''}\delta(t'-t'')dt'' = e^{\beta' t'} \qquad 0 < t' < t \tag{71}$$

and therefore

$$\left\langle v(t)F(t')\right\rangle = \begin{cases} 2De^{-\beta(t-t')} & 0 < t' < t \\ 0 & \text{otherwise} \end{cases} \tag{72}$$

Consider now $\langle v(t)F(t+\tau)\rangle$ with τ positive. Accordingly

$$\langle v(t)F(t+\tau)\rangle = 0 \qquad \tau > 0 \tag{73}$$

This a reflection of causality in that a future force has no effect on the present velocity. Now consider

$$\langle v(t)F(t-\tau)\rangle = \begin{cases} 2De^{-\beta\tau} & 0 < \tau < t \\ 0 & \text{otherwise} \end{cases} \tag{74}$$

which shows that as long as the force acts at a time earlier than t, the velocity is affected by it.

The equal time case: For the equal time case, it is better to redo the calculation. In Eq. (67), take $t = t'$,

$$v(t)F(t) = e^{-\beta t}v(0)F(t) + e^{-\beta t}\int_0^t e^{\beta t'}F(t')F(t)dt' \tag{75}$$

and therefore

$$\langle v(t)F(t)\rangle = e^{-\beta t}\int_0^t e^{\beta t'}\langle F(t')F(t)\rangle dt' \tag{76}$$

$$= 2De^{-\beta t}\int_0^t e^{\beta t'}\delta(t-t')dt' \tag{77}$$

We take

$$\int_0^t e^{\beta t'}\delta(t-t')dt' = \frac{1}{2}e^{\beta t} \tag{78}$$

to yield,

$$\langle v(t)F(t)\rangle = D \tag{79}$$

4.6 Energy Balance

For deterministic systems that obey Newton's law, one attempts to obtain a conservation law or an equation for energy flow by the following procedure, as applied to the Langevin equation. Multiply the Langevin equation by $v(t)$ to obtain

$$v(t)\frac{dv(t)}{dt} = -\beta v^2(t) + v(t)F(t) \tag{80}$$

and rewrite it as

$$\frac{1}{2}\frac{dv^2(t)}{dt} = -\beta v^2(t) + v(t)F(t) \tag{81}$$

The kinetic energy (per unit mass) is defined by

$$T = \frac{1}{2}v^2(t) \tag{82}$$

The change in T is therefore

$$\frac{d}{dt}T = -2\beta T + v(t)F(t) \tag{83}$$

Taking expectation values of both sides we have

$$\frac{d}{dt}\langle T\rangle = -2\beta\langle T\rangle + \langle v(t)F(t)\rangle \tag{84}$$

This shows that the change in $\langle T\rangle$ is governed by two terms. The term $\langle v(t)F(t)\rangle$ increases it and the term $-2\beta\langle T\rangle$ decreases it. One says that $\langle v(t)F(t)\rangle$ is the average work done and $-2\beta\langle T\rangle$ is the dissipation.

If we use Eq. (79)

$$\langle v(t)F(t)\rangle = D \tag{85}$$

then we have that

$$\frac{d}{dt}\langle T\rangle = -2\beta\langle T\rangle + D \tag{86}$$

Now if we assume that

$$\frac{d}{dt}\langle T\rangle = 0 \qquad t = \infty \tag{87}$$

then

$$\langle T\rangle = \frac{D}{2\beta} \qquad t = \infty \tag{88}$$

Equation (87) was *assumed* by Langevin. It is reasonable on statistical mechanics grounds. However, Eq. (87) follows directly from the Langevin equation as was seen in Sect. 4.2 and was first shown by Uhlenbeck and Ornstein.

5 Statistical Properties of Position

We now obtain the statistical properties of position, $x(t)$, that is governed by

$$\frac{dx(t)}{dt} = v(t) \tag{89}$$

There are three approaches one may take:

Approach 1. Solving symbolically Eq. (89) we have,

$$x(t) = x(0) + \int_0^t v(t')\,dt' \tag{90}$$

Since we have derived the statistical properties of $v(t)$, we can obtain the statistical properties of $x(t)$. Note that once we have the statistical properties of $v(t)$, we do not need the statistical properties of $F(t)$.

Approach 2. Combing Eqs. (22) and (89) we have the differential equation

$$\frac{d^2x(t)}{dt^2} = -\beta\frac{dx(t)}{dt} + F(t) \tag{91}$$

This a second order differential equation whose formal solution is given by (see Appendix 1)

$$x(t) = x(0) + \frac{v(0)}{\beta}(1 - e^{-\beta t}) + \frac{1}{\beta}\int_0^t \left(1 - e^{\beta(t'-t)}\right) F(t')dt' \tag{92}$$

We can now use the same methods we used for the Langevin equation for velocity but of course we need the statistical properties of $F(t')$.

Approach 3. It is also of interest to write $x(t)$ in terms of $v(t)$ directly. In Appendix A we show that

$$x(t) = x(0) - \frac{1}{\beta}(v(t) - v(0)) + \frac{1}{\beta}\int_0^t F(t') \tag{93}$$

This explicitly expresses $x(t)$ in terms of $v(t)$ and hence the statistical properties of the two can be directly related.

5.1 Average Position

Taking the expectation value of Eq. (90)

$$\langle x\rangle_t = \langle x\rangle_0 + \int_0^t \left\langle v(t')\right\rangle dt' \tag{94}$$

and using Eq. (24)

$$\langle v\rangle_t = \langle v\rangle_0\, e^{-\beta t} \tag{95}$$

we have

$$\langle x\rangle_t = \langle x\rangle_0 + \langle v\rangle_0 \int_0^t e^{-\beta t'}\, dt' \tag{96}$$

giving the expectation value of position at time t,

$$\langle x\rangle_t = \langle x\rangle_0 + \frac{\langle v\rangle_0}{\beta}(1 - e^{-\beta t}) \tag{97}$$

Limits. The limits at infinity and zero are

$$\langle x \rangle_{t\to\infty} \to \langle x \rangle_0 + \frac{\langle v \rangle_0}{\beta} \tag{98}$$

$$\langle x \rangle_{t\to 0} \to \langle x \rangle_0 + \langle v \rangle_0 \, t \tag{99}$$

5.2 *Standard Deviation of Position*

We shall first evaluate the standard deviation of position at time t, defined by

$$\sigma_x^2(t) = \left\langle (x(t) - \langle x \rangle_t)^2 \right\rangle \tag{100}$$

In Appendix 4 we show that

$$\sigma_x^2(t) = \sigma_x^2(0) + \int_0^t \int_0^t \frac{D}{\beta} \left(e^{-\beta |t'-t''|} - e^{-\beta(t'+t'')} \right) dt' dt'' \tag{101}$$

and we further show that

$$\int_0^t \int_0^t \frac{D}{\beta} \left(e^{-\beta(|t'-t''|} - e^{-\beta(t'+t'')} \right) dt' dt'' = \frac{2D}{\beta^2} t + \frac{D}{\beta^3} \left(4e^{-\beta t} - 3 - e^{-2\beta t} \right) \tag{102}$$

Therefore

$$\sigma_x^2(t) = \sigma_x^2(0) + \frac{2D}{\beta^2} t + \frac{D}{\beta^3} \left(4e^{-\beta t} - 3 - e^{-2\beta t} \right) \tag{103}$$

5.3 *Second Moment of Position*

Writing Eq. (103) explicitly

$$\left\langle x^2 \right\rangle_t - \langle x \rangle_t^2 = \left\langle x^2 \right\rangle_0 - \langle x \rangle_0^2 + \frac{2D}{\beta^2} t + \frac{D}{\beta^3} \left(4e^{-\beta t} - 3 - e^{-2\beta t} \right) \tag{104}$$

and squaring Eq. (97)

$$\langle x \rangle_t^2 = \langle x \rangle_0^2 + \frac{\langle v \rangle_0^2}{\beta^2} (1 - e^{-\beta t})^2 + 2 \frac{\langle x \rangle_0 \langle v \rangle_0}{\beta} (1 - e^{-\beta t}) \tag{105}$$

we have that

$$\left\langle x^2\right\rangle_t = \left\langle x^2\right\rangle_0 - \frac{\langle v\rangle_0^2}{\beta^2}(1-e^{-\beta t})^2 - 2\frac{\langle x\rangle_0 \langle v\rangle_0}{\beta}(1-e^{-\beta t}) \tag{106}$$

$$+\frac{2D}{\beta^2}t + \frac{D}{\beta^3}\left(4e^{-\beta t} - 3 - e^{-2\beta t}\right) \tag{107}$$

5.4 *Limits*

For the limits of Eq. (103) we have

$$\sigma_x^2(t) \sim \frac{2D}{\beta^2}t \qquad t \to \infty \tag{108}$$

which is Einstein' result. Also

$$\sigma_x(t) \sim \sqrt{\frac{2D}{\beta^2}t} \qquad t \to \infty \tag{109}$$

which is not differentiable at *zero*, but of course it does not hold at zero. Historically, Eq. (109) was derived by different methods and the fact that the derivative does not exist at zero was taken as a criticism. But of course Einstein was aware that his result only held for large times.

For small times one obtains that

$$\sigma_x^2(t \to 0) \sim \frac{D}{3}t^3 \tag{110}$$

which is differentiable at zero.

5.5 *Deviation from Initial Position*

It is also of interest to calculate the deviation from the initial position

$$\lambda_{x_0}^2(t) = \left\langle (x(t) - x(0))^2 \right\rangle \tag{111}$$

Starting with Eq. (92)

$$x(t) = x(0) + \frac{v(0)}{\beta}(1-e^{-\beta t}) + \frac{1}{\beta}\int_0^t \left(1 - e^{\beta(t'-t)}\right) F(t')dt' \tag{112}$$

we have

$$(x(t)-x(0))^2 = \frac{v^2(0)}{\beta^2}(1-e^{-\beta t})^2 + \frac{1}{\beta^2}\int_0^t\int_0^t \left(1-e^{\beta(t'-t)}\right)\left(1-e^{\beta(t''-t)}\right)F(t')F(t'')dt'dt''$$
$$+2\frac{v(0)}{\beta^2}(1-e^{-\beta t})\int_0^t \left(1-e^{\beta(t'-t)}\right)F(t')dt' \tag{113}$$

Taking expectation values

$$\lambda_{x_0}^2(t) = \frac{\langle v^2\rangle_0}{\beta^2}(1-e^{-\beta t})^2 + \frac{1}{\beta^2}\int_0^t\int_0^t \left(1-e^{\beta(t'-t)}\right)\left(1-e^{\beta(t''-t)}\right)\langle F(t')F(t'')\rangle dt'dt'' \tag{114}$$

In Appendix 5 we show that

$$\frac{1}{\beta^2}\int_0^t\int_0^t \left(1-e^{\beta(t'-t)}\right)\left(1-e^{\beta(t''-t)}\right)F(t')F(t'')dt'dt'' = \frac{D}{\beta^3}\left[2\beta t - 3 + 4e^{-\beta t} - e^{-2\beta t}\right] \tag{115}$$

and therefore

$$\lambda_{x_0}^2(t) = \frac{\langle v^2\rangle_0}{\beta^2}(1-e^{-\beta t})^2 + \frac{2D}{\beta^2}\left[t - \frac{3-4e^{-\beta t}+e^{-2\beta t}}{2\beta}\right] \tag{116}$$

Using Eqs. (103) and (116), we obtain the relation between $\sigma_x^2(t)$ and the deviation from the initial position $\lambda_{x_0}^2(t)$,

$$\sigma_x^2(t) = \lambda_{x_0}^2(t) - \frac{\langle v^2\rangle_t}{\beta^2}(1-e^{-\beta t})^2 \tag{117}$$

5.6 *Correlation of Position and Force*

We now consider the two-time cross-correlation between position and force at different times, $\langle x(t)F(t')\rangle$. Starting with Eq. (92),

$$x(t) = x(0) + \frac{v(0)}{\beta}(1-e^{-\beta t}) + \frac{1}{\beta}\int_0^t \left(1-e^{\beta(t''-t)}\right)F(t'')dt'' \tag{118}$$

and multiplying by the force at time t', we have

$$x(t)F(t') = x(0)F(t') + \frac{v(0)}{\beta}(1-e^{-\beta t})F(t') + \frac{1}{\beta}\int_0^t \left(1-e^{\beta(t''-t)}\right)F(t'')F(t')dt'' \tag{119}$$

Taking expectation values

$$\left\langle x(t)F(t')\right\rangle = \frac{2D}{\beta}\int_0^t \left(1 - e^{\beta(t''-t)}\right)\delta(t''-t')dt'' \tag{120}$$

which evaluates to

$$\left\langle x(t)F(t')\right\rangle = \frac{2D}{\beta}\begin{cases} 1 - e^{-\beta(t-t')} & 0 < t' < t \\ 0 & \text{otherwise} \end{cases} \tag{121}$$

Note that the correlation between position and the force at a later time is zero,

$$\langle x(t)F(t+\tau)\rangle = 0 \qquad \tau > 0 \tag{122}$$

The reason for this is that if the force acts at a time later than the position time, it will have had no affect on the position. However

$$\langle x(t)F(t-\tau)\rangle = \frac{2D}{\beta}\begin{cases} 1 - e^{-\beta\tau} & 0 < \tau < t \\ 0 & \text{otherwise} \end{cases} \tag{123}$$

For equal times,

$$\langle x(t)F(t)\rangle = 0 \tag{124}$$

6 Correlation of Velocity and Position

We now calculate the two time cross correlation function of position and velocity. Starting with Eq. (90)

$$x(t) = x(0) + \int_0^t v(t')\,dt' \tag{125}$$

and multiplying by $v(t')$, we have

$$x(t)v(t') = x(0)v(t') + \int_0^t v(t')v(t'')\,dt'' \tag{126}$$

Taking expectation values

$$\left\langle x(t)v(t')\right\rangle = \left\langle x(0)v(t')\right\rangle + \int_0^t \left\langle v(t')v(s)\right\rangle ds \tag{127}$$

$$= \langle x\rangle_0 \left\langle v(t')\right\rangle + \int_0^t \left\langle v(t')v(s)\right\rangle ds \tag{128}$$

Consider the integral term in Eq. (128). Using Eq. (55) we have

$$\int_0^t \left\langle v(t')v(s)\right\rangle ds = \langle v\rangle_0^2 \int_0^t e^{-\beta(t'+s)}ds + \frac{D}{\beta}\int_0^t \left(e^{-\beta(|t'-s|} - e^{-\beta(t'+s)}\right) ds \tag{129}$$

But

$$\frac{D}{\beta}\int_0^t \left(e^{-\beta(|t'-s|} - e^{-\beta(t'+s)}\right) ds = \frac{D}{\beta^2}\left(2 - 2e^{-\beta t'} - e^{-\beta(t-t')} + e^{-\beta(t+t')}\right) \tag{130}$$

and also

$$\langle v\rangle_0^2 \int_0^t e^{-\beta(t'+s)}ds = \frac{1}{\beta}\langle v\rangle_0^2 e^{-\beta t'}(1 - e^{-\beta t}) \tag{131}$$

Therefore

$$\int_0^t \left\langle v(t')v(s)\right\rangle ds = \frac{1}{\beta}\langle v\rangle_0^2 e^{-\beta t'}(1 - e^{-\beta t}) + \frac{D}{\beta^2}\left(2 - 2e^{-\beta t'} - e^{-\beta(t-t')} + e^{-\beta(t+t')}\right) \tag{132}$$

and hence

$$\begin{aligned}\left\langle x(t)v(t')\right\rangle &= \left\langle x(0)v(t')\right\rangle + \langle v\rangle_0^2 \frac{1}{\beta}e^{-\beta t'}(1 - e^{-\beta t}) \\ &\quad + \frac{D}{\beta^2}\left(2 - 2e^{-\beta t'} - e^{-\beta(t-t')} + e^{-\beta(t+t')}\right)\end{aligned} \tag{133}$$

or

$$\begin{aligned}\left\langle x(t)v(t')\right\rangle &= \langle x(0)v(0)\rangle e^{-\beta t'} + \frac{2D}{\beta^2} + e^{-\beta t'}\left(\frac{\langle v\rangle_0^2}{\beta} - \frac{2D}{\beta^2}\right) \\ &\quad + \left(\frac{D}{\beta^2} - \frac{\langle v\rangle_0^2}{\beta}\right)e^{-\beta(t+t')} - \frac{D}{\beta^2}e^{-\beta(t'-t)}\end{aligned} \tag{134}$$

For $t = t'$,

$$\begin{aligned}\langle x(t)v(t)\rangle &= \langle x(0)v(0)\rangle e^{-\beta t} + \frac{2D}{\beta^2} + e^{-\beta t}\left(\frac{\langle v\rangle_0^2}{\beta} - \frac{2D}{\beta^2}\right) \\ &\quad + \left(\frac{D}{\beta^2} - \frac{\langle v\rangle_0^2}{\beta}\right)e^{-2\beta t} - \frac{D}{\beta^2}\end{aligned} \tag{135}$$

7 How Did Langevin Solve His Equation?

Langevin's insight was that the important Einstein result, the expectation value of the second moment

$$\langle x^2 \rangle_t \sim t \qquad t \to \infty \tag{136}$$

can be obtained without the differential equation for the probability density. He aims to obtain a ordinary differential equations for $\langle x^2 \rangle_t$. Starting with

$$\frac{dv(t)}{dt} = -\beta v(t) + F(t) \tag{137}$$

he multiplies by x, giving

$$x(t)\frac{dv(t)}{dt} = -\beta x(t)v(t) + x(t)F(t) \tag{138}$$

Simple manipulation gives

$$\frac{1}{2}\frac{d^2x^2}{dt^2} = v^2 - \frac{\beta}{2}\frac{dx^2}{dt} + F(t)x \tag{139}$$

This is an equation for x^2 which is the quantity he wants. Unfortunately, it is not solely an equation for x^2 because there is a v^2 term. Nonetheless he takes expectation values

$$\frac{1}{2}\frac{d^2}{dt^2}\langle x^2 \rangle = \langle v^2 \rangle - \frac{\beta}{2}\frac{d}{dt}\langle x^2 \rangle + \langle F(t)x(t) \rangle \tag{140}$$

Again, this equation cannot be solved because $\langle v^2 \rangle$ and $\langle F(t)x(t) \rangle$ are unknown. But on physical grounds, he argues that for large times

$$\langle v^2 \rangle = \text{constant} = \text{c} \qquad t \to \infty \tag{141}$$

Of course, we know that is correct since it was derived in Sect. 4.2. However, Langevin knew it had to be the case because of a theorem in statistical mechanics called the equipartition theorem which states that in equilibrium, the temperature of two constituents of a gas are the same. In this case, the Brownian particle would reach the same temperature as the atoms surrounding it, and hence have constant velocity squared. He takes $\langle F(t)x \rangle$ it to be zero

$$\langle F(t)x(t) \rangle = 0 \tag{142}$$

with the explanation "evidently null by reason of the irregularity of".

Putting Eqs. (141) and (142) into Eq. (140), we have

$$\frac{1}{2}\frac{d^2}{dt^2}\langle x^2 \rangle = c - \frac{\beta}{2}\frac{d}{dt}\langle x^2 \rangle \tag{143}$$

or

$$\frac{d^2}{dt^2}\langle x^2 \rangle + \beta\frac{d}{dt}\langle x^2 \rangle = 2\text{c} \tag{144}$$

Therefore he obtains an equation containing *only* $\langle x^2 \rangle$. It follows that

$$\frac{d}{dt}\langle x^2 \rangle = \frac{2c}{\beta} + e^{-\beta t} \tag{145}$$

He now further argues that for long times $e^{-\beta t} \sim 0$ in which case

$$\frac{d}{dt}\langle x^2 \rangle \sim \frac{2c}{\beta} \tag{146}$$

and therefore (with $\langle x^2 \rangle_0 = 0$) we have

$$\langle x^2 \rangle_t = \frac{2c}{\beta} t \tag{147}$$

This is Einstein's result, derived from a random differential equation and without knowing the probability density of x. This is the first time that a random differential equation was solved.

It is of some interest to see if we can use Eq. (140) to obtain the exact answer for $\langle x^2 \rangle$ assuming that we know $\langle v^2 \rangle_t$ and $\langle F(t)x(t) \rangle$ exactly. Considering zero initial conditions, we know from Eqs. (124) and (33) that (exactly)

$$\langle F(t)x(t) \rangle = 0 \tag{148}$$

and

$$\langle v^2 \rangle_t = \frac{D}{\beta}\left(1 - e^{-2\beta t}\right) \tag{149}$$

Putting these values into Eq. (140) we have that

$$\frac{1}{2}\frac{d^2}{dt^2}\langle x^2 \rangle + \frac{\beta}{2}\frac{d}{dt}\langle x^2 \rangle = \frac{D}{\beta}\left(1 - e^{-2\beta t}\right) \tag{150}$$

The solution is

$$\langle x^2 \rangle_t = \frac{2D}{\beta^2} t + \frac{D}{\beta^3}\left(4e^{-\beta t} - 3 - e^{-2\beta t}\right) \tag{151}$$

which is Eq. (107) with zero initial conditions

8 Non-stationary Power Spectrum of Velocity

The power spectrum gives an indication of the intensities of frequency that exist in a time function. Wang and Uhlenbeck were the first to consider the power spectrum of the velocity by way of the Langevin equation. However, the power spectrum is defined only for a wide sense stationary processes which means that the autocorrelation function

$$\langle X(t)X(t+\tau)\rangle = R(\tau) \tag{152}$$

is only a function of the difference of the two times. For such a situations, the power spectrum of the stochastic process $X(t)$ is

$$S(\omega) = \frac{1}{\sqrt{2\pi}} \int R(\tau) e^{-i\tau\omega} d\tau \tag{153}$$

Can we apply this theorem to velocity? Examining Eq. (59), which we repeat here,

$$\langle v(t)v(t+\tau)\rangle = e^{-\beta(2t+\tau)} \left\langle v^2 \right\rangle_0 + \frac{D}{\beta} \left(e^{-\beta|\tau|} - e^{-\beta(2t+\tau)} \right) \tag{154}$$

we see that it is not stationary, that is, it is not only a function of τ only. Therefore, we cannot use Eq. (153). However for large times

$$\langle v(t)v(t+\tau)\rangle_{t\to\infty} = \frac{D}{\beta} \begin{cases} e^{\beta\tau} & \tau < 0 \\ e^{-\beta\tau} & \tau > 0 \end{cases} \tag{155}$$

which shows that for large times, the autocorrelation function is just a function of τ. Therefore in some sense, Eq. (153) can be used. Wang and Uhlenbeck obtained the following power spectrum for $v(t)$

$$S_v(\omega) = \frac{1}{\sqrt{2\pi}} \frac{N_0}{\beta^2 + \omega^2} \tag{156}$$

which holds only for large times.

Galleani and Cohen obtained the time dependent spectrum for velocity [18, 19]. The time dependent spectrum is defined by way of the Wigner spectrum, $\overline{W}(t,\omega)$, which is commonly called the instantaneous spectrum,

$$\overline{W}(t,\omega) = \frac{1}{2\pi} \int \langle v(t-\tau/2)v(t+\tau/2)\rangle \, e^{-i\tau\omega} d\tau \tag{157}$$

It is the ensemble average of the Wigner distribution [9, 10, 44]. The Wigner spectrum is a function that describes the intensity in *both* time and frequency. The exact $\overline{W}(t,\omega)$ has been calculated and is given by

$$\overline{W}_v(t,\omega) = \frac{1}{\pi} \left(\left\langle v_0^2 \right\rangle - \frac{D}{\beta} \right) e^{-2\beta t} \frac{\sin 2\omega t}{\omega} + \frac{D}{\pi} \frac{1}{\beta^2 + \omega^2}$$
$$- \frac{D}{\pi} \frac{e^{-2\beta t}}{\beta^2 + \omega^2} (\cos 2\omega t - \omega/\beta \sin 2\omega t) \qquad t \geq 0$$

For the infinite time limit,

$$\lim_{t\to 0} \overline{W}_v(t,\omega) = \frac{2D}{\beta^2+\omega^2} \tag{158}$$

which is the classical Wang-Uhlenbeck result.

Appendix 1: Formal Solutions

Treating the Langevin equation as an ordinary differential equation with a driving force $F(t)$, the solution, with initial condition $v(0)$ is

$$v(t) = e^{-\beta t}v(0) + \int_0^t e^{\beta(t'-t)}F(t')dt' \tag{159}$$

This can be verified by direct substitution. Since

$$x(t) = x(0) + \int_0^t v(t')\,dt' \tag{160}$$

we have

$$x(t) = x(0) + \int_0^t \left[e^{-\beta t'}v(0) + \int_0^{t'} e^{\beta(t''-t')}F(t'')dt''\right]dt' \tag{161}$$

$$= x(0) + \frac{v(0)}{\beta}(1-e^{-\beta t}) + \int_0^t\int_0^{t'} e^{\beta(t''-t')}F(t'')dt''dt' \tag{162}$$

But

$$\int_0^t\int_0^{t'} e^{\beta(t''-t')}F(t'')dt''dt' = \int_0^t e^{-\beta t'}\int_0^{t'} e^{\beta t''}F(t'')dt''\,dt' \tag{163}$$

$$= -\frac{1}{\beta}e^{-\beta t}\int_0^t e^{\beta s}F(t')dt' + \frac{1}{\beta}\int_0^t F(t')dt' \tag{164}$$

$$= -\frac{1}{\beta}\int_0^t \left(e^{\beta(t'-t)} - 1\right)F(t')dt' \tag{165}$$

Therefore

$$x(t) = x(0) + \frac{v(0)}{\beta}(1-e^{-\beta t}) + \frac{1}{\beta}\int_0^t \left(1 - e^{\beta(t'-t)}\right)F(t')dt' \tag{166}$$

Also, rewrite Eq. (159) as

$$\int_0^t e^{\beta(t'-t)}F(t')dt' = v(t) - e^{-\beta t}v(0) \tag{167}$$

and substitute it into Eq. (166) to obtain

$$x(t) = x(0) + \frac{v(0) - v(t)}{\beta} + \frac{1}{\beta}\int_0^t F(t')dt' \tag{168}$$

This gives a direct relation between x and v.

In addition

$$(x(t) - x(0))\beta + (v(t) - v(0)) = \int_0^t F(t')dt' \tag{169}$$

and squaring gives

$$(x(t) - x(0))^2\beta^2 + (v(t) - v(0))^2 + (x(t) - x(0))(v(t) - v(0))2\beta = \left(\int_0^t F(t')dt'\right)^2 \tag{170}$$

and using

$$\beta(x(t) - x(0)) = -(v(t) - v(0) + \int_0^t F(t')dt' \tag{171}$$

gives

$$(x(t) - x(0))^2\beta^2 = (v(t) - v(0))^2 - 2(v(t) - v(0))\int_0^t F(t')dt' + \left(\int_0^t F(t')dt'\right)^2 \tag{172}$$

This gives a direct relation between $x^2(t)$ and $v^2(t)$.

Appendix 2: Evaluation of $\int_0^t \int_0^t e^{\beta(t'+t'')} \langle F(t')F(t'')\rangle dt'dt''$

Consider the evaluation of

$$\int_0^t \int_0^t e^{\beta(t'+t'')} \langle F(t')F(t'')\rangle dt'dt'' \tag{173}$$

Using

$$\langle F(t')F(t'')\rangle = 2D\delta(t' - t'') \tag{174}$$

we have

$$\int_0^t \int_0^t e^{\beta(t'+t'')} \langle F(t')F(t'')\rangle dt''dt' = 2D\int_0^t \int_0^t e^{\beta(t'+t'')}\delta(t' - t'')dt''dt' \tag{175}$$

The inner integral gives

$$\int_0^t e^{\beta(t'+t'')}\delta(t' - t'')dt'' = \begin{cases} 2De^{2\beta t'} & 0 < t' < t \\ 0 & \text{otherwise} \end{cases} \tag{176}$$

Hence

$$\int_0^t \int_0^t e^{\beta(t'+t'')} \langle F(t')F(t'')\rangle dt'dt'' = 2D \int_0^t e^{2\beta t'} dt' = \frac{D}{\beta}(e^{-2\beta t} - 1) \tag{177}$$

Multiplying both sides by $e^{-2\beta t}$ we have

$$e^{-2\beta t} \int_0^t \int_0^t e^{\beta(t'+t'')} \langle F(t')F(t'')\rangle dt'dt'' = \frac{D}{\beta}(1 - e^{-2\beta t}) \tag{178}$$

which is Eq. (31) of the main text.

Appendix 3: Evaluation of $\int_0^t \int_0^s e^{\beta t'} e^{\beta t''} \langle F(t')F(t'')\rangle dt'dt''$

We evaluate

$$\int_0^t \int_0^s e^{\beta t'} e^{\beta t''} \langle F(t')F(t'')\rangle dt'dt'' \tag{179}$$

which enters in the valuation of $\langle v(t)v(s)\rangle$ as per Eq. (51). We have

$$\int_0^t \int_0^s e^{\beta t'} e^{\beta t''} \langle F(t')F(t'')\rangle dt'dt'' = \int_0^t \int_0^s e^{\beta t'} e^{\beta t''} \delta(t' - t'')dt'dt'' \tag{180}$$

$$= 2D \int_0^t e^{2\beta t'} dt' \qquad 0 < t' < s \tag{181}$$

Imposing the constraint $0 < t' < s$ on the remaining integral we have that

$$\int_0^t e^{2\beta t'} dt' = \begin{cases} \frac{1}{2\beta}\left(e^{2\beta s} - 1\right) & t > s \\ \frac{1}{2\beta}\left(e^{2\beta t} - 1\right) & t < s \end{cases} \tag{182}$$

Therefore

$$\int_0^t \int_0^s e^{\beta t'} e^{\beta t''} \langle F(t')F(t'')\rangle dt'dt'' = \frac{D}{\beta} \begin{cases} \left(e^{2\beta s} - 1\right) & t > s \\ \left(e^{2\beta t} - 1\right) & t < s \end{cases} \tag{183}$$

Appendix 4: Standard Deviation of $x(t)$

We obtain the standard deviation of position,

$$\sigma_x^2(t) = \left\langle \left(x(t) - \langle x\rangle_t\right)^2 \right\rangle = \sigma_x^2(0) + \frac{2D}{\beta^2} t + \frac{D}{\beta^3}\left(4e^{-\beta t} - 3 - e^{-2\beta t}\right) \tag{184}$$

Staring with

$$x(t) = x(0) + \int_0^t v(t')\,dt' \tag{185}$$

and

$$\langle x\rangle_t = \langle x\rangle_0 + \frac{1}{\beta}\langle v\rangle_0 \left(1 - e^{-\beta t}\right) \tag{186}$$

and subtracting Eq. (186) from Eq. (185) we have

$$x(t) - \langle x\rangle_t = x(0) - \langle x\rangle_0 - \frac{1}{\beta}\langle v\rangle_0 \left(1 - e^{-\beta t}\right) + x(0) + \int_0^t v(t')\,dt' \tag{187}$$

$$= \int_0^t v(t')\,dt' - \frac{1}{\beta}\langle v\rangle_0 \left(1 - e^{-\beta t}\right) \tag{188}$$

Squaring and taking expectation values yields

$$\sigma_x^2(t) = \sigma_x^2(0) + \int_0^t\int_0^t \left\langle v(t')v(t'')\right\rangle dt'dt'' + \frac{1}{\beta^2}\langle v\rangle_0^2 \left(1 - e^{-\beta t}\right)^2 - \frac{2}{\beta}\langle v\rangle_0 \left(1 - e^{-\beta t}\right)\int_0^t \left\langle v(t')\right\rangle dt' \tag{189}$$

Using Eq. (24), we have that

$$\int_0^t \left\langle v(t')\right\rangle dt' = \int_0^t \langle v\rangle_0 e^{-\beta t'}\,dt' = \frac{1}{\beta}\langle v\rangle_0 \left(1 - e^{-\beta t}\right) \tag{190}$$

Therefore

$$\sigma_x^2(t) = \sigma_x^2(0) + \int_0^t\int_0^t \left\langle v(t')v(t'')\right\rangle dt'dt'' - \frac{1}{\beta^2}\langle v\rangle_0^2 \left(1 - e^{-\beta t}\right)^2 \tag{191}$$

But we know from Eq. (55) that

$$\langle v(t)v(s)\rangle = e^{-\beta(t+s)}\left\langle v^2\right\rangle_0 + \frac{D}{\beta}\left(e^{-\beta(|t-s|} - e^{-\beta(t+s)}\right) \tag{192}$$

and therefore

$$\sigma_x^2(t) = \sigma_x^2(0) + \left\langle v^2\right\rangle_0 \int_0^t\int_0^t e^{-\beta(t'+t'')}dt'dt'' + \frac{D}{\beta}\int_0^t\int_0^t \left(e^{-\beta(|t'-t''|} - e^{-\beta(t'+t'')}\right) dt'dt'' \tag{193}$$

$$- \frac{1}{\beta^2}\langle v\rangle_0^2 \left(1 - e^{-\beta t}\right)^2 \tag{194}$$

But clearly

$$\left\langle v^2\right\rangle_0 \int_0^t\int_0^t e^{-\beta(t'+t'')}dt'dt'' = \frac{1}{\beta^2}\langle v\rangle_0^2 \left(1 - e^{-\beta t}\right)^2 \tag{195}$$

and therefore

$$\sigma_x^2(t) = \sigma_x^2(0) + \frac{D}{\beta}\int_0^t\int_0^t \left(e^{-\beta(|t'-t''|} - e^{-\beta(t'+t'')}\right) dt'dt'' \tag{196}$$

We now evaluate the integral in Eq. (196). First consider

$$\int_0^t \left(e^{-\beta(|t'-t''|}\right) dt' = \int_0^{t''} \left(e^{-\beta(t''-t')}\right) dt' + \int_{t''}^t \left(e^{-\beta(t'-t'')}\right) dt' \tag{197}$$

$$= \frac{1}{\beta}e^{-\beta t''}(e^{\beta t''} - 1) - e^{\beta t''}\frac{1}{\beta}(e^{-\beta t} - e^{-\beta t''}) \tag{198}$$

$$= \frac{1}{\beta}\left(1 - e^{-\beta t''} - e^{-\beta t}e^{\beta t''} + 1\right) \tag{199}$$

or

$$\int_0^t \left(e^{-\beta(|t'-t''|}\right) dt' = \frac{1}{\beta}\left(2 - e^{-\beta t''} - e^{-\beta t}e^{\beta t''}\right) \tag{200}$$

Integrating over t'' we obtain

$$\int_0^t \frac{1}{\beta}\left(2 - e^{-\beta t''} - e^{-\beta t}e^{\beta t''}\right) dt'' = \frac{1}{\beta}\frac{1}{\beta}\left(2t + \frac{1}{\beta}(e^{-\beta t} - 1) - \frac{1}{\beta}e^{-\beta t}(e^{\beta t} - 1)\right) \tag{201}$$

$$= \frac{1}{\beta^2}\left(2t + \frac{2}{\beta}(e^{-\beta t} - 1)\right) \tag{202}$$

Giving

$$\int_0^t\int_0^t \left(e^{-\beta(|t'-t''|}\right) dt'dt'' = \frac{1}{\beta^2}\left(2t + \frac{2}{\beta}(e^{-\beta t} - 1)\right) \tag{203}$$

The second integral is

$$\int_0^t\int_0^t e^{-\beta(t'+t'')}\, dt'dt'' = \frac{1}{\beta^2}(e^{-\beta t} - 1)^2 \tag{204}$$

Subtracting Eq. (203) from Eq. (204) and multiplying by $\frac{D}{\beta}$ we have

$$\frac{D}{\beta}\int_0^t\int_0^t \left(e^{-\beta(|t'-t''|} - e^{-\beta(t'+t'')}\right) dt'dt'' = \frac{D}{\beta}\left[\frac{1}{\beta^2}\left(2t + \frac{2}{\beta}(e^{-\beta t} - 1)\right) - \frac{1}{\beta^2}(e^{-\beta t} - 1)^2\right] \tag{205}$$

which simplifies to

$$\frac{D}{\beta}\int_0^t\int_0^t \left(e^{-\beta(|t'-t''|} - e^{-\beta(t'+t'')}\right) dt'dt'' = \frac{2D}{\beta^2}t + \frac{D}{\beta^3}\left(4e^{-\beta t} - 3 - e^{-2\beta t}\right) \tag{206}$$

Appendix 5: Evaluation of Eq. (115)

We show Eq. (115) of the text. We have that

$$\frac{1}{\beta^2}\int_0^t\int_0^t\left(1-e^{\beta(t'-t)}\right)\left(1-e^{\beta(t''-t)}\right)F(t')F(t'')dt'dt'' \tag{207}$$

$$=\frac{1}{\beta^2}\int_0^t\int_0^t\left(1-e^{\beta(t'-t)}\right)\left(1-e^{\beta(t''-t)}\right)2D\delta(t'-t'')dt'dt'' \tag{208}$$

$$=\frac{2D}{\beta^2}\int_0^t\left(1-e^{\beta(t'-t)}\right)^2dt' \tag{209}$$

$$=\frac{2D}{\beta^2}\int_0^t\left[1-2e^{\beta(t'-t)}+e^{2\beta(t'-t)}\right]dt' \tag{210}$$

$$=\frac{2D}{\beta^2}\left[t-\frac{2}{\beta}e^{\beta(t'-t)}+\frac{1}{2\beta}e^{2\beta(t'-t)}\right]|_0^t \tag{211}$$

$$=\frac{2D}{\beta^2}\left[t-\frac{2}{\beta}+\frac{1}{2\beta}+\frac{2}{\beta}e^{-\beta t}-\frac{1}{2\beta}e^{-2\beta t}\right] \tag{212}$$

Therefore

$$\frac{1}{\beta^2}\int_0^t\int_0^t\left(1-e^{\beta(t'-t)}\right)\left(1-e^{\beta(t''-t)}\right)F(t')F(t'')dt'dt''=\frac{D}{\beta^3}\left[2\beta t-3+4e^{-\beta t}-e^{-2\beta t}\right] \tag{213}$$

References

1. Ahmad, A., Cohen, L.: Random force in gravitation systems. Astrophys. J. **179**, 885 (1973)
2. Ahmad, A., Cohen, L.: Dynamical friction in gravitational systems. Astrophys. J. **188**, 469 (1974)
3. Brown, R.: A brief account of microscopical observations made in the months of june, july, and august, 1827 on the particles contained in the pollen of plants; and on the general existence of active molecules in organic and inorganic bodies. Philos. Mag. n.s **4**, 161–173 (1828)
4. Chandrasekhar, S.: A statistical theory of stellar encounters. Astrophys. J. **94**, 511–525 (1941)
5. Chandrasekhar, S., von Neumann, J.: The statistics of the gravitational field arising from a random distribution of stars, I, the speed of fluctuations. Astrophys. J. **95**, 53–489 (1942). II, The speed or fluctuations; dynamical friction; spatial correlations. Astrophys. J. **97**, 1–27 (1943)
6. Chandrasekhar, S.: Stochastic problems in physics and astronomy. Rev. Mod. Phys. **15**, 1–89 (1943)
7. Coffey, W.T., Kalmykov, Y.P., Waldron, T.: The Langevin Equation. World Scientific (1996)
8. Cohen, L., Ahmad, A.: The two-time autocorrelation function for force in bounded gravitational systems. Astrophys. J. **197**, 667 (1975)
9. Cohen, L.: Time-frequency distributions—a review. Proc. IEEE **77**, 941–981 (1989)
10. Cohen, L.: Time-Frequency Analysis. Prentice-Hall (1995)
11. Cohen, L.: The history of noise. IEEE Sign. Process. Mag. **22**(6), 20–45 (2005)
12. Doob, J.L.: The brownian movement and stochastic equations. Ann. Math. **43**, 351 (1942)

13. Einstein, A.: Über die von der molekularkinetischen Theorie der Wärme geforderte Bewegung von in ruhenden Flüssigkeiten suspendierten Teilchen, (On the Movement of Small Particles Suspended in Stationary Liquids Required by the Molecular-Kinetic Theory of Heat). Annalen der Physik **17**, 549–560 (1905)
14. Einstein, A.: Méthode pour la détermination de valeurs statistiques d'observations concermant des grandeurs soumises à des fluctuations irrégulières (Method for the Determination of Statistical Values of Observations Regarding Quantities Subject to Irregular Fluctuations). Archives des sciences physiques et naturelles **37**, 254–256 (1914)
15. Ford, G.W., Kac, M., Mazur, P.: Statistical mechanics of assemblies of coupled oscillators. J. Math. Phys. **6**(4), 504–515 (1965)
16. Furth, R.: Investigations on the Theory of the Brownian Movement. Dover (1956)
17. Gardiner, W.: Handbook of Stochastic Methods. Springer (1983)
18. Galleani, L., Cohen, L.: Direct time-frequency characterization of linear systems governed by differential equations. Sign. Process. Lett. **11**, 721–724 (2004)
19. Galleani, L., Cohen, L.: Nonstationary stochastic differential equations. In: Marshall, S., Sicuranza, G. (eds.) Advances of nonlinear signal and image processing, pp. 1–13. Hindowi Publishing (2006)
20. Galleani, L., Cohen, L.: Nonstationary and stationary noise. In: Proceedings of the SPIE, vol. 6234, id. 623416 (2006)
21. Gitterman, M.: The Noisy Oscillator. World Scientific (2005)
22. Huang, K.: Statistical Physics. CRC Press (2001)
23. Kubo, R.: The fluctuation-dissipation theorem. Rep. Prog. Phys. **29** Part I, 255 (1966)
24. Kac, M.: Random walk and the theory of brownian motion. Am. Math. Mon. **54**(7), 369–391 (1947)
25. Kramers, H.: Brownian motion in a field of force and the diffusion model of chemical reactions. Physica **7**, 284 (1940)
26. Langevin, P.: Sur la theorie du mouvement brownien, (On the Theory of Brownian Motion). C. R. Acad. Sci. (Paris) **146**, 530–533 (1908)
27. MacDonald, D.K.C.: Noise and Fluctuations. Wiley (1962)
28. Middleton, D.: Introduction to Statistical Communication Theory. McGraw-Hill (1960)
29. Manoliu, A., Kittel, C.: Correlation in the Langevin theory of Brownian motion. Am. J. Phys. **47**, 678–680 (1979)
30. O'Connell, R.F.: Fluctuations and noise: a general model with applications. Proc. SPIE **5842**, 206 (2005)
31. Perrin, J.: Les atoms, librairie felix alcan, Paris (1913) (Atoms, (trans: Hammick, D.L.). Ox Bow Press, Woodbridge, 1990)
32. Papoulis, A., Pillai, S.U.: Probability, Random Variables and Stochastic Processes. McGraw-Hill (2002)
33. Rice, S.O.: Mathematical analysis of random noise. Bell Syst. Tech. J. **23** & **24**,1–162 (1944 and 1945)
34. Risken, H.: The Fokker-Planck Equation. Springer (1996)
35. Schleich, W.P.: Quantum Optics in Phase Space. Wiley (2001)
36. Scully, M.O., Zubairy, M.S.: Quantum Optics. Cambridge University Press (1997)
37. Smoluchowski, M.V.: Annalen der Physik **21**, 756 (1906)
38. Uhlenbeck, G.E., Ornstein, L.S.: On the theory of Brownian Motion. Phys. Rev. **36**, 823–841 (1930)
39. van Kampen, N.: Stochastic Processes in Physics and Chemistry. North Holland (1981)
40. van Lear Jr., G.A., Uhlenbeck, G.E.: The brownian motion of strings and elastic rods. Phys. Rev. **38**, 1583–1598 (1931)
41. Wang, M.C., Uhlenbeck, G.E.: On the theory of the Brownian Motion II. Rev. Mod. Phys. **17**, 323–342 (1945)
42. Wax, N.: Selected Papers on Noise and Stochastic Processes. Dover (1954)
43. Wiener, N.: Differential space. J. Math. Phys. **2**, 131–174 (1923)
44. Wigner, E.P.: On the quantum correction for thermodynamic equilibrium. Phys. Rev. **40**, 749–759 (1932)
45. Zwanzig, R.: Nonequilibrium Statistical Mechanics. Oxford University Press (2001)

Geometry-Fitted Fourier-Mellin Transform Pairs

Darren Crowdy

Abstract The construction of novel Fourier/Mellin-type transform pairs that are tailor-made for given planar regions within the special class of circular domains is surveyed. Circular domains are those having boundary components that are either circular arcs or straight lines. The new transform pairs generalize the classical Fourier and Mellin transforms. These geometry-fitted transform pairs can be used to great advantage in solving boundary value problems defined in these domains.

Keywords Fourier transform · Mellin transform · Geometric function theory

1 Introduction

This article surveys some of the mathematical ideas laid out in the author's plenary lecture at 10th International ISAAC Meeting in Macau in 2015. The topic is the construction of novel Fourier-Mellin type transform pairs that are "tailor-made" for given planar domains within a special class.

The class of domains amenable to the construction—at the time of writing at least—is the class of *circular domains*, either simply or multiply connected, having boundary components made up of straight lines, arcs of circles, or a mixture of both. Figure 1 shows examples: a simply connected convex quadrilateral (a polygon), a simply connected lens-shaped domain (a circular polygon), and the "disc-in-channel" geometry (a doubly connected circular domain) that arises in many applications.

The author's recent results in this area have been inspired by the extensive body of mathematical work over the last few decades pioneered by A.S. Fokas and collaborators, and now commonly referred to as the *Fokas method* [11, 12]. That work describes a unified transform approach to initial and boundary value problems for both linear and nonlinear integrable partial differential equations. In one strand of

D. Crowdy (✉)
Department of Mathematics, Imperial College London, 180 Queen's Gate, London SW7 2AZ, UK
e-mail: d.crowdy@imperial.ac.uk

T. Qian and L.G. Rodino (eds.), *Mathematical Analysis, Probability and Applications – Plenary Lectures*, Springer Proceedings in Mathematics & Statistics 177, DOI 10.1007/978-3-319-41945-9_2

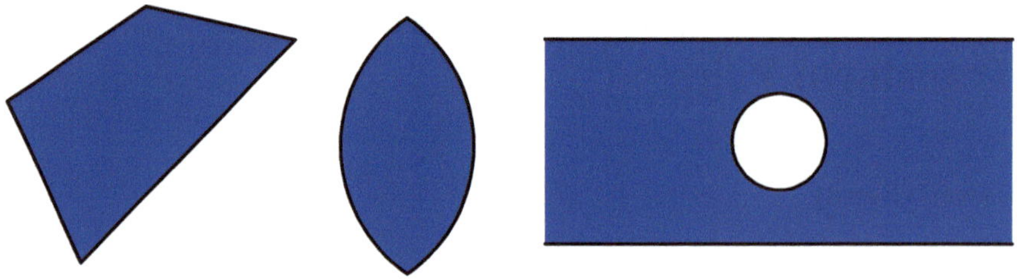

Fig. 1 Example *circular* domains: a convex polygon, a lens-shaped circular-arc domain, and the "disc-in-channel" geometry

this work a new constructive approach to the solution of boundary value problems for Laplace's equation in convex polygons has been described by Fokas and Kapaev [10] with the extension to biharmonic fields made by Crowdy and Fokas [9]. That work was based on an analytical formulation involving the spectral analysis of a Lax pair and use of Riemann-Hilbert methods. The new approach outlined here—and described in more detail in [2, 3]—has a more geometrical flavour and has led the way to generalization of results previously pertaining only to simply connected convex polygons to the much broader class of circular domains, including multiply connected cases.

We first review the results described in [2, 3]. Given a bounded N-sided convex polygon with straight line edges $\{S_n | n = 1, ..., N\}$ inclined at angles $\{\chi_n | n = 1, ..., N\}$ to the positive real axis (e.g., the quadrilateral shown in Fig. 1) we can derive the following transform pair to represent a function $f(z)$ analytic in the polygon:

$$f(z) = \frac{1}{2\pi} \sum_{j=1}^{N} \int_L \rho_{jj}(k) e^{-\mathrm{i}\chi_j} e^{\mathrm{i}e^{-\mathrm{i}\chi_j} kz} dk, \quad \rho_{mn}(k) = \int_{S_n} f(z') e^{-\mathrm{i}e^{-\mathrm{i}\chi_m} kz'} dz', \quad (1)$$

where L is the ray along the positive real axis in the k-plane (see Fig. 2) and where the *spectral functions* (or "transforms") satisfy the so-called *global relations* [2, 11, 12]

$$\sum_{n=1}^{N} \rho_{mn}(k) = 0, \qquad k \in \mathbb{C}, \qquad m = 1, ..., N. \quad (2)$$

Only the diagonal elements of what we call the *spectral matrix* $\rho_{mn}(k)$ appear in the inverse transform formula for $f(z)$.

In precise analogy, given a bounded convex N-sided circular polygon with edges that are arcs of circles with centres $\{\delta_n | n = 1, ..., N\}$ and radii $\{q_n | n = 1, ..., N\}$ (e.g., the lens-shaped domain of Fig. 1) we can derive the following transform pair to represent a function $f(z)$ analytic in it:

$$f(z) = \frac{1}{2\pi \mathrm{i}} \sum_{j=1}^{N} \left\{ \int_{L_1} \frac{\rho_{jj}(k)}{1 - e^{2\pi \mathrm{i}k}} \left[\frac{z - \delta_j}{q_j} \right]^k dk + \int_{L_2} \rho_{jj}(k) \left[\frac{z - \delta_j}{q_j} \right]^k dk \right.$$
$$\left. + \int_{L_3} \frac{\rho_{jj}(k) e^{2\pi \mathrm{i}k}}{1 - e^{2\pi \mathrm{i}k}} \left[\frac{z - \delta_j}{q_j} \right]^k dk \right\},$$
$$\rho_{mn}(k) = \frac{1}{q_m} \int_{C_n} \left[\frac{z' - \delta_m}{q_m} \right]^{-k-1} f(z') dz', \tag{3}$$

where L_1, L_2 and L_3 are the contours in the k-plane (see Fig. 2) and where the spectral functions satisfy the global relations

$$\sum_{n=1}^{N} \rho_{mn}(k) = 0, \qquad k \in -\mathbb{N}, \qquad m = 1, ..., N. \tag{4}$$

Again, only the diagonal elements of spectral matrix $\rho_{mn}(k)$ appear in the inverse transform formula for $f(z)$.

The plan of the article is as follows. First, in Sects. 2 and 3, we discuss the geometrical construction of the transform pairs (1) and (3). In Sect. 4 we combine those general ideas to construct a useful Fourier-Mellin type transform pair for the doubly connected disc-in-channel geometry of Fig. 1. Finally, Sect. 5 illustrates how the transform pairs can be used in practice by solving an accessory parameter problem in conformal mapping theory and finding a useful conformal mapping function associated with the disc-in-channel geometry of Fig. 1.

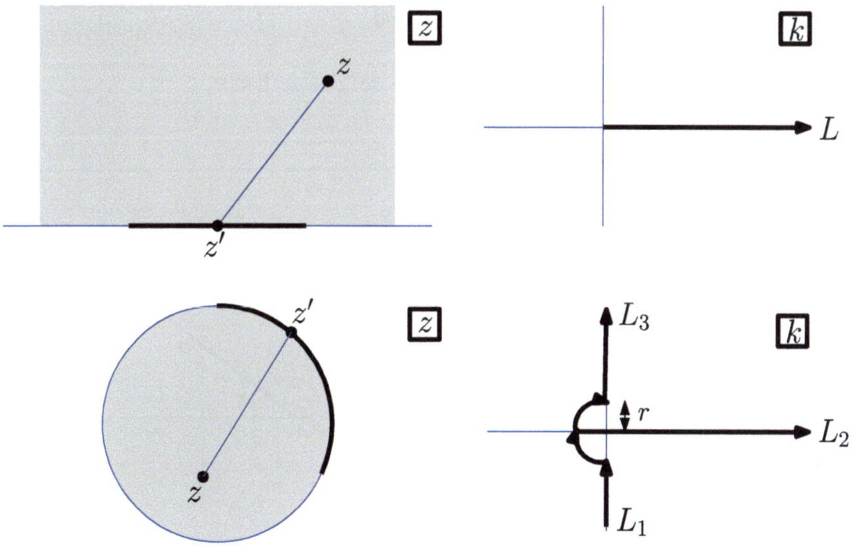

Fig. 2 The basic geometrical units in the z-plane, shown *left*, with the corresponding integration contours in the k-plane shown to the *right* ($0 < r < 1$)

2 Geometrical Approach to Transform Pairs

Suppose a point z' lies on some finite length slit on the real axis and z is in the upper-half plane (the left schematic of Fig. 3) then

$$0 < \arg[z - z'] < \pi. \tag{5}$$

It follows that

$$\int_L e^{\mathrm{i}k(z-z')} dk = \left[\frac{e^{\mathrm{i}k(z-z')}}{\mathrm{i}(z-z')} \right]_0^\infty = \frac{1}{\mathrm{i}(z'-z)} \tag{6}$$

or,

$$\frac{1}{z'-z} = \mathrm{i} \int_L e^{\mathrm{i}k(z-z')} dk, \qquad 0 < \arg[z - z'] < \pi. \tag{7}$$

It is easy to check that the contribution from the upper limit of integration vanishes for the particular choices of z' and z to which we have restricted consideration.

On the other hand, suppose z' lies on some other finite length slit making angle χ with the positive real axis and suppose that z is in the slanted half plane shown shaded in Fig. 3 (the half plane "to the left" of the slit as one follows its tangent with uniform inclination angle χ). Now the transformation

$$z' \mapsto e^{-\mathrm{i}\chi}(z' - \alpha), \qquad z \mapsto e^{-\mathrm{i}\chi}(z - \alpha), \tag{8}$$

for example, where the (unimportant) constant α is shown in Fig. 3, takes the slit to the real axis, and z to the upper-half plane, and

$$0 < \arg[e^{-\mathrm{i}\chi}(z - \alpha) - e^{-\mathrm{i}\chi}(z' - \alpha)] < \pi. \tag{9}$$

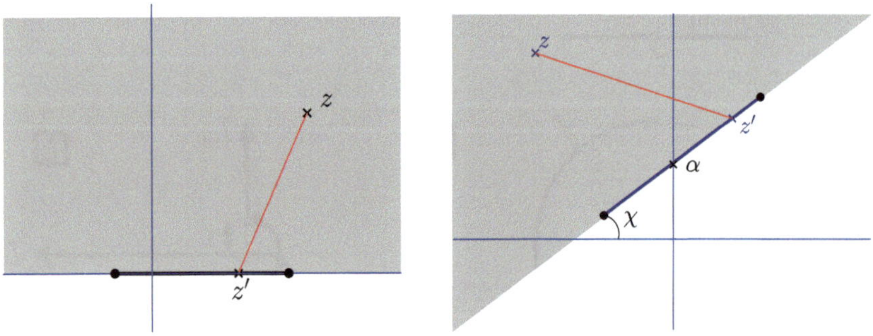

Fig. 3 Geometrical positioning of z and z' for the validity of (6) and (11)

Hence, on use of (7) with the substitutions (8), we can write

$$\frac{1}{e^{-\mathrm{i}\chi}(z'-\alpha)-e^{-\mathrm{i}\chi}(z-\alpha)} = \mathrm{i}\int_L e^{\mathrm{i}k(e^{-\mathrm{i}\chi}(z-\alpha)-e^{-\mathrm{i}\chi}(z'-\alpha))}dk, \qquad (10)$$

or, on cancellation of α and rearrangement,

$$\frac{1}{z'-z} = \mathrm{i}\int_L e^{\mathrm{i}e^{-\mathrm{i}\chi}k(z-z')}e^{-\mathrm{i}\chi}dk. \qquad (11)$$

Now consider a bounded convex polygon P with N sides $\{S_j | j = 1, \ldots, N\}$. Figure 4 shows an example with $N = 3$. For a function $f(z)$ analytic in P, Cauchy's integral formula provides that for $z \in P$,

$$f(z) = \frac{1}{2\pi\mathrm{i}}\oint_{\partial P}\frac{f(z')dz'}{z'-z} \qquad (12)$$

or, on separating the boundary integral into a sum over the N sides,

$$f(z) = \frac{1}{2\pi\mathrm{i}}\sum_{j=1}^{N}\int_{S_j} f(z')\,\frac{1}{(z'-z)}\,dz'. \qquad (13)$$

But if side S_j has inclination χ_j then (11) can be used, with $\chi \mapsto \chi_j$, to reexpress the Cauchy kernel, that is $1/(z'-z)$, uniformly for all $z \in P$ and z' on the respective sides

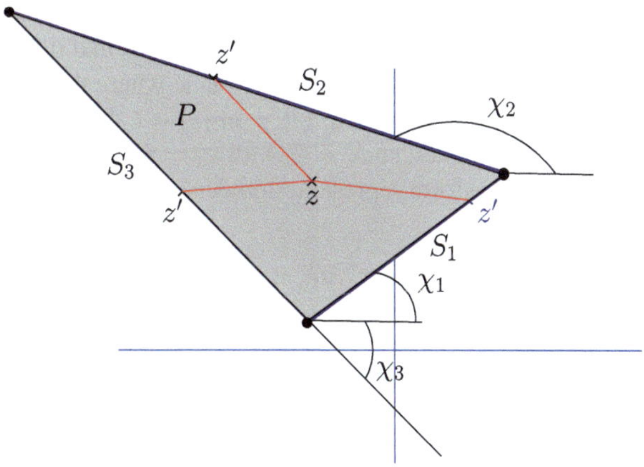

Fig. 4 A convex polygon P as an intersection of $N = 3$ half planes with N angles $\{\chi_j | j = 1, 2, 3\}$. Formula (11) can be used in the Cauchy integral formula with $\chi = \chi_j$ when z' is on side S_j (for $j = 1, 2, 3$)

$$f(z) = \frac{1}{2\pi \mathrm{i}} \sum_{j=1}^{N} \int_{S_j} f(z') \left\{ \mathrm{i} \int_L e^{\mathrm{i}e^{-\mathrm{i}\chi_j} k(z-z')} e^{-\mathrm{i}\chi_j} dk \right\} dz'. \tag{14}$$

On reversing the order of integration we can write

$$f(z) = \frac{1}{2\pi} \sum_{j=1}^{N} \int_L \rho_{jj}(k) e^{-\mathrm{i}\chi_j} e^{\mathrm{i}e^{-\mathrm{i}\chi_j} kz} dk, \tag{15}$$

where, for integers m, n between 1 and N, we define the *spectral matrix* [2] to be

$$\rho_{mn}(k) \equiv \int_{S_n} f(z') e^{-\mathrm{i}e^{-\mathrm{i}\chi_m} kz'} dz', \tag{16}$$

and where $L = [0, \infty)$ is the *fundamental contour* [2] for straight line edges shown in Fig. 2. We have then arrived at the transform pair (1).

The spectral matrix elements have their own analytical structure. Observe that, for any $k \in \mathbb{C}$, and for any $m = 1, ..., N$,

$$\sum_{n=1}^{N} \rho_{mn}(k) = \sum_{n=1}^{N} \int_{S_n} f(z') e^{-\mathrm{i}e^{-\mathrm{i}\chi_m} kz'} dz' = \int_{\partial P} f(z') e^{-\mathrm{i}e^{-\mathrm{i}\chi_m} kz'} dz' = 0, \tag{17}$$

where we have used Cauchy's theorem and the fact that $f(z') e^{-\mathrm{i}e^{-\mathrm{i}\chi_m} kz'}$ (for $m = 1, ..., N$) is analytic inside P.

Special case: How does the traditional Fourier transform pair fit into this geometrical view? Transform pairs for unbounded polygons, such as strips and semi-strips, can be derived with minor modifications: the only difference is that the global relations are now valid in restricted parts of the spectral k-plane where the spectral functions are well defined. If P is the infinite strip $-l < \mathrm{Im}[z] < l$ then, geometrically, it is the intersection of two half planes, so $N = 2$, with $\chi_1 = 0$ and $\chi_2 = \pi$. For any $f(z)$ analytic in this strip the transform representation derived above is

$$f(z) = \frac{1}{2\pi} \int_L \rho_{11}(k) e^{\mathrm{i}kz} dk - \frac{1}{2\pi} \int_L \rho_{22}(k) e^{-\mathrm{i}kz} dk, \tag{18}$$

where the spectral functions are

$$\rho_{11}(k) = \int_{-\infty}^{\infty} f(z) e^{-\mathrm{i}kz} dz, \qquad \rho_{12}(k) = \int_{\infty+\mathrm{i}l}^{-\infty+\mathrm{i}l} f(z) e^{-\mathrm{i}kz} dz, \tag{19}$$
$$\rho_{21}(k) = \int_{-\infty}^{\infty} f(z) e^{\mathrm{i}kz} dz \qquad \rho_{22}(k) = \int_{+\infty+\mathrm{i}l}^{-\infty+\mathrm{i}l} f(z) e^{\mathrm{i}kz} dz.$$

The global relations in this case are

$$\rho_{11}(k) + \rho_{12}(k) = 0, \qquad \rho_{21}(k) + \rho_{22}(k) = 0, \qquad k \in \mathbb{R} \tag{20}$$

which, in contrast to the case of a bounded convex polygon (where the global relations are valid for all $k \in \mathbb{C}$), are only valid for $k \in \mathbb{R}$. It is clear from their definitions that

$$\rho_{22}(-k) = \rho_{12}(k) \tag{21}$$

implying, after a change of variable $k \mapsto -k$ in the second integral of (18), that we can write

$$f(z) = \frac{1}{2\pi}\int_L \rho_{11}(k)e^{ikz}dk + \frac{1}{2\pi}\int_0^{-\infty} \rho_{12}(k)e^{ikz}dk. \tag{22}$$

On use of the first global relation in (20) we can eliminate $\rho_{12}(k)$:

$$f(z) = \frac{1}{2\pi}\int_L \rho_{11}(k)e^{ikz}dk - \frac{1}{2\pi}\int_0^{-\infty} \rho_{11}(k)e^{ikz}dk = \frac{1}{2\pi}\int_{-\infty}^{\infty} \rho_{11}(k)e^{ikz}dk. \tag{23}$$

Dropping the (now unnecessary) subscripts on $\rho_{11}(k)$, we arrive at the well-known Fourier transform pair

$$f(z) = \frac{1}{2\pi}\int_{-\infty}^{\infty} \rho(k)e^{ikz}dk, \qquad \rho(k) = \int_{-\infty}^{\infty} f(z)e^{-ikz}dz. \tag{24}$$

In retrieving the classical Fourier transform in this way we see how our derivation generalizes it to produce "geometry-fitted" transform pairs for any simply connected convex polygon.

3 Transform Pairs for Circular Polygons

It is natural to ask if the construction extends to other domains beyond convex polygons. The answer is in the affirmative, and the author [2] has recently shown how to extend the construction to the much broader class of so-called *circular domains*, or circular polygons, including multiply connected ones [3]. The simple convex polygons just considered are a subset of this more general class.

A key step is to establish [2] the following formula valid for $|z| < 1$:

$$\frac{1}{1-z} = \int_{L_1} \frac{1}{1-e^{2\pi ik}}z^k dk + \int_{L_2} z^k dk + \int_{L_3} \frac{e^{2\pi ik}}{1-e^{2\pi ik}}z^k dk. \tag{25}$$

This is the basic identity that replaces the result (7) in the construction of transform pairs for circular polygons. The *fundamental contour* (for circular-arc edges [2]) is now made up of three components labelled L_1, L_2 and L_3 and shown in Fig. 2. The parameter r is arbitrary but must be chosen so that $0 < r < 1$. In an appendix to [2] the author shows how to derive (25) in a natural way from the results of the previous section. We omit details here noting only that the appearance of the expressions z^k in (25) remind us of the classical Mellin transform.

Suppose, more generally, that z is a point inside some circle C_j with centre $\delta_j \in \mathbb{C}$ and radius $q_j \in \mathbb{R}$. Suppose too that z' is a point on the circle C_j. Then $|z - \delta_j| < |z' - \delta_j|$ and, on use of (25), the Cauchy kernel for z' on C_j and z inside C_j has the spectral representation

$$\begin{aligned}\frac{1}{z'-z} &= \frac{1}{(z'-\delta_j)-(z-\delta_j)} \\ &= \frac{1}{(z'-\delta_j)}\frac{1}{[1-(z-\delta_j)/(z'-\delta_j)]} \\ &= \int_{L_1}\frac{1}{1-e^{2\pi ik}}\frac{(z-\delta_j)^k}{(z'-\delta_j)^{k+1}}dk + \int_{L_2}\frac{(z-\delta_j)^k}{(z'-\delta_j)^{k+1}}dk \\ &\quad + \int_{L_3}\frac{e^{2\pi ik}}{1-e^{2\pi ik}}\frac{(z-\delta_j)^k}{(z'-\delta_j)^{k+1}}dk. \end{aligned} \tag{26}$$

It is important, especially for numerical implementations, to write this as

$$\begin{aligned}\frac{1}{z'-z} &= \int_{L_1}\frac{1}{1-e^{2\pi ik}}\frac{1}{q_j}\left(\frac{z-\delta_j}{q_j}\right)^k\left[\frac{z'-\delta_j}{q_j}\right]^{-k-1}dk \\ &\quad + \int_{L_2}\frac{1}{q_j}\left(\frac{z-\delta_j}{q_j}\right)^k\left[\frac{z'-\delta_j}{q_j}\right]^{-k-1}dk \\ &\quad + \int_{L_3}\frac{e^{2\pi ik}}{1-e^{2\pi ik}}\frac{1}{q_j}\left(\frac{z-\delta_j}{q_j}\right)^k\left[\frac{z'-\delta_j}{q_j}\right]^{-k-1}dk. \end{aligned} \tag{27}$$

Just as a convex N-sided polygon was interpreted geometrically as the intersection of N half plane regions, a convex N-sided *circular polygon* can be viewed as the intersection of N circular discs. Figure 5 shows an example circular polygon D with $N = 3$ bounded by circular arcs denoted by C_1, C_2 and C_3. The Cauchy integral formula for a function $f(z)$ analytic in this region is

$$f(z) = \frac{1}{2\pi \mathrm{i}}\oint_{\partial D}\frac{f(z')}{z'-z}dz' = \frac{1}{2\pi \mathrm{i}}\sum_{j=1}^{N}\int_{C_j}\frac{f(z')}{z'-z}dz', \tag{28}$$

where ∂D denotes the boundary of D and where, in the second equality, the integral around ∂D has been separated into the N separate integrals around the individual circular arcs $\{C_j | j = 1, ..., N\}$.

Now for $z \in D$ we can substitute (27) into the Cauchy integral formula (28) when z' sits on each of the separate boundary arcs $\{C_j | j = 1, ..., N\}$ to find

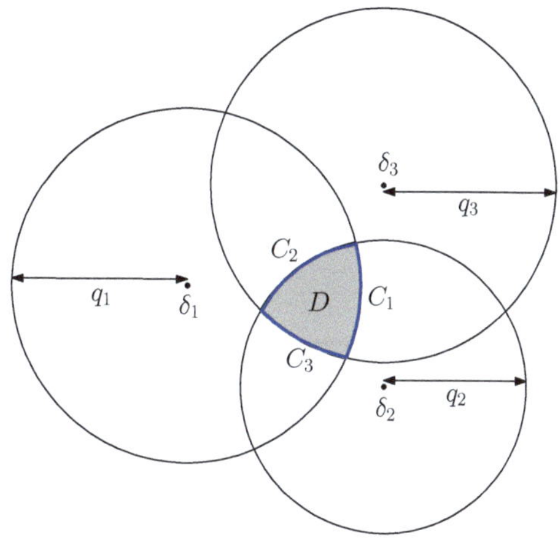

Fig. 5 A circular polygon D with $N = 3$ sides denoted by C_1, C_2 and C_3

$$f(z) = \frac{1}{2\pi \mathrm{i}} \sum_{j=1}^{N} \left\{ \int_{L_1} \frac{\rho_{jj}(k)}{1 - e^{2\pi \mathrm{i}k}} \left[\frac{z - \delta_j}{q_j} \right]^k dk + \int_{L_2} \rho_{jj}(k) \left[\frac{z - \delta_j}{q_j} \right]^k dk \right.$$
$$\left. + \int_{L_3} \frac{\rho_{jj}(k) e^{2\pi \mathrm{i}k}}{1 - e^{2\pi \mathrm{i}k}} \left[\frac{z - \delta_j}{q_j} \right]^k dk \right\}, \tag{29}$$

where we have swapped the order of integration and introduced the N-by-N *spectral matrix*

$$\rho_{mn}(k) \equiv \frac{1}{q_m} \int_{C_n} \left[\frac{z' - \delta_m}{q_m} \right]^{-k-1} f(z') dz'. \tag{30}$$

Global relations for this system are

$$\sum_{n=1}^{N} \rho_{mn}(k) = 0, \qquad k \in -\mathbb{N} \tag{31}$$

for any $m = 1, 2, \ldots, N$. There are N such global relations but each is an equivalent statement of the analyticity of $f(z)$ in the domain D. In this way, we have constructed the "tailor-made" transform pair (3) for a circular polygon.

4 Disc-in-Channel Geometry

The geometrical construction can be extended to *multiply connected* circular domains [3], and to domains whose boundaries are a combination of straight line and circular-arc edges. An example geometry, important in applications [4, 5, 13, 15, 16], is the disc-in-channel geometry of Fig. 2. The transform representation of a function $\hat{\chi}(z)$ that is analytic *and single-valued* in such a domain can be shown [3] to be

$$\hat{\chi}(z) = \underbrace{\frac{1}{2\pi}\int_0^{\infty} \rho_{11}(k)e^{ikz}dk + \frac{1}{2\pi}\int_0^{-\infty} \rho_{33}(k)e^{ikz}dk}_{\text{Fourier-type transform}}$$
$$\underbrace{-\frac{1}{2\pi i}\left\{\int_{L_1} \frac{\rho_{22}(k)}{1-e^{2\pi ik}}\frac{1}{z^{k+1}}dk + \int_{L_2} \rho_{22}(k)\frac{1}{z^{k+1}}dk + \int_{L_3} \frac{\rho_{22}(k)e^{2\pi ik}}{1-e^{2\pi ik}}\frac{1}{z^{k+1}}dk\right\}}_{\text{Mellin-type transform}}, \tag{32}$$

where the simultaneous appearance of both "Fourier-type" and "Mellin-type" contributions naturally reflects the hybrid geometry of the domain (and motivates the designation "Fourier-Mellin transforms" [2]). The elements of the spectral matrix are defined as follows:

$$\rho_{11}(k) = \int_{-\infty-il}^{+\infty-il} \hat{\chi}(z)e^{-ikz}dz = \rho_{31}(k) \qquad \rho_{22}(k) = -\oint_{|z|=1} \hat{\chi}(z)z^k dz, \tag{33}$$

and

$$\rho_{33}(k) = \int_{\infty+il}^{-\infty+il} \hat{\chi}(z)e^{-ikz}dz = \rho_{13}(k), \tag{34}$$

with

$$\rho_{21}(k) = \int_{-\infty-il}^{+\infty-il} \hat{\chi}(z)z^k dz, \qquad \rho_{23}(k) = \int_{\infty+il}^{-\infty+il} \hat{\chi}(z)z^k dz, \tag{35}$$

and

$$\rho_{12}(k) = \rho_{32}(k) = -\oint_{|z|=1} \hat{\chi}(z)e^{-ikz}dz. \tag{36}$$

The functions appearing in the spectral matrix satisfy the *global relations*

$$\begin{aligned}\rho_{11}(k) + \rho_{12}(k) + \rho_{13}(k) &= 0, \qquad k \in \mathbb{R},\\ \rho_{31}(k) + \rho_{32}(k) + \rho_{33}(k) &= 0, \qquad k \in \mathbb{R},\end{aligned} \tag{37}$$

which are equivalent, and

$$\rho_{21}(k) + \rho_{22}(k) + \rho_{23}(k) = 0, \qquad k \in -\mathbb{N}. \tag{38}$$

As discussed in [3], the doubly connected nature of the domain means that both (37) and (38) must be analysed to find the unknown spectral functions.

5 Application to Conformal Mapping

Suppose an application demands the solution for a harmonic field satisfying some boundary value problem in the disc-in-channel geometry of Fig. 1. If that boundary value problem is conformally invariant one might think to make use of conformal mapping. Even so, this is more easily said than done. The domain is doubly connected so a suitable pre-image domain is some annulus $\rho < |\zeta| < 1$ in a parametric complex ζ-plane (see Fig. 6); the conformal modulus ρ must be found as part of the construction of the conformal mapping. The domain is also a circular-arc domain and the construction of conformal mappings from the unit disc, say, to simply connected circular-arc domains is treated in standard texts [1, 14]. The extension of that theory to doubly connected domains (relevant to this example) has been presented more recently by Crowdy and Fokas [6], with the extension to arbitrary multiply connected domains given by Crowdy, Fokas and Green [7]. A well-known difficulty in all these conformal mapping constructions is solving for the *accessory parameters* [1, 6, 7, 14]. In this example ρ is one such accessory parameter.

We now show how that accessory parameter problem can be conveniently solved—linearized, in fact—by the generalized Fourier-Mellin transform pairs derived earlier. The main idea is to use the transform method to construct not the conformal mapping $z = z(\zeta)$, say, from the annulus $\rho < |\zeta| < 1$ to the given disc-in-channel geometry, but its inverse, which we denote by $\zeta = \zeta(z)$. Actually, the latter function is often more useful in applications since if a conformally invariant boundary value problem can be solved in the more convenient annulus geometry $\rho < |\zeta| < 1$ then knowledge of the transformation $\zeta = \zeta(z)$ allows immediate solution of the boundary value problem in the original disc-in-channel geometry.

To proceed with the construction we define the subsidiary function

$$\chi(z) \equiv \log \zeta(z). \tag{39}$$

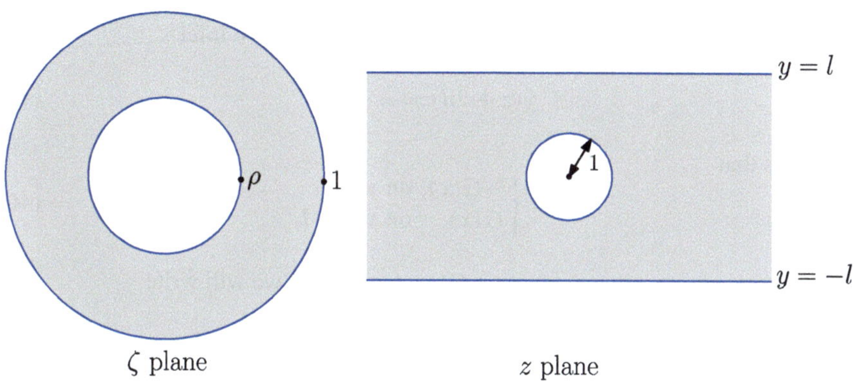

Fig. 6 Conformal mapping problem: to construct the mapping $\zeta = \zeta(z)$ from the disc-in-channel geometry to the concentric annulus $\rho < |\zeta| < 1$

It is reasonable to assume, on grounds of symmetry, that $\zeta = 1$ maps to the end of the channel as $x \to \infty$ and $\zeta = -1$ maps to $x \to -\infty$. We can write

$$\chi(z) = \log \tanh\left(\frac{\pi z}{4l}\right) + \hat{\chi}(z), \tag{40}$$

where $\hat{\chi}(z) \to 0$ as $x \to \pm\infty$. The first term is just the logarithm of the conformal mapping of the channel (without the circular hole) to a unit disc; the latter is easily derived using elementary considerations (e.g., the classical Schwarz-Christoffel formula [1, 8]). Since the boundary conditions on $\chi(z)$ are

$$\text{Re}[\chi(z)] = \begin{cases} 0, & \text{on } y = \pm l, \\ \log\rho, & \text{on } |z| = 1, \end{cases} \tag{41}$$

then, on use of (40), the following boundary conditions on $\hat{\chi}(z)$ pertain:

$$\text{Re}[\hat{\chi}(z)] = \begin{cases} \log\left|\coth\left(\frac{\pi z}{4l}\right)\right|, & \text{on } y = \pm l, \\ \\ \log\rho + \log\left|\coth\left(\frac{\pi z}{4l}\right)\right|, & \text{on } |z| = 1. \end{cases} \tag{42}$$

Recall that ρ is not known in advance and must be found.

By the symmetries of the proposed mapping between regions we expect that if a point $\zeta = e^{\mathrm{i}\theta}$ on the upper-half unit circle corresponds to $z = x + \mathrm{i}l$ then the point $\overline{\zeta}$ will correspond to $z = x - \mathrm{i}l$. This means that, for each x,

$$\chi(x + \mathrm{i}l) = \log\zeta = \mathrm{i}\theta = -\chi(x - \mathrm{i}l), \tag{43}$$

implying the relation

$$\chi(z + 2\mathrm{i}l) = -\chi(z). \tag{44}$$

Since the first term in (40) also satisfies this identity then we infer

$$\hat{\chi}(z + 2\mathrm{i}l) = -\hat{\chi}(z). \tag{45}$$

It follows that

$$\hat{\chi} = \begin{cases} -G(x), & \text{on } y = l, \\ G(x), & \text{on } y = -l, \end{cases} \tag{46}$$

for some (purely imaginary) function $G(x)$. On $|z| = 1$ we will write

$$\hat{\chi}(z) = r(z) + \mathrm{i}H(z) \tag{47}$$

where

$$r(z) = \log \rho + \log \left| \coth \left(\frac{\pi z}{4l} \right) \right|, \qquad H(z) = a_0 + \sum_{m \geq 1} a_m z^m + \frac{\overline{a_m}}{z^m}, \tag{48}$$

and where the coefficients $a_0 \in \mathbb{R}$, $\{a_m \in \mathbb{C} | m \geq 1\}$ are to be found. We also expect, on grounds of symmetry,

$$\overline{\chi(\overline{z})} = \chi(z), \qquad \text{on } \overline{z} = z \tag{49}$$

implying that

$$\overline{r}(z) - \mathrm{i}\overline{H}(z) = r(z) + \mathrm{i}H(z), \quad \text{or} \quad \overline{H}(z) = -H(z). \tag{50}$$

This condition implies $a_0 = 0$ and

$$a_n = \mathrm{i}b_n, \tag{51}$$

for some real set $\{b_n\}$. We will make use of these facts later.

The key observation is this: the function $\hat{\chi}(z)$ is single-valued and analytic in the fluid region D. It therefore has a transform representation of the form (32). To find it, we must determine the unknown spectral functions. This can be done by analysis of the global relations (37) and (38).

Now (37) and (38) give, respectively,

$$\int_{-\infty}^{\infty} G(x) e^{-\mathrm{i}kx} \left[e^{-kl} + e^{kl} \right] dx - \oint_{|z|=1} e^{-\mathrm{i}kz} \hat{\chi}(z) dz = 0, \quad k \in \mathbb{R}, \tag{52}$$

$$\int_{-\infty}^{\infty} G(x) \left[\frac{1}{(x - \mathrm{i}l)^n} + \frac{1}{(x + \mathrm{i}l)^n} \right] dx - \oint_{|z|=1} \hat{\chi}(z) \frac{dz}{z^n} = 0, \quad n \in \mathbb{N}. \tag{53}$$

Equation (52) implies that

$$2 \cosh(kl) \mathcal{G}(k) = B(k) + R_1(k), \tag{54}$$

where

$$\mathcal{G}(k) \equiv \int_{-\infty}^{\infty} G(x) e^{-\mathrm{i}kx} dx \tag{55}$$

and

$$B(k) \equiv \oint_{|z|=1} \mathrm{i}H(z) e^{-\mathrm{i}kz} dz, \qquad R_1(k) \equiv \oint_{|z|=1} r(z) e^{-\mathrm{i}kz} dz. \tag{56}$$

It follows that

$$\mathcal{G}(k) = \frac{B(k) + R_1(k)}{2\cosh(kl)} \tag{57}$$

and the inverse Fourier transform provides

$$G(x) = \frac{1}{2\pi}\int_{-\infty}^{\infty}\left[\frac{B(k) + R_1(k)}{2\cosh(kl)}\right]e^{\mathrm{i}kx}dk. \tag{58}$$

The second global relation (53) implies that, for $n \in \mathbb{N}$,

$$\oint_{|z|=1}\frac{\mathrm{i}H(z)}{z^n}dz + R_2(n-1) = \int_{-\infty}^{\infty}G(x)\left[\frac{1}{(x-\mathrm{i}l)^n} + \frac{1}{(x+\mathrm{i}l)^n}\right]dx, \tag{59}$$

where we define

$$R_2(n) \equiv \oint_{|z|=1}\frac{r(z)}{z^{n+1}}dz. \tag{60}$$

It is easy to show that for $n \geq 1$,

$$\oint_{|z|=1}\frac{\mathrm{i}H(z)}{z^n}dz = -2\pi a_{n-1}. \tag{61}$$

Equation (59) then implies that

$$a_{n-1} = -\frac{1}{2\pi}\int_{-\infty}^{\infty}G(x)\left[\frac{1}{(x-\mathrm{i}l)^n} + \frac{1}{(x+\mathrm{i}l)^n}\right]dx + \frac{R_2(n-1)}{2\pi}. \tag{62}$$

On substitution of (58) for $G(x)$, we find

$$a_{n-1} = \int_{-\infty}^{\infty}J(k,n-1)\left[\frac{B(k) + R_1(k)}{2\cosh(kl)}\right]dk + \frac{R_2(n-1)}{2\pi}, \qquad n \geq 1, \tag{63}$$

where we define

$$J(k,n) \equiv -\frac{1}{4\pi^2}\int_{-\infty}^{\infty}e^{\mathrm{i}kx}\left[\frac{1}{(x-\mathrm{i}l)^{n+1}} + \frac{1}{(x+\mathrm{i}l)^{n+1}}\right]dx. \tag{64}$$

Some residue calculus reveals that for $n \geq 1$,

$$J(k,n) = \begin{cases} -\dfrac{\mathrm{i}e^{-kl}(\mathrm{i}k)^n}{2\pi n!}, & k \geq 0, \\[2ex] \dfrac{\mathrm{i}e^{kl}(\mathrm{i}k)^n}{2\pi n!}, & k < 0, \end{cases} \tag{65}$$

and, for $n = 0$,

$$J(k,0) = \begin{cases} -\dfrac{\mathrm{i}e^{-kl}}{2\pi}, & k > 0, \\ 0, & k = 0, \\ \dfrac{\mathrm{i}e^{kl}}{2\pi}, & k < 0. \end{cases} \tag{66}$$

It follows that, for $n \geq 1$,

$$a_{n-1} = \int_{-\infty}^{\infty} \frac{J(k,n-1)B(k)}{2\cosh(kl)} dk + \int_{-\infty}^{\infty} \frac{J(k,n-1)R_1(k)}{2\cosh(kl)} dk + \frac{R_2(n-1)}{2\pi}. \tag{67}$$

But

$$\begin{aligned} B(k) &= \oint_{|z|=1} e^{-\mathrm{i}kz}\mathrm{i}H(z)dz = \oint_{|z|=1} \mathrm{i}e^{-\mathrm{i}kz} \left[a_0 + \sum_{m\geq 1} a_m z^m + \frac{\overline{a_m}}{z^m} \right] dz \\ &= -2\pi \sum_{m\geq 1} \frac{\overline{a_m}(-\mathrm{i}k)^{m-1}}{(m-1)!}. \end{aligned} \tag{68}$$

Hence

$$a_{n-1} + \sum_{m\geq 1} A_{n-1,m}\overline{a_m} = E_{n-1}, \qquad n \geq 1, \tag{69}$$

where

$$\begin{aligned} A_{n,m} &= \pi \int_{-\infty}^{\infty} \frac{J(k,n)(-\mathrm{i}k)^{m-1}}{(m-1)!\cosh(kl)} dk, \\ E_n &= \int_{-\infty}^{\infty} \frac{J(k,n)R_1(k)}{2\cosh(kl)} dk + \frac{R_2(n)}{2\pi}. \end{aligned} \tag{70}$$

Finally, on use of (51), system (69) becomes the system of real equations

$$\sum_{m\geq 1} A_{0m} b_m = \mathrm{i}E_0, \tag{71}$$

$$b_n - \sum_{m\geq 1} A_{nm} b_m = -\mathrm{i}E_n, \qquad n \geq 1. \tag{72}$$

It can be checked easily that E_0 is the only quantity that depends on the unknown $\log \rho$. We can therefore solve (72) for the set of coefficients $\{b_n | n \geq 1\}$ and then use (71) *a posteriori* to determine $\log \rho$. With the coefficients determined from this simple linear system all the spectral functions needed in the representation (32) of the required function $\hat{\chi}(z)$ can be found. The required inverse conformal mapping $\zeta(z)$ then follows.

The construction above was originally presented in an appendix to [5] where it was used as a check on a solution given there.

6 Summary

It is hoped that this article provides a useful overview of recent developments concerning these "geometry-fitted" Fourier-Mellin transform pairs, and how to make constructive use of them. The author has recently employed the new transform pairs described here in a number of different applications [2–5] with much earlier work on solving various PDEs in convex polygons carried out by other authors [11, 12]. We believe that the full scope and implications of the method for applications, and its various extensions, have yet to be explored.

Acknowledgments The author acknowledges the support of an Established Career Fellowship from the Engineering and Physical Sciences Research Council in the UK (EP/K019430/1) and a Wolfson Research Merit Award from the Royal Society.

References

1. Ablowitz, M., Fokas, A.S.: Complex Variables and Applications. Cambridge University Press (1997)
2. Crowdy, D.G.: Fourier-Mellin transforms for circular domains. Comput. Meth. Funct. Theory **15**(4), 655–687 (2015)
3. Crowdy, D.G.: A transform method for Laplace's equation in multiply connected circular domains. IMA J. Appl. Math. **80**(6), 1902–1931 (2015)
4. Crowdy, D.G.: Effective slip lengths for longitudinal shear flow over partial-slip circular bubble mattresses. Fluid Dyn. Res. **47**, 065507 (2015)
5. Crowdy, D.G.: Uniform flow past a periodic array of cylinders. Eur. J. Mech. B/Fluids **56**, 120–129 (2015)
6. Crowdy, D.G., Fokas, A.S.: Conformal mappings to a doubly connected polycircular arc domain. Proc. R. Soc. A **463**, 1885–1907 (2007)
7. Crowdy, D.G., Fokas, A.S., Green, C.C.: Conformal mappings to multiply connected polycircular arc domains. Comput. Meth. Funct. Theory **11**(2), 685–706 (2011)
8. Crowdy, D.G.: The Schwarz-Christoffel mapping to bounded multiply connected polygonal domains. Proc. R. Soc. A **461**, 2653–2678 (2005)
9. Crowdy, D.G., Fokas, A.S.: Explicit integral solutions for the plane elastostatic semi-strip. Proc. R. Soc. A **460**, 1285–1310 (2004)
10. Fokas, A.S., Kapaev, A.A.: On a transform method for the Laplace equation in a polygon. IMA J. Appl. Math. **68**, 355–408 (2003)
11. Fokas, A.S.: A unified approach to boundary value problems. In: CBMS-NSF Regional Conference Series in Applied Mathematics, No 78. SIAM, Philadelphia (2008)
12. Fokas, A.S., Pelloni, B. (eds.): Unified Transform for Boundary Value Problems: Applications and Advances. SIAM Monographs, Philadelphia (2015)

13. Martin, P.A., Dalrymple, R.A.: Scattering of long waves by cylindrical obstacles and gratings using matched asymptotic expansions. J. Fluid Mech. **188**, 465–490 (1988)
14. Nehari, Z.: Conformal Mapping. Dover Publications, New York (2011)
15. Poritsky, H.: Potential of a charged cylinder between two parallel grounded planes. J. Math. Phys. **39**, 35–48 (1960)
16. Richmond, H.W.: On the electrostatic field of a plane or circular grating formed of thick rounded bars. Proc. Lond. Math. Soc. **22**(2), 389–403 (1923)

First Order Approach to L^p Estimates for the Stokes Operator on Lipschitz Domains

Alan McIntosh and Sylvie Monniaux

Abstract This paper concerns Hodge-Dirac operators $D_H = d + \delta$ acting in $L^p(\Omega, \Lambda)$ where Ω is a bounded open subset of $\mathbb{R}^n$ satisfying some kind of Lipschitz condition, Λ is the exterior algebra of $\mathbb{R}^n$, d is the exterior derivative acting on the de Rham complex of differential forms on Ω, and δ is the interior derivative with tangential boundary conditions. In $L^2(\Omega, \Lambda)$, $d' = \delta$ and D_H is self-adjoint, thus having bounded resolvent $\{(I + \mathrm{it}D_H)\}_{\{t\in\mathbb{R}\}}$ as well as a bounded functional calculus in $L^2(\Omega, \Lambda)$. We investigate the range of values $p_H < p < p^H$ about $p = 2$ for which D_H has bounded resolvents and a bounded holomorphic functional calculus in $L^p(\Omega, \Lambda)$.

Keywords Hodge-Dirac operator · Lipschitz domains · Stokes operator · First order approach · Hodge boundary conditions

1 Introduction

At the ISAAC meeting in Macau, the first author discussed the harmonic analysis of first order systems on bounded domains, with particular reference to his current joint research with the second author concerning the L^p theory of Hodge-Dirac operators on Lipschitz domains, with implications for the Stokes' operator on such domains with Hodge boundary conditions. In this article, we present an overview of this material, staying with the three dimensional situation. Full definitions and proofs in higher dimensions can be found in [14]. In other papers with Marius Mitrea, the second author has pursued applications to the Navier-Stokes equation on Lipschitz

A. McIntosh
Mathematical Sciences Institute, Australian National University,
Canberra ACT 2601, Australia
e-mail: alan.mcintosh@anu.edu.au

S. Monniaux (✉)
I2M, UMR 7373, Centrale Marseille,
Aix-Marseille Université, CNRS, 13453, Marseille, France
e-mail: sylvie.monniaux@univ-amu.fr

T. Qian and L.G. Rodino (eds.), *Mathematical Analysis, Probability and Applications – Plenary Lectures*, Springer Proceedings in Mathematics & Statistics 177, DOI 10.1007/978-3-319-41945-9_3

domains. We will not comment further on that here, except to mention that the non-linear applications depend on having results for the linear Stokes operator in the case $p = 3$ or possibly $p = 3/2$ (the dual exponent to 3).

Harmonic analysis $\Longrightarrow$ *First order systems*

Hodge-Dirac systems on bounded domains

$\Downarrow$

Potential Application $\Longleftarrow$ *Second order equations*

Navier-Stokes equation with Hodge boundary conditions

Hodge-Laplacian; Hodge-Stokes Operator

2 Hodge-Dirac operators

Our aim is to investigate the L^p theory of the first order *Hodge-Dirac operator*

$$D_H = d_\Omega + \delta_{\overline{\Omega}}$$

acting on a *bounded domain* $\Omega \subset \mathbb{R}^3$ satisfying some kind of *Lipschitz condition*.

Here d_Ω is the *exterior derivative* acting on *differential forms* in $L^p(\Omega, \Lambda)$, and $\delta_{\overline{\Omega}}$ is the adjoint operator which includes the *tangential boundary condition*

$$\nu \lrcorner u|_{\partial\Omega} = 0$$

i.e. the *normal component* of u at the boundary $\partial\Omega$ is zero, at least on that part of the boundary where it is well defined. This is effectively half a boundary condition for D_H, which is what is expected for a first order system.

Let us now define our terms.

3 Lipschitz domains

Henceforth Ω denotes a *bounded connected open subset* of $\mathbb{R}^3$, and B denotes the unit ball in $\mathbb{R}^3$. We say that

- Ω is *very weakly Lipschitz* if $\Omega = \cup_{j=1}^N (\rho_j B)$ for some natural number N, where each map $\rho_j : B \to \rho_j B \subset \mathbb{R}^3$ is uniformly locally bilipschitz, and

$1 = \sum_{j=1}^{N} \chi_j$ on Ω, where each $\chi_j : \Omega \to [0, 1]$ is a Lipschitz function with $\mathrm{sppt}_\Omega(\chi_j) \subset \rho_j B$;

- Ω is *strongly Lipschitz* if, locally, the boundary $\partial\Omega$ of Ω is a portion of the graph of a Lipschitz function $g : \mathbb{R}^2 \to \mathbb{R}$ (with respect to some rotated coordinate system), with Ω being to one side of the graph;
- Ω is *smooth* if each such function g is smooth.

In the above, $\mathrm{sppt}_\Omega(\chi_j)$ denotes the closure of $\{x \in \Omega\,;\ \chi_j(x) \neq 0\}$ in Ω.

Every strongly Lipschitz domain is weakly Lipschitz (which we shall not discuss further, but refer the reader to [5]) and every weakly Lipschitz domain is very weakly Lipschitz. A weakly Lipschitz domain which is not strongly Lipschitz is the well known two brick domain (consisting of one brick on top of another, pointing in orthogonal directions), and a very weakly Lipschitz domain which is not weakly Lipschitz is the unit ball with the half-disk $\{(x_1, x_2, x_3) \in B\,;\ x_3 = 0, x_1 > 0\}$ removed.

In a strongly Lipschitz domain (and indeed in a weakly Lipschitz domain), there is a well-defined outward-pointing unit normal $\nu(y)$ for almost every $y \in \partial\Omega$. In fact $\nu \in L^\infty(\partial\Omega; \mathbb{R}^3)$. As can be seen from the above example, the unit normal is not necessarily defined on the whole boundary of a very weakly Lipschitz domain.

4 Exterior Algebra

- The *exterior algebra* on $\mathbb{R}^3$ with basis e_1, e_2, e_3 is

$$
\begin{aligned}
&\Lambda = \Lambda^0 \oplus \Lambda^1 \oplus \Lambda^2 \oplus \Lambda^3 \approx \mathbb{C} \oplus \mathbb{C}^3 \oplus \mathbb{C}^3 \oplus \mathbb{C} \\
&u = u^0 + u^1 + u^2 + u^3 \qquad \text{where} \\
&\Lambda^0 = \mathbb{C} \\
&\Lambda^1 = \mathbb{C}^3 : \qquad u^1 = u^1_1 e_1 + u^1_2 e_2 + u^1_3 e_3 \\
&\Lambda^2 \approx \mathbb{C}^3 : \qquad u^2 = u^2_{2,3}\, e_2 \wedge e_3 + u^2_{3,1} e_3 \wedge e_1 + u^2_{1,2}\, e_1 \wedge e_2 \\
&\Lambda^3 \approx \mathbb{C} : \qquad u^3 = u^3_{1,2,3}\, e_1 \wedge e_2 \wedge e_3 \qquad (e_k \wedge e_j = -e_j \wedge e_k)
\end{aligned}
$$

- $L^p(\Omega, \Lambda) = L^p(\Omega, \mathbb{C}) \oplus L^p(\Omega, \mathbb{C}^3) \oplus L^p(\Omega, \mathbb{C}^3) \oplus L^p(\Omega, \mathbb{C})$
- If $a = \sum_j a_j e_j \in \mathbb{R}^3$, $u \in \Lambda^\ell$, then $a \wedge u = \sum_j a_j e_j \wedge u \in \Lambda^{\ell+1}$
- If also $v \in \Lambda^{\ell+1}$ then $a \lrcorner v \in \Lambda^\ell$ and $\langle a \wedge u\,,\ v\rangle = \langle u\,,\ a \lrcorner v\rangle$
- $du = \nabla \wedge u = \sum_j e_j \wedge \partial_j u$, $\delta u = -\nabla \lrcorner u = -\sum_j e_j \lrcorner \partial_j u$
- The *exterior product* $\wedge$ and the *contraction* $\lrcorner$ can be represented by scalar multiplication, dot products and cross products.

5 The de Rham Complex on $\Omega \subset \mathbb{R}^3$

Suppose that Ω denotes a bounded open subset of $\mathbb{R}^3$ and $1 < p < \infty$.

The *exterior derivative* d_Ω defined on Ω can be expressed as follows:

$$d_\Omega : 0 \to L^p(\Omega, \mathbb{C}) \xrightarrow{\nabla_\Omega} L^p(\Omega, \mathbb{C}^3) \xrightarrow{\mathrm{curl}_\Omega} L^p(\Omega, \mathbb{C}^3) \xrightarrow{\mathrm{div}_\Omega} L^p(\Omega, \mathbb{C}) \to 0$$

(noting that curl is sometimes written as rot or $\nabla \times$, and div as $\nabla.$).

As an operator, $d_\Omega : \mathcal{D}^p(d_\Omega) \to L^p(\Omega, \Lambda)$ is an unbounded operator with *domain* $\mathcal{D}^p(d_\Omega) = \{u \in L^p(\Omega, \Lambda)\,;\ d_\Omega u \in L^p(\Omega, \Lambda)\}$.

Note that ${d_\Omega}^2 = 0$ because $\mathrm{curl}_\Omega \nabla_\Omega = 0$ and $\mathrm{div}_\Omega \mathrm{curl}_\Omega = 0$, or as we can see directly, ${d_\Omega}^2 u = \sum_{j,k} e_j \wedge e_k \partial_j \partial_k u = 0$ by the skew-symmetry of the wedge product.

Hence the *range* of d_Ω is contained in the *null-space* of d_Ω, i.e. $\mathcal{R}^p(d_\Omega) \subset \mathcal{N}^p(d_\Omega)$ where $\mathcal{R}^p(d_\Omega) = \{v \in L^p(\Omega, \Lambda)\,;\ v = d_\Omega u$ for some $u \in \mathcal{D}^p(d_\Omega)\}$ and $\mathcal{N}^p(d_\Omega) = \{u \in \mathcal{D}^p(d_\Omega)\,;\ d_\Omega u = 0\}$.

If Ω is *very weakly Lipschitz*, then $\overline{\mathcal{R}^p(d_\Omega)} = \mathcal{R}^p(d_\Omega)$ and the codimension of $\mathcal{R}^p(d_\Omega)$ in $\mathcal{N}^p(d_\Omega)$ is finite dimensional. We return to these facts in Sect. 19.

6 The Dual de Rham Complex

With Ω and p as above, let $q = p'$ (i.e. $\frac{1}{p} + \frac{1}{q} = 1$).

The dual of the exterior derivative $d_\Omega : \mathcal{D}^q(d_\Omega) \to L^q(\Omega, \Lambda)$:

$$d_\Omega : 0 \to L^q(\Omega, \mathbb{C}) \xrightarrow{\nabla_\Omega} L^q(\Omega, \mathbb{C}^3) \xrightarrow{\mathrm{curl}_\Omega} L^q(\Omega, \mathbb{C}^3) \xrightarrow{\mathrm{div}_\Omega} L^q(\Omega, \mathbb{C}) \to 0$$

is $\delta_{\overline{\Omega}} : \mathcal{D}^p(\delta_{\overline{\Omega}}) \to L^p(\Omega, \Lambda)$:

$$0 \leftarrow L^p(\Omega, \mathbb{C}) \xleftarrow{-\mathrm{div}_{\overline{\Omega}}} L^p(\Omega, \mathbb{C}^3) \xleftarrow{\mathrm{curl}_{\overline{\Omega}}} L^p(\Omega, \mathbb{C}^3) \xleftarrow{-\nabla_{\overline{\Omega}}} L^p(\Omega, \mathbb{C}) \leftarrow 0 : \delta_{\overline{\Omega}}$$

where the domain $\mathcal{D}^p(\delta_{\overline{\Omega}})$ is the completion of $C_c^\infty(\Omega, \Lambda)$ in the graph norm $\|u\|_p + \|\delta_{\overline{\Omega}} u\|_p$.

Again, ${\delta_{\overline{\Omega}}}^2 = 0$, i.e. $\mathcal{R}^p(\delta_{\overline{\Omega}}) \subset \mathcal{N}^p(\delta_{\overline{\Omega}})$.

If Ω is *very weakly Lipschitz*, then $\overline{\mathcal{R}^p(\delta_{\overline{\Omega}})} = \mathcal{R}^p(\delta_{\overline{\Omega}})$, with finite codimension in $\mathcal{N}^p(\delta_{\overline{\Omega}})$.

If Ω is *strongly Lipschitz*, then the *normal component* of $u \in \mathcal{D}^p(\delta_{\overline{\Omega}})$ at the boundary is *zero*, i.e.

$$\mathcal{D}^p(\delta_{\overline{\Omega}}) = \{u \in L^p(\Omega, \Lambda)\,;\ \delta_\Omega u \in L^p(\Omega, \Lambda), \nu \lrcorner u|_{\partial\Omega} = 0\}\,.$$

Remark 6.1 The condition $\nu \lrcorner u|_{\partial\Omega} = 0$ is to be understood in the following sense: for $u \in L^p(\Omega, \Lambda)$ such that $\delta_\Omega u \in L^p(\Omega, \Lambda)$ in a strongly Lipschitz domain, the normal component at the boundary $\nu \lrcorner u|_{\partial\Omega}$ is defined as a functional on traces of differential forms $v \in W^{1,p'}(\Omega, \Lambda)$ (where $\frac{1}{p'} + \frac{1}{p} = 1$) by the integration by parts formula:

$$\langle \nu \lrcorner u, v\rangle_{\partial\Omega} = \langle u, dv\rangle_\Omega - \langle \delta u, v\rangle_\Omega.$$

Since $Tr_{|\partial\Omega}\big(W^{1,p'}(\Omega, \Lambda)\big) \subseteq B^{p',p'}_{1/p}(\partial\Omega, \Lambda)$, we obtain that $\nu \lrcorner u \in B^{p,p}_{-1/p}(\partial\Omega, \Lambda)$. For more details, we refer to [16, Sect. 2.3].

Remark 6.2 Some care needs to be taken when consulting references, in that different authors use different sign conventions for δ and Δ.

Remark 6.3 The definitions and results concerning very weakly Lipschitz domains in $\mathbb{R}^3$ can be adapted to domains in a Riemannian manifold with very little effort.

7 Hypothesis

For the rest of this article, Ω denotes a *very weakly Lipschitz* **domain in $\mathbb{R}^3$.**

8 The Hodge-Dirac Operator $D_H = d_\Omega + \delta_{\overline{\Omega}}$ in $L^2(\Omega, \Lambda)$

First we consider the case $p = 2$. Then the exterior derivative d_Ω and adjoint interior derivative $\delta_{\overline{\Omega}}$ are unbounded operators in $L^2(\Omega, \Lambda)$ which satisfy

- ${d_\Omega}^2 = 0\,,\quad {\delta_{\overline{\Omega}}}^2 = 0\,,\quad {d_\Omega}^* = \delta_{\overline{\Omega}}\,,\quad {\delta_{\overline{\Omega}}}^* = d_\Omega.$

In $L^2(\Omega, \Lambda)$, define the *Hodge-Dirac operator with tangential boundary condition* $D_H := d_\Omega + \delta_{\overline{\Omega}}$ with $\mathcal{D}^2(D_H) = \mathcal{D}^2(d_\Omega) \cap \mathcal{D}^2(\delta_{\overline{\Omega}})$. It is straightforward to check the following properties (using the properties of d_Ω and $\delta_{\overline{\Omega}}$ just described)

- The *Hodge-Dirac operator* $D_H = d_\Omega + \delta_{\overline{\Omega}}$ is self-adjoint in $L^2(\Omega, \Lambda)$;
- $\mathcal{N}^2(D_H) = \mathcal{N}^2(d_\Omega) \cap \mathcal{N}^2(\delta_{\overline{\Omega}})$ is finite dimensional;
- The *Hodge decomposition* of $L^2(\Omega, \Lambda)$ takes the form

$$\begin{array}{rcl} L^2(\Omega, \Lambda) &=& \mathcal{N}^2(d_\Omega) \overset{\perp}{\oplus} \mathcal{R}^2(\delta_{\overline{\Omega}}) \\ && \quad\;\;\cup \qquad\quad \cap \\ L^2(\Omega, \Lambda) &=& \mathcal{R}^2(d_\Omega) \overset{\perp}{\oplus} \mathcal{N}^2(\delta_{\overline{\Omega}}) \qquad \text{and so} \\ L^2(\Omega, \Lambda) &=& \mathcal{R}^2(d_\Omega) \overset{\perp}{\oplus} \mathcal{R}^2(\delta_{\overline{\Omega}}) \overset{\perp}{\oplus} \mathcal{N}^2(D_H)\,. \end{array}$$

- In particular, on restricting to the space of square integrable vector fields, $L^2(\Omega, \Lambda^1) = L^2(\Omega, \mathbb{C}^3)$, we have

$$
\begin{array}{l}
L^2(\Omega,\Lambda^1) = \mathcal{N}^2(\mathrm{curl}_\Omega) \overset{\perp}{\oplus} \mathcal{R}^2(\mathrm{curl}_{\overline{\Omega}}) \\
\qquad\qquad\qquad\quad \cup \qquad\qquad \cap \\
L^2(\Omega,\Lambda^1) = \mathcal{R}^2(\nabla_\Omega) \overset{\perp}{\oplus} \boxed{\mathcal{N}^2(\mathrm{div}_{\overline{\Omega}}) = \mathcal{H}^2} \qquad \text{and so} \\
L^2(\Omega,\Lambda^1) = \mathcal{R}^2(\nabla_\Omega) \overset{\perp}{\oplus} \mathcal{R}^2(\mathrm{curl}_{\overline{\Omega}}) \overset{\perp}{\oplus} \mathcal{N}^2(D_H)
\end{array}
$$

where $\mathcal{H}^2 := \mathcal{N}^2(\mathrm{div}_{\overline{\Omega}}) \subset L^2(\Omega, \Lambda^1)$.

In the case when Ω is strongly Lipschitz, $\mathcal{H}^2$ is the space of divergence-free square integrable vector fields which satisfy the tangential boundary condition $\nu.u|_{\partial\Omega} = 0$.

9 The Hodge-Laplacian $-\Delta_H = {D_H}^2$

In $L^2(\Omega, \Lambda)$, define the *Hodge-Laplacian* $-\Delta_H := {D_H}^2 = d_\Omega\delta_{\overline{\Omega}} + \delta_{\overline{\Omega}} d_\Omega$ with $\mathcal{D}^2(\Delta_H) = \mathcal{D}^2(d_\Omega\delta_{\overline{\Omega}}) \cap \mathcal{D}^2(\delta_{\overline{\Omega}} d_\Omega)$. This is called the Hodge-Laplacian with absolute or generalised boundary conditions. We remark that Δ_H has the sign convention $\Delta_H u = {\partial_1}^2 u + {\partial_2}^2 u + {\partial_3}^2 u$, $u \in \mathcal{D}^2(\Delta_H)$.

It is straightforward to check the following properties:

- The *Hodge-Laplacian* $-\Delta_H = d_\Omega\delta_{\overline{\Omega}} + \delta_{\overline{\Omega}} d_\Omega$ is nonnegative self-adjoint in $L^2(\Omega, \Lambda)$;
- $\mathcal{N}^2(\Delta_H) = \mathcal{N}^2(D_H) = \mathcal{N}^2(d_\Omega) \cap \mathcal{N}^2(\delta_{\overline{\Omega}})$;
- The Hodge-Laplacian preserves each of the spaces $L^2(\Omega, \Lambda^k)$, $0 \leq k \leq 3$, and so splits as a direct sum of its restrictions to these spaces, as can be seen from the expression $-\Delta_H = d_\Omega\delta_{\overline{\Omega}} + \delta_{\overline{\Omega}} d_\Omega$ with

$$
d_\Omega : 0 \underset{\leftarrow}{\rightarrow} L^2(\Omega,\mathbb{C}) \underset{\underset{-\mathrm{div}_{\overline{\Omega}}}{\leftarrow}}{\overset{\nabla_\Omega}{\rightarrow}} \underset{\underset{\boxed{\mathcal{H}^2}}{\cup}}{L^2(\Omega,\mathbb{C}^3)} \underset{\underset{\mathrm{curl}_{\overline{\Omega}}}{\leftarrow}}{\overset{\mathrm{curl}_\Omega}{\rightarrow}} L^2(\Omega,\mathbb{C}^3) \underset{\underset{-\nabla_{\overline{\Omega}}}{\leftarrow}}{\overset{\mathrm{div}_\Omega}{\rightarrow}} L^2(\Omega,\mathbb{C}) \underset{\leftarrow}{\rightarrow} 0 : \delta_{\overline{\Omega}}
$$

$$
-\Delta_H = \underset{(= -\Delta_{\mathrm{Neumann}})}{-\mathrm{div}_{\overline{\Omega}}\nabla_\Omega} \oplus (-\nabla_\Omega \mathrm{div}_{\overline{\Omega}} + \mathrm{curl}_{\overline{\Omega}}\mathrm{curl}_\Omega) \oplus (\mathrm{curl}_\Omega \mathrm{curl}_{\overline{\Omega}} - \nabla_{\overline{\Omega}}\mathrm{div}_\Omega) \oplus \underset{(= -\Delta_{\mathrm{Dirichlet}})}{-\mathrm{div}_\Omega \nabla_{\overline{\Omega}}}
$$

Indeed it also preserves each component of the Hodge decomposition, in particular $\mathcal{H}^2 = L^2(\Omega, \Lambda^1) \cap \mathcal{N}^2(\delta_{\overline{\Omega}}) = \mathcal{N}^2(\mathrm{div}_{\overline{\Omega}})$.

10 The Hodge-Stokes Operator $S_H = -\Delta_H|_{\mathcal{H}^2}$

In $\mathcal{H}^2$, define the *Stokes operator with Hodge boundary conditions* by $S_H u = -\Delta_H u = \mathrm{curl}_{\overline{\Omega}}\mathrm{curl}_\Omega u$, $u \in \mathcal{H}^2$ (i.e. $\mathrm{div}_{\overline{\Omega}} u = 0$) with $\mathcal{D}^2(S_H) = \{u \in L^2(\Omega, \Lambda^1)$; $\mathrm{div}_{\overline{\Omega}} u = 0$, $\mathrm{curl}_{\overline{\Omega}}\mathrm{curl}_\Omega u \in L^2(\Omega, \Lambda^1)\}$. It is straightforward to check the following properties:

- The *Hodge-Stokes* operator $S_H = \mathrm{curl}_{\overline{\Omega}}\mathrm{curl}_\Omega$ is nonnegative self-adjoint in $\mathcal{H}^2$;
- $\mathcal{N}^2(S_H) = \mathcal{N}^2(D_H) \cap L^2(\Omega, \Lambda^1) = \mathcal{N}^2(\mathrm{curl}_\Omega) \cap \mathcal{N}^2(\mathrm{div}_{\overline{\Omega}})$ is finite dimensional;

If Ω is *strongly Lipschitz* and $u \in \mathcal{D}^2(S_H)$, then the *tangential boundary conditions* $\nu.u|_{\partial\Omega} = 0$; $\nu \times \mathrm{curl} u|_{\partial\Omega} = 0$ hold. See, e.g., [17, Sect. 3].

11 L^2 Results for D_H, Δ_H and S_H

To summarise, we have the following properties:

- $L^2(\Omega, \Lambda) = \mathcal{R}^2(d_\Omega) \overset{\perp}{\oplus} \mathcal{R}^2(\delta_{\overline{\Omega}}) \overset{\perp}{\oplus} \mathcal{N}^2(D_H)$;
- Hodge-Dirac operator $D_H = d_\Omega + \delta_{\overline{\Omega}}$ is self-adjoint in $L^2(\Omega, \Lambda)$;
- Hodge-Laplacian $-\Delta_H = {D_H}^2 = d_\Omega \delta_{\overline{\Omega}} + \delta_{\overline{\Omega}} d_\Omega$ is nonnegative self-adjoint in $L^2(\Omega, \Lambda)$;
- Hodge-Stokes operator $S_H = -\Delta_H|_{\mathcal{H}^2}$ is nonnegative self-adjoint in $\mathcal{H}^2 = \mathcal{N}^2(\mathrm{div}_{\overline{\Omega}})$.

So D_H, Δ_H, S_H all have *resolvent bounds*, e.g.

$$\begin{aligned}
\|(I + itD_H)^{-1}u\|_2 &\le \|u\|_2 \quad \forall u \in L^2(\Omega, \Lambda)\,, \quad \forall\, t \in \mathbb{R} \setminus \{0\}\\
\|(I - t^2\Delta_H)^{-1}u\|_2 &\le \|u\|_2 \quad \forall u \in L^2(\Omega, \Lambda)\,, \quad \forall\, t > 0\\
\|(I + t^2 S_H)^{-1}u\|_2 &\le \|u\|_2 \quad \forall u \in \mathcal{H}^2 \quad\quad\quad\;\; , \quad \forall\, t > 0
\end{aligned}$$

and all have *functional calculi* of self-adjoint operators, in particular

$$\|D_H u\|_2 = \|\,\mathrm{sgn}(D_H)\sqrt{-\Delta_H}\,u\|_2 = \|\sqrt{-\Delta_H}\,u\|_2 \quad \forall u \in \mathcal{D}^2(D_H) = \mathcal{D}^2(\sqrt{-\Delta_H})\,.$$

12 L^p Questions for D_H, Δ_H and S_H, $1 < p < \infty$

Whether or not the L^p versions of these properties hold, depends on Ω and p. Of course, we no longer have orthogonality of the Hodge decomposition, and the constants in the resolvent bounds and the functional calculi may depend on p. Allowing for this, when Ω is smooth, all of the properties hold for all $p \in (1, \infty)$.

In our situation, namely when Ω is a very weakly Lipschitz domain, we list the main properties and then discuss their relationship with one another, and conditions under which they hold.

(H_p) D_H has an L^p Hodge decomposition: $L^p(\Omega,\Lambda) = \mathcal{R}^p(d_\Omega) \oplus \mathcal{R}^p(\delta_{\overline{\Omega}}) \oplus \mathcal{N}^p(D_H)$;

(R_p) D_H is bisectorial in L^p, in particular $\|(I+itD_H)^{-1}u\|_p \le C\|u\|_p \quad \forall\, t \in \mathbb{R}\setminus\{0\}$;

(F_p) D_H has a bounded $H^\infty(S^o_\mu)$ functional calculus in $L^p(\Omega,\Lambda)$ for all $\mu > 0$: $\|f(D_H)u\|_p \le C_\mu \|f\|_\infty \|u\|_p \quad \forall f \in H^\infty(S^o_\mu)$, in particular, $\|D_H u\|_p \approx \|\sqrt{-\Delta_H}\, u\|_p$.

Here $S^o_\mu = \{z \in \mathbb{C};\ |\arg z| < \mu \text{ or } |\arg(-z)| < \mu\}$, $0 < \mu < \pi/2$.
Let us note that:

- $(F_p) \Longrightarrow (H_p)$: Exercise.
- $(F_p) \Longrightarrow (R_p) \Longrightarrow$ Hodge-Laplacian is sectorial in $L^p(\Omega,\Lambda)$, in particular Δ_H has the L^p resolvent bounds

$$\|(I - t^2\Delta_H)^{-1}u\|_p = \|(I+itD_H)^{-1}(I-itD_H)^{-1}u\|_p \le C^2\|u\|_p \quad \forall\, t > 0\,.$$

- $(F_p) \Longrightarrow$ Hodge-Laplacian has a bounded $H^\infty(S^o_\mu)$ functional calculus $\forall \mu > 0$ $\Longrightarrow$ *maximal regularity* results for the parabolic equation (see Sect. 13)

$$\begin{aligned}\partial_t F(t,.) - \Delta_H F(t,.) &= h(t,.) \in L^q((0,T);L^p(\Omega,\Lambda))\,, t > 0\\ F(0,.) &= 0\,.\end{aligned}$$

13 Background on Bisectorial Operators and Holomophic Functional Calculus

If the reader maintains attention on the resolvent bounds stated for the Hodge-Dirac operator, the Hodge-Laplacian and the Hodge-Stokes operator, then this material is not needed. But we will briefly describe the above-mentioned concepts for those who are interested.

Let $0 \le \omega < \mu < \frac{\pi}{2}$. Define *closed and open sectors and double sectors* in the complex plane by

$$\begin{aligned} S_{\omega+} &:= \{z \in \mathbb{C} : |\arg z| \le \omega\} \cup \{0\}\,, \qquad S_{\omega-} := -S_{\omega+}\,,\\ S^o_{\mu+} &:= \{z \in \mathbb{C} : z \ne 0,\ |\arg z| < \mu\}\,, \qquad S^o_{\mu-} := -S^o_{\mu+}\,,\\ S_\omega &:= S_{\omega+} \cup S_{\omega-}\,, \qquad S^o_\mu := S^o_{\mu+} \cup S^o_{\mu-}\,.\end{aligned}$$

Let $0 \le \omega < \frac{\pi}{2}$. A closed operator D acting on a closed subspace $\mathcal{X}^p$ of $L^p(\Omega, \Lambda)$ is called *bisectorial with angle* ω if its spectrum $\sigma(D) \subset S_\omega$, and for all $\theta \in (\omega, \frac{\pi}{2})$ there exists $C_\theta > 0$ such that

$$\|\lambda(\lambda I - D)^{-1}u\|_p \le C_\theta \|u\|_p \quad \forall \lambda \in \mathbb{C} \setminus S_\theta \,, \forall u \in \mathcal{X}^p \,.$$

In (R_p), we really mean that D_H is *bisectorial with angle* 0, and present the particular resolvent bounds for $\lambda = i/t$ with t real.

Let $0 \le \omega < \pi$. A closed operator D acting on $\mathcal{X}^p$ is called *sectorial with angle* ω if $\sigma(D) \subset S_{\omega+}$, and for all $\theta \in (\omega, \pi)$ there exists $C_\theta > 0$ such that

$$\|\lambda(\lambda I - D)^{-1}u\|_p \le C_\theta \|u\|_p \quad \forall \lambda \in \mathbb{C} \setminus S_{\theta+} \,, \forall u \in \mathcal{X}^p \,.$$

For the Hodge-Laplacian, we really mean *sectorial with angle* 0, and present the particular resolvent bounds for $\lambda = -1/t^2$ with $t > 0$.

Denote by $H^\infty(S^o_\mu)$ the space of all bounded holomorphic functions on S^o_μ, and by $\Psi(S^o_\mu)$ the subspace of those functions ψ which satisfy $|\psi(z)| \le C \min\{|z|^\alpha, |z|^{-\alpha}\}$ for some $\alpha > 0$. Similarly define $H^\infty(S^o_{\mu+})$ and $\Psi(S^o_{\mu+})$.

For D bisectorial with angle ω in $\mathcal{X}^p$ and $\psi \in \Psi(S^o_\mu)$, $\omega < \mu < \frac{\pi}{2}$ (or sectorial with angle ω and $\psi \in \Psi(S^o_{\mu+})$, $\omega < \mu < \pi$) define $\psi(D)$ through the Cauchy integral

$$\psi(D)u = \frac{1}{2\pi i} \int_\gamma \psi(z)(zI - D)^{-1}u \, dz, \qquad u \in \mathcal{X}^p \,,$$

where γ denotes the boundary of S_θ (or $S_{\theta+}$) for some $\theta \in (\omega, \mu)$, oriented counter-clockwise. Then D is said to have a *bounded holomorphic functional calculus with angle* μ, or a *bounded* $H^\infty(S^o_\mu)$ (or $H^\infty(S^o_{\mu+})$) *functional calculus* in $\mathcal{X}^p$ if there exists $C > 0$ such that

$$\|\psi(D)u\|_p \le C_p \|\psi\|_\infty \|u\|_p \qquad \forall u \in \mathcal{X}^p \,, \forall \psi \in \Psi(S^o_\mu) \text{ (or } \Psi(S^o_{\mu+})) \,.$$

For such an operator, the functional calculus extends to all $f \in H^\infty(S^o_\mu)$ (or $H^\infty(S^o_{\mu+})$) on defining

$$f(D)u = \lim_{n \to \infty} \psi_n(D)u, \qquad u \in \mathcal{X}^p,$$

where the functions $\psi_n \in \Psi(S^o_\mu)$ are uniformly bounded and tend locally uniformly to f. (We are implicitly taking $f(0) = 0$ here.)

We list some properties.

- If D is bisectorial of angle $\omega < \pi/2$, then D^2 is sectorial of angle $2\omega < \pi$.
- If D has a bounded $H^\infty(S^o_\mu)$ functional calculus, then D^2 has a bounded $H^\infty(S^o_{2\mu+})$ functional calculus.

- If D is a bisectorial operator with a bounded holomorphic functional calculus in $\mathcal{X}^p$, then $\|\operatorname{sgn}(D)u\|_p \leq C_p\|u\|_p$ for all $u \in \mathcal{X}^p$ where

$$\operatorname{sgn}(z) = \begin{cases} -1 & z \in S^o_{\mu-} \\ 0 & z = 0 \\ +1 & z \in S^o_{\mu+} \end{cases}$$

and so D has Riesz transform bounds in $\mathcal{X}^p$:

$$\|Du\|_p = \|\operatorname{sgn}(D)\sqrt{D^2}\,u\|_p \leq C_p\|\sqrt{D^2}\,u\|_p$$
$$\|\sqrt{D^2}\,u\|_p = \|\operatorname{sgn}(D)Du\|_p \leq C_p\|Du\|_p\,, \quad u \in \mathcal{D}(D) = \mathcal{D}(\sqrt{D^2})\,.$$

- If S is a sectorial operator with a bounded holomorphic functional calculus of angle $< \pi/2$ in $\mathcal{X}^p$, and $1 < q < \infty$, $0 < T \leq \infty$, then the parabolic equation

$$\partial_t F(t,\,.) + SF(t,\,.) = h(t,\,.) \in L^q((0,T);\mathcal{X}^p)\,, t > 0$$
$$F(0,\,.) = 0$$

has maximal regularity in the sense that

$$\Big\{\int_0^T \|F(t,\,.)\|_p{}^q\,dt\Big\}^{1/q} + \Big\{\int_0^T \|SF(t,\,.)\|_p{}^q\,dt\Big\}^{1/q} \leq C_{p,q}\Big\{\int_0^T \|h(t,\,.)\|_p{}^q\,dt\Big\}^{1/q}.$$

For further details on the above material, see [2, 8, 13] or the lecture notes [1, 12].

Solution to Exercise. Show that (H_p) is a consequence of $\|D_H u\|_p \approx \|\sqrt{-\Delta_H}\,u\|_p$.
We need $\|D_H u\|_p \approx \|d_\Omega u\|_p + \|\delta_{\overline{\Omega}} u\|_p$, or equivalently $\|d_\Omega u\|_p \lesssim \|D_H u\|_p$.
Write $u = \sum_{k=0}^3 u^k$, $u^k \in L^p(\Omega, \Lambda^k)$, then

$$\|d_\Omega u\|_p \approx \sum_{\ell=0}^{3} \|(d_\Omega u)^\ell\|_p = \sum_{k=0}^{3} \|d_\Omega(u^k)\|_p \leq \sum_{k=0}^{3} \|D_H(u^k)\|_p \approx \sum_{k=0}^{3} \|\sqrt{-\Delta_H}(u^k)\|_p$$
$$= \sum_{k=0}^{3} \|(\sqrt{-\Delta_H}\,u)^k\|_p \approx \|\sqrt{-\Delta_H}\,u\|_p \approx \|D_H u\|_p\,.$$

(The bound $\|d_\Omega(u^k)\|_p \leq \|d_\Omega(u^k) + \delta_{\overline{\Omega}}(u^k)\|_p$ holds because $d_\Omega(u^k) \in L^p(\Omega, \Lambda^{k+1})$ and $\delta_{\overline{\Omega}}(u^k) \in L^p(\Omega, \Lambda^{k-1})$.) The idea for this result comes from [3, Sect. 5]. □

14 L^p Hodge Decomposition

It is a consequence of the interpolation properties of the spaces $\mathcal{R}^p(d_\Omega)$ and $\mathcal{R}^p(\delta_{\underline{\Omega}})$ (see Remark 19.2) that property (H_p) is stable in p in the following sense.

Theorem 14.1 *There exist Hodge exponents p_H, $p^H = {p_H}'$ with $1 \le p_H < 2 < p^H \le \infty$ such that the Hodge decomposition (H_p)*

$$L^p(\Omega, \Lambda) = \mathcal{R}^p(d_\Omega) \oplus \mathcal{R}^p(\delta_{\underline{\Omega}}) \oplus \mathcal{N}^p(D_H)$$

holds in the L^p norm if and only if $p_H < p < p^H$.

This is proved in [14, Sect. 4], following a similar proof in [11, Sect. 3.2].

It is well known that, when Ω has *smooth* boundary, then $p_H = 1$ and $p^H = \infty$. See, e.g. [18, Theorems 2.4.2–2.4.14] for the general case of smooth compact Riemannian manifolds with boundary.

If Ω is a *strongly Lipschitz* domain in $\mathbb{R}^3$, then $p_H < 3/2 < 3 < p^H$. See, e.g., [15, Theorem 1.1]. In [14] we reprove this result, with the new techniques having the advantage of providing a new result in higher dimensions, namely that $p_H < 2n/(n+1) < 2n/(n-1) < p^H$ when Ω is a bounded strongly Lipschitz domain in $\mathbb{R}^n$. In fact we show that D_H has a bounded holomorphic functional calculus in $L^p(\Omega, \Lambda)$ for some $p < 2n/(n+1)$ (and hence, by duality, in $L^{p'}(\Omega, \Lambda)$), and apply the Exercise in Sect. 12.

15 L^p Results for D_H, Δ_H and S_H, $p_H < p < p^H$

In [14], we prove that for all p in the Hodge range, the Hodge-Dirac operator has a bounded holomorphic functional calculus. We do not include a proof here, but say a little more in Sect. 23.

Theorem 15.1 *Suppose that Ω is a very weakly Lipschitz domain in $\mathbb{R}^3$, and that $p_H < p < p^H$, i.e. (H_p) $L^p(\Omega, \Lambda) = \mathcal{R}^p(d_\Omega) \oplus \mathcal{R}^p(\delta_{\underline{\Omega}}) \oplus \mathcal{N}^p(D_H)$. Then*

- (R_p) *The Hodge-Dirac operator D_H is bisectorial in $L^p(\Omega, \Lambda)$, in particular $\|(I + itD_H)^{-1}u\|_p \le C\|u\|_p \quad \forall t \in \mathbb{R} \setminus \{0\}, \forall u \in L^p(\Omega, \Lambda)$;*
- (F_p) *D_H has a bounded $H^\infty(S^o_\mu)$ functional calculus in $L^p(\Omega, \Lambda)$ for all $\mu > 0$, in particular, $\|D_H u\|_p \approx \|\sqrt{-\Delta_H}\, u\|_p$ for all $u \in \mathcal{D}^p(D_H) = \mathcal{D}^p(\sqrt{-\Delta_H})$.*

Corollary 15.2 *(i) The Hodge-Laplacian $-\Delta_H = {D_H}^2 = d_\Omega \delta_{\underline{\Omega}} + \delta_{\underline{\Omega}} d_\Omega$ is L^p sectorial with a bounded holomorphic functional calculus, in particular,*
$\|(I - t^2\Delta_H)^{-1}u\|_p \le C^2\|u\|_p \quad \forall t > 0, \forall u \in L^p(\Omega, \Lambda)$.
(ii) The Hodge-Stokes operator $S_H = -\Delta_H|_{\mathcal{H}^p}$ is sectorial with a bounded holomorphic functional calculus in $\mathcal{H}^p := \{u \in L^p(\Omega, \Lambda^1)\,;\ \mathrm{div}_{\underline{\Omega}} u = 0\}$, in particular,
$\|(I + t^2 S_H)^{-1}u\|_p \le C^2\|u\|_p \quad \forall t > 0, \forall u \in \mathcal{H}^p$.

In the case of a bounded strongly Lipschitz domain, it was shown in [17] that $-\Delta_H$ and S_H are L^p sectorial for p in an open interval containing $[\frac{3}{2}, 3]$ in dimension 3. To our knowledge, the fact that they have a functional calculus is new, due to [14]. It was proved in [10] that for the same range of p the Riesz transforms $d_\Omega(-\Delta_H)^{-\frac{1}{2}}$ and $\delta_{\overline{\Omega}}(-\Delta_H)^{-\frac{1}{2}}$ are bounded in $L^p(\Omega, \Lambda)$, again in the case of a bounded strongly Lipschitz domain.

Again: If Ω is a *very weakly Lipschitz* domain in $\mathbb{R}^3$, and $p_H < p < p^H$, then D_H, Δ_H, S_H all have L^p resolvent bounds,

$$\|(I + itD_H)^{-1}u\|_p \leq C\|u\|_p \quad \forall u \in L^p(\Omega, \Lambda), \ \forall\, t \in \mathbb{R} \setminus \{0\}$$
$$\|(I - t^2\Delta_H)^{-1}u\|_p \leq C^2\|u\|_p \quad \forall u \in L^p(\Omega, \Lambda), \ \forall\, t > 0$$
$$\|(I + t^2 S_H)^{-1}u\|_p \leq C^2\|u\|_p \quad \forall u \in \mathcal{H}^p, \ \forall\, t > 0$$

and all have corresponding holomorphic functional calculi.

In fact, D_H cannot have a functional calculus in $L^p(\Omega, \Lambda)$ for p outside the interval (p_H, p^H), as shown in the Exercise in Sect. 12.

But S_H CAN, and DOES, at least for $\max\{1, p_{H_S}\} < p \leq p_H$ where p_{H_S} is the Sobolev exponent below p_H i.e. $\frac{1}{p_{H_S}} = \frac{1}{p_H} + \frac{1}{3}$.

Note: (i) Since $p_H < 2$, it is easily computed that $p_{H_S} < 6/5$.

(ii) If Ω is *strongly Lipschitz*, then $p_H < 3/2$, and so $p_{H_S} < 1$.

16 L^p Result for Hodge-Stokes Operator S_H, $p_{H_S} < p < p^H$

Theorem 16.1 *Suppose Ω is a very weakly Lipschitz domain in $\mathbb{R}^3$, and* max $\{1, p_{H_S}\} < p < p^H$. *Then the Hodge-Stokes operator $S_H = -\Delta_H|_{\mathcal{H}^p}$ is sectorial with a bounded holomorphic functional calculus in $\mathcal{H}^p = \{u \in L^p(\Omega, \Lambda^1)\,;$* $\mathrm{div}_{\overline{\Omega}} u = 0\}$. *In particular,*

$$\|(I + t^2 S_H)^{-1}u\|_p \leq C^2\|u\|_p\,, \quad \forall u \in \mathcal{H}^p, \ \forall t > 0\,.$$

Corollary 16.2 *Suppose Ω is a strongly Lipschitz domain in $\mathbb{R}^3$, and $1 < p < p^H$. Then S_H is sectorial with a bounded holomorphic functional calculus in $\mathcal{H}^p$.*

These results are proved in [14]. Here we will not look further into functional calculi, but will indicate how to apply the fact that the Hodge-Dirac operator has L^q resolvent bounds when $p_H < q < p^H$, to derive L^p resolvent bounds for the Hodge-Stokes operator when $p_{H_S} < p \leq p_H$.

The proofs depend on the theory of regularised Poincaré and Bogovskiĭ potential operators as developed in [7, 16] for the case when Ω is starlike or strongly Lipschitz. Here we start with the special case of the unit ball $B \subset \mathbb{R}^3$, and then derive what we need for very weakly Lipschitz domains.

17 Potential Operator on the Unit Ball

Let

- $B = B(0, 1)$, the unit ball in $\mathbb{R}^3$, centred at the origin;
- $\theta \in C_c^\infty(\frac{1}{2}B, \mathbb{R})$ with $\int \theta = 1$;
- $R_B : L^p(B, \Lambda) \to W^{1,p}(B, \Lambda)$, the *regularised Poincaré potential operator* defined by $R_B u = \sum_{k=1}^3 R_B u^k$,

$$R_B u^k(x) = \int_B \theta(a)(x-a) \lrcorner \int_0^1 t^{k-1} u^k(a + t(x-a))\, dt\, da \qquad (k = 1, 2, 3)\,,$$

$u = \sum_{k=0}^3 u^k \in L^p(B, \Lambda) = \oplus_{k=0}^3 L^p(B, \Lambda^k)$.

$$\begin{array}{ccccccccc} & u^0 & & u^1 & & u^2 & & u^3 & \\ & \in & & \in & & \in & & \in & \\ d_B : 0 \rightleftarrows & L^p(B, \mathbb{C}) & \underset{R_B}{\overset{\nabla_B}{\rightleftarrows}} & L^p(B, \mathbb{C}^3) & \underset{R_B}{\overset{\text{curl}_B}{\rightleftarrows}} & L^p(B, \mathbb{C}^3) & \underset{R_B}{\overset{\text{div}_B}{\rightleftarrows}} & L^p(B, \mathbb{C}) & \rightleftarrows 0 \end{array}$$

Then $R_B : L^p(B, \Lambda) \to W^{1,p}(B, \Lambda)$ is bounded, $R_B : L^p(B, \Lambda) \to L^p(B, \Lambda)$ is compact, and

$$d_B R_B u + R_B d_B u + \left(\int \theta u^0 \right) 1 = u \qquad \forall\, u \in L^p(B, \Lambda)$$

(where 1 denotes the constant function $1 \in L^p(\Omega, \Lambda^0)$). We write this as

$$d_B R_B u + R_B d_B u + K_B u = u$$

where $K_B u = (\int \theta u^0)1$ and note that $K_B : L^p(B, \Lambda) \to L^\infty(B, \Lambda^0)$ is bounded, and $K_B : L^p(B, \Lambda) \to L^p(B, \Lambda^0)$ is compact. The operator K_B compensates for the fact that the above sequence for d_B misses out on being exact, due to the gradient map ∇_B having a one dimensional null-space consisting of constant functions in $L^p(B, \Lambda^0)$.

Moreover, if $1 < p = q_S < q < \infty$, where $p = q_S$ is the Sobolev exponent below q, i.e.

$$\tfrac{1}{p} = \tfrac{1}{q} + \tfrac{1}{3}$$

then the potential map $R_B : L^p(B, \Lambda) \to L^q(B, \Lambda)$ is bounded.

18 Potential Operator on Bilipschitz Transformation of the Unit Ball

Suppose $\rho : B \to \rho B \subset \mathbb{R}^3$ is a uniformly locally bilipschitz transformation. Then the pull-back $\rho^* : L^p(\rho B, \Lambda) \to L^p(B, \Lambda)$ is bounded, and

$$d_{\rho B} = (\rho^*)^{-1} d_B \rho^*;$$

recall that $(\rho^* u)(x) = (\underline{\rho}_x)^* u(\rho(x))$ where $\underline{\rho}_x$ is the Jacobian matrix of ρ at x.

Define $R_{\rho B} : L^p(\rho B, \Lambda) \to L^q(\rho B, \Lambda)$ and $K_{\rho B} : L^p(\rho B, \Lambda) \to L^\infty(\rho B, \Lambda)$ by

$$R_{\rho B} = (\rho^*)^{-1} R_B \rho^* \qquad \text{and} \qquad K_{\rho B} = (\rho^*)^{-1} K_B \rho^*$$

so that

$$d_{\rho B} R_{\rho B} u + R_{\rho B} d_{\rho B} u + K_{\rho B} u = u\ .$$

$$d_{\rho B} : 0 \rightleftarrows L^p(\rho B, \mathbb{C}) \underset{R_{\rho B}}{\overset{\nabla_{\rho B}}{\rightleftarrows}} L^p(\rho B, \mathbb{C}^3) \underset{R_{\rho B}}{\overset{\mathrm{curl}_{\rho B}}{\rightleftarrows}} L^p(\rho B, \mathbb{C}^3) \underset{R_{\rho B}}{\overset{\mathrm{div}_{\rho B}}{\rightleftarrows}} L^p(\rho B, \mathbb{C}) \rightleftarrows 0$$

The operators $R_{\rho B}$ and $K_{\rho B}$ have the same boundedness and compactness properties as R_B and K_B.

19 Potential Operators on *Very Weakly Lipschitz* Domains

- $1 < p < q < \infty \quad (\frac{1}{p} = \frac{1}{q} + \frac{1}{3})$.
- Ω is *very weakly Lipschitz*, i.e. $\Omega = \cup_{j=1}^N (\rho_j B)$ where each $\rho_j : B \to \rho_j B \subset \mathbb{R}^3$ is uniformly locally bilipschitz, and
- $1 = \sum_{j=1}^N \chi_j$ on Ω, where each $\chi_j : \Omega \to [0, 1]$ is a Lipschitz function with $\mathrm{sppt}_\Omega(\chi_j) \subset \rho_j B$.
- Define $R_\Omega = \sum_{j=1}^N \chi_j R_{\rho_j B}$ and $K_\Omega u = \sum_{j=1}^N (\chi_j K_{\rho_j B} u - (\nabla \chi_j) \wedge R_{\rho_j B} u)$.

$$\begin{array}{ccccccccc} & u^0 & & u^1 & & u^2 & & u^3 & \\ & \in & & \in & & \in & & \in & \\ d_\Omega : 0 \rightleftarrows & L^p(\Omega, \mathbb{C}) & \underset{R_\Omega}{\overset{\nabla_\Omega}{\rightleftarrows}} & L^p(\Omega, \mathbb{C}^3) & \underset{R_\Omega}{\overset{\mathrm{curl}_\Omega}{\rightleftarrows}} & L^p(\Omega, \mathbb{C}^3) & \underset{R_\Omega}{\overset{\mathrm{div}_\Omega}{\rightleftarrows}} & L^p(\Omega, \mathbb{C}) & \rightleftarrows 0 \end{array}$$

It is straightforward to apply the properties mentioned in the previous two sections to prove the following result.

Theorem 19.1 *The exterior derivative d_Ω has a potential map $R_\Omega : L^p(\Omega, \Lambda) \to L^q(\Omega, \Lambda)$ satisfying*

$$d_\Omega R_\Omega u + R_\Omega d_\Omega u + K_\Omega u = u \qquad \forall u \in L^p(\Omega, \Lambda) ,$$

where $K_\Omega : L^p(\Omega, \Lambda) \to L^q(\Omega, \Lambda)$. Moreover K_Ω and R_Ω are compact operators in $L^p(\Omega, \Lambda)$.

Remark 19.2 Although we will not use this fact in the coming sections, we remark than R_Ω can be modified in such a way that $d_\Omega R_\Omega u = u$ for all $u \in \mathcal{R}^p(d_\Omega)$.

Using this modification, we have that $d_\Omega R_\Omega : L^p(\Omega, \Lambda) \to \mathcal{R}^p(d_\Omega)$ is a *bounded projection* for all p, $1 < p < \infty$, and as a corollary, the spaces $\mathcal{R}^p(d_\Omega)$ $(1 < p < \infty)$ are closed subspaces of $L^p(\Omega, \Lambda)$ which interpolate by the complex method.

In this case, R_Ω is a true potential operator. For example, if u^1 is a gradient vector field, then $w_0 = R_\Omega u^1 \in L^q(\Omega, \mathbb{C})$ is its potential, because $\nabla_\Omega w_0 = d_\Omega R_\Omega u^1 = u^1$.

Remark 19.3 With a modified R_Ω as in Remark 19.2, define $\mathcal{Z}^p = K_\Omega(\mathcal{N}^p(d_\Omega))$. Then $\mathcal{N}^p(d_\Omega) = \mathcal{R}^p(d_\Omega) \oplus \mathcal{Z}^p$ with decomposition $u = d_\Omega R_\Omega u + K_\Omega u$ for all $u \in \mathcal{N}^p(d_\Omega)$. So the spaces in the decomposition are closed, and $\mathcal{Z}^p$ is finite dimensional, on account of the compactness of K_Ω. Thus $\mathcal{R}^p(d_\Omega)$ has finite codimension in $\mathcal{N}^p(d_\Omega)$, as claimed in Sect. 5.

In the following section $T_{\overline{\Omega}}$ could be similarly modified to give $u = \delta_{\overline{\Omega}} T_{\overline{\Omega}} u$ for all $u \in \mathcal{R}^p(\delta_{\overline{\Omega}})$.

20 Dual Potential Operators

- $1 < p < q < \infty \qquad (\frac{1}{p} = \frac{1}{q} + \frac{1}{3})$;
- $T_{\overline{\Omega}} : L^p(\Omega, \Lambda) \to L^q(\Omega, \Lambda)$ is dual to $R_\Omega : L^{q'}(\Omega, \Lambda) \to L^{p'}(\Omega, \Lambda)$;
- $L_{\overline{\Omega}} : L^p(\Omega, \Lambda) \to L^q(\Omega, \Lambda)$ is dual to $K_\Omega : L^{q'}(\Omega, \Lambda) \to L^{p'}(\Omega, \Lambda)$.

Then, dual to the equation $d_\Omega R_\Omega u + R_\Omega d_\Omega u + K_\Omega u = u$, is

$$\boxed{u = \delta_{\overline{\Omega}} T_{\overline{\Omega}} u + T_{\overline{\Omega}} \delta_{\overline{\Omega}} u + L_{\overline{\Omega}} u}$$

so that $T_{\overline{\Omega}}$ is a potential operator for $\delta_{\overline{\Omega}}$, called the Bogovskiĭ operator

$$0 \leftrightarrows L^p(\Omega, \mathbb{C}) \underset{T_{\overline{\Omega}}}{\overset{-\nabla_{\overline{\Omega}}}{\leftrightarrows}} L^p(\Omega, \mathbb{C}^3) \underset{T_{\overline{\Omega}}}{\overset{\mathrm{curl}_{\overline{\Omega}}}{\leftrightarrows}} L^p(\Omega, \mathbb{C}^3) \underset{T_{\overline{\Omega}}}{\overset{-\mathrm{div}_{\overline{\Omega}}}{\leftrightarrows}} L^p(\Omega, \mathbb{C}) \leftrightarrows 0 : \delta_{\overline{\Omega}}$$

21 L^p Results for Δ_H on $\mathcal{N}^p(\delta_{\overline{\Omega}})$, $p_{H_S} < p < p^H$

Suppose that Ω is a *very weakly Lipschitz* domain. We have stated in Theorem 15.1 that when $p_H < q < p^H$, the Hodge-Dirac operator $D_H = d_\Omega + \delta_{\overline{\Omega}}$ is bisectorial with a bounded holomorphic funtional calculus in $L^q(\Omega, \Lambda)$. Our aim now is to extend this result as follows.

Theorem 21.1 *Suppose that*

- $p_H < q < p^H$;
- $\max\{1, q_S\} \le p \le q$ *where* q_S *is the lower Sobolev exponent of* q*, i.e.* $\frac{1}{q_S} = \frac{1}{q} + \frac{1}{3}$.

Then the Hodge-Laplacian $-\Delta_H$ *is sectorial with a bounded holomorphic functional calculus in* $\mathcal{N}^p(\delta_{\overline{\Omega}}) = \{u \in L^p(\Omega, \Lambda) ; \delta_{\overline{\Omega}} u = 0\}$. *In particular,*

$$\|(I - t^2\Delta_H)^{-1}u\|_p \le C^2\|u\|_p\,, \quad \forall u \in \mathcal{N}^p(\delta_{\overline{\Omega}}),\ \forall t > 0\,. \tag{1}$$

Similar resolvent bounds also holds on $\mathcal{N}(d_\Omega)$ and hence on $\mathcal{R}(\delta_{\overline{\Omega}})$ and on $\mathcal{R}(d_\Omega)$. On restricting to $L^p(\Omega, \Lambda^1)$, we obtain Theorem 16.1 as a corollary.

For the results on functional calculi, we refer the reader to [14]. We do not fully prove the resolvent bounds either, but give the spirit of the method by outlining the estimates in the case when $p = q_S$.

22 L^p Resolvent Bounds for Δ_H on $\mathcal{N}^p(\delta_{\overline{\Omega}})$, $p = q_S$, $p_H < q < p^H$

- Ω is *very weakly Lipschitz* and $p_H < q < p^H$, $p = q_S > 1$.
- The idea is to modify the techniques of Blunck-Kunstmann [6], but there is still quite a bit to do, because we are working on the subspace $\mathcal{N}^p(\delta_{\overline{\Omega}})$. We will not consider the functional calculus here, but will outline a proof of resolvent bounds.
- The easy part: When $t \ge 1$, and $\delta_{\overline{\Omega}} u = 0$, then

$$\begin{aligned}
\|(I - t^2\Delta_H)^{-1}u\|_p &\lesssim \|(I - t^2\Delta_H)^{-1}u\|_q \qquad \text{(because } \Omega \text{ is bounded)}\\
&= \|(I - t^2\Delta_H)^{-1}(\delta_{\overline{\Omega}} T_{\overline{\Omega}} + L_{\overline{\Omega}})u\|_q\\
&\le t\|\delta_{\overline{\Omega}}(I - t^2\Delta_H)^{-1}T_{\overline{\Omega}}u\|_q + \|(I - t^2\Delta_H)^{-1}L_{\overline{\Omega}}u\|_q\\
&\lesssim \|tD_H(I + t^2{D_H}^2)^{-1}T_{\overline{\Omega}}u\|_q + \|(I + t^2{D_H}^2)^{-1}L_{\overline{\Omega}}u\|_q\\
&\lesssim \|T_{\overline{\Omega}}u\|_q + \|L_{\overline{\Omega}}u\|_q\\
&\lesssim \|u\|_p
\end{aligned}$$

  (using Hodge decomposition in $L^q(\Omega, \Lambda)$ in line 4, and resolvent bounds for D_H in $L^q(\Omega, \Lambda)$ in line 5).
- Henceforth take $0 < t < 1$.

- Cover Ω: Let $\underline{Q}^t_j$ $(j \in J)$ be the cubes in $\mathbb{R}^3$ with side-length t and corners at points in $t\mathbb{Z}^3$, which intersect Ω. Let $Q^t_j = 4\underline{Q}^t_j \cap \Omega$. Then $\Omega = \cup Q^t_j$. Write $1 = \sum_{j\in J} \eta_j^2$ on Ω, where $\eta_j \in C^1_c(4\underline{Q}^t_j, [0,1])$ and $\|\nabla \eta_j\|_\infty \leq 1/t$. The "cubes" Q^t_j have finite overlap, in fact $\sum_{j\in J} 1_{Q^t_j} \leq 64$. (Here $1_{Q^t_j}$ denotes the function with value 1 on Q^t_j and zero elsewhere on Q.)
- L^q *off-diagonal bounds* in $\mathrm{dist}(Q^t_j, Q^t_k) = \inf\{|x-y|; x \in Q^t_j, y \in Q^t_k\}$ are a consequence of the L^q resolvent bounds. See [14, Sect. 5], or adapt the L^2 proofs in [4]. We need the following two bounds.

  For each $N \in \mathbb{N}$, there exists C_N such that, when $\mathrm{sppt}(f) \in Q^t_k$, then

$$\|1_{Q^t_j}(I - t^2\Delta_H)^{-1} f\|_q \leq C_N \left(\tfrac{t}{t+\mathrm{dist}(Q^t_j, Q^t_k)}\right)^N \|f\|_q \qquad \text{and}$$
$$t\|1_{Q^t_j}(I - t^2\Delta_H)^{-1}\delta_{\overline{\Omega}} f\|_q \leq C_N \left(\tfrac{t}{t+\mathrm{dist}(Q^t_j, Q^t_k)}\right)^N \|f\|_q \,.$$

- Decompose $u \in \mathcal{N}^p(\delta_{\overline{\Omega}})$ (using $\delta_{\overline{\Omega}}(\eta_k f) - \eta_k \delta_{\overline{\Omega}} f = (\nabla \eta_k) \lrcorner f$):

$$\begin{aligned}
u &= \sum_{k\in J} \eta_k{}^2 u = \sum_{k\in J} \eta_k\, I\, \eta_k u \\
&= \sum_{k\in J} (\eta_k\, \delta_{\overline{\Omega}} T_{\overline{\Omega}}\, \eta_k u + \eta_k\, T_{\overline{\Omega}}\, \delta_{\overline{\Omega}}\, \eta_k u + \eta_k\, L_{\overline{\Omega}}\, \eta_k u) \\
&= \sum_{k\in J} (\delta_{\overline{\Omega}}(\eta_k T_{\overline{\Omega}}\, \eta_k u) - (\nabla \eta_k) \lrcorner T_{\overline{\Omega}}\, \eta_k u + \eta_k T_{\overline{\Omega}}(\nabla \eta_k) \lrcorner u + \eta_k L_{\overline{\Omega}}\, \eta_k u) \\
&= \sum_{k\in J} (\delta_{\overline{\Omega}}\, w_k + \tfrac{1}{t} v_k) \qquad \text{where}
\end{aligned}$$

$$\begin{aligned}
w_k &= \eta_k T_{\overline{\Omega}}\, \eta_k u \qquad \text{and} \\
v_k &= -(t\nabla \eta_k) \lrcorner T_{\overline{\Omega}}\, \eta_k u + \eta_k T_{\overline{\Omega}}(t\nabla \eta_k) \lrcorner u + t\, \eta_k L_{\overline{\Omega}}\, \eta_k u \,.
\end{aligned}$$

- On using the $L^p - L^q$ bounds on $T_{\overline{\Omega}}$ and $L_{\overline{\Omega}}$, we obtain

$$\|w_k\|_q \lesssim \|\eta_k u\|_p \lesssim \|1_{Q^t_k} u\|_p \quad \text{with } \mathrm{sppt}(w_k) \subset Q^t_k \qquad \text{and}$$
$$\|v_k\|_q \lesssim (1+t)\|1_{Q^t_k} u\|_p \lesssim \|1_{Q^t_k} u\|_p \quad \text{with } \mathrm{sppt}(v_k) \subset Q^t_k \,.$$

Here now is the resolvent estimate. Suppose $\delta_{\overline{\Omega}} u = 0$. Then

$$
\begin{aligned}
\|(I-t^2\Delta_H)^{-1}u\|_p &\le \Big[\sum_{j\in J}\int_{Q_j^t}|(I-t^2\Delta_H)^{-1}u|^p\Big]^{\frac1p}\\
&= \Big[\sum_{j\in J}(\|1_{Q_j^t}(I-t^2\Delta_H)^{-1}u\|_p)^p\Big]^{\frac1p}\\
&\le \Big[\sum_{j\in J}(\|1_{Q_j^t}(I-t^2\Delta_H)^{-1}u\|_q|Q_j^t|^{\frac13})^p\Big]^{\frac1p} \qquad (\tfrac1p=\tfrac1q+\tfrac13)\\
&\lesssim \Big[\sum_{j\in J}(\sum_{k\in J}\|1_{Q_j^t}(I-t^2\Delta_H)^{-1}(\delta_{\overline{\Omega}}w_k+\tfrac1t v_k)\|_q t)^p\Big]^{\frac1p}\\
&\lesssim \Big[\sum_{j\in J}(\sum_{k\in J}\big(\tfrac{t}{t+\mathrm{dist}(Q_j^t,Q_k^t)}\big)^4(\|w_k\|_q+\|v_k\|_q))^p\Big]^{\frac1p} \qquad (^*)\\
&\lesssim \Big[\sum_{j\in J}(\sum_{k\in J}\big(\tfrac{t}{t+\mathrm{dist}(Q_j^t,Q_k^t)}\big)^4\|1_{Q_k^t}u\|_p)^p\Big]^{\frac1p}\\
&\lesssim \Big(\sup_j\sum_{k\in J}\big(\tfrac{t}{t+\mathrm{dist}(Q_j^t,Q_k^t)}\big)^4\Big)\Big[\sum_{k\in J}\|1_{Q_k^t}u\|_p{}^p\Big]^{\frac1p} \qquad (^{**})\\
&\lesssim \Big[\sum_{k\in J}\|1_{Q_k^t}u\|_p{}^p\Big]^{\frac1p} = \Big[\sum_{k\in J}\int_{Q_k^t}|u|^p\Big]^{\frac1p} \qquad (^{***})\\
&= \Big(\int_\Omega\sum_{k\in J}1_{Q_k^t}|u|^p\Big)^{\frac1p} \lesssim \|u\|_p \qquad (^{****})
\end{aligned}
$$

as claimed.

- In (*) we used the off-diagonal bounds with $N=4$;
- In (**) we used the *Schur estimate* in $\ell^p(J)$, with $A_{j,k}=\big(\frac{t}{t+\mathrm{dist}(Q_j^t,Q_k^t)}\big)^4$ and $\beta_k=\|1_{Q_k^t}u\|_p$:

$$
\begin{aligned}
\Big[\sum_j|\sum_k A_{j,k}\beta_k|^p\Big]^{\frac1p} &\le \big(\sup_j\sum_k|A_{j,k}|\big)^{\frac1{p'}}\big(\sup_k\sum_j|A_{j,k}|\big)^{\frac1p}\big(\sum_k|\beta_k|^p\big)^{\frac1p}\\
&= \big(\sup_j\sum_k|A_{j,k}|\big)\big(\sum_k|\beta_k|^p\big)^{\frac1p} \quad \text{when } A_{j,k}=A_{k,j}\ ;
\end{aligned}
$$

- In (***) we used that, given Q_j^t,

$$\sum_k \left(\tfrac{t}{t+\operatorname{dist}(Q_j^t,Q_k^t)}\right)^4 \lesssim C_0 + \sum_{M=0}^{\infty} \sum_{\{k;2^M t \le \operatorname{dist}(Q_j^t,Q_k^t)<2^{(M+1)}t\}} \tfrac{1}{2^{4M}}$$
$$\lesssim C_0 + \sum_{M=0}^{\infty} 2^{3M} \tfrac{1}{2^{4M}} = C_0 + \sum_{M=0}^{\infty} \tfrac{1}{2^M} \le C\,;$$

- In (****) we used the finite overlap of the cubes.

This completes the proof of (1) in the case when $p = q_S$. The proof of L^p sectoriality when $q_S \le p < q$ requires minor modification. To show that S_H has a bounded holomorphic functional calculus requires further work, using a Calderón–Zygmund decomposition of Ω. For this, the reader is referred to [14].

23 Remarks on Obtaining Resolvent Bounds in the Hodge Range

In the previous section we applied Theorem 15.1. But suppose we just start with the L^2 resolvent bounds. Then a similar procedure to that described above, can be used to obtain resolvent bounds for D_H on $\mathcal{N}^p(\delta_{\overline{\Omega}})$ when $6/5 = 2_S \le p \le 2$. Moreover, use of the potential operators R_Ω will lead to resolvent bounds on $\mathcal{N}^p(d_\Omega)$, also when $6/5 \le p \le 2$. Now, if p is also in the Hodge range, we then obtain resolvent bounds on all of $L^p(\Omega, \Lambda)$, i.e. we obtain resolvent bounds for D_H on $L^p(\Omega, \Lambda)$ when $\max\{6/5, p_H\} < p \le 2$. Repeating this procedure once more if necessary, we obtain resolvent bounds on $L^p(\Omega, \Lambda)$ for $p_H < p \le 2$ (as $(6/5)_S < 1$). A duality argument then gives resolvent bounds when $2 \le p < p^H$. In this way, the statement (R_p) can be proved when $p_H < p < p^H$, as stated in Theorem 15.1. See [14] for details.

We remark that such an iteration method has been used previously in [9] in the study of more general first order systems on $\mathbb{R}^n$. A similar iteration procedure has been used also in [10, 17].

24 Parabolic Equations

As mentioned in Sect. 13, operators with a bounded holomorphic functional calculus on a closed subspace $\mathcal{X}^p$ of $L^p(\Omega, \Lambda)$, also satisfy maximal regularity. So, on taking $\mathcal{X}^p = \mathcal{H}^p$, we obtain:

Theorem 24.1 *Suppose that Ω is a very weakly Lipschitz domain in $\mathbb{R}^3$, that* $\max\{1, p_{HS}\} < p < p^H$, *and that $1 < q < \infty$, $0 < T \le \infty$. Suppose also that*

$$\partial_t F(t,\,.) + S_H F(t,\,.) = h(t,\,.) \in L^q((0,T);\mathcal{H}^p)\,,\, t>0$$
$$F(0,\,.) = 0\,.$$

Then

$$\Big\{\int_0^T \|F(t,\,.)\|_p{}^q\,dt\Big\}^{1/q} + \Big\{\int_0^T \|SF(t,\,.)\|_p{}^q\,dt\Big\}^{1/q} \le C_{p,q}\Big\{\int_0^T \|h(t,\,.)\|_p{}^q\,dt\Big\}^{1/q}.$$

Acknowledgments The first author would like to thank the organisers of the ISAAC meeting in Macau for arranging such an interesting conference, and in particular Tao Qian for his kind hospitality. The authors appreciate the support of the Mathematical Sciences Institute at the Australian National University, Canberra, where much of the collaboration took place, as well as the Laboratoire International Associé "Analysis and Geometry" and the Mathematical Institute in Marseille (I2M). Both authors were supported by the Australian Research Council.

References

1. Albrecht, D., Duong, X., McIntosh, A.: Operator theory and harmonic analysis. In: Instructional Workshop on Analysis and Geometry, Part III. Canberra (1995). Proc. Centre Math. Appl. Austral. Nat. Univ. **34**, 77–136 (1996)
2. Auscher, P., McIntosh, A., Nahmod, A.: Holomorphic functional calculi of operators, quadratic estimates and interpolation. Indiana Univ. Math. J. **46**, 375–403 (1997)
3. Auscher, P., McIntosh, A., Russ, E.: Hardy spaces of differential forms on riemannian manifolds. J. Geom. Anal. **18**, 192–248 (2008)
4. Axelsson, A., Keith, S., McIntosh, A.: Quadratic estimates and functional calculi of perturbed Dirac operators. Invent. Math. **163**, 455–497 (2006)
5. Axelsson, A., McIntosh, A.: Hodge decompositions on weakly Lipschitz domains. Advances in Analysis and Geometry, Trends Mathematics, pp. 3–29. Birkhäuser, Basel (2004)
6. Blunck, S., Kunstmann, P.: Calderón-Zygmund theory for non-integral operators and the H^∞ functional calculus. Rev. Mat. Iberoamericana **19**, 919–942 (2003)
7. Costabel, M., McIntosh, A.: On Bogovskiĭ and regularised Poincaré operators for de Rham complexes on Lipschitz domains. Math. Z. **265**, 297–320 (2010)
8. Cowling, M., Doust, I., McIntosh, A., Yagi, A.: Banach space operators with a bounded H^∞ functional calculus. J. Austral. Math. Soc. Ser. A **60**, 51–89 (1996)
9. Frey, D., McIntosh, A., Portal, P.: Conical square function estimates and functional calculi for perturbed Hodge-Dirac operators in L^p. J. d'Analyse Mathématique. arXiv:1407.4774
10. Hofmann, S., Mitrea, M., Monniaux, S.: Riesz transforms associated with the Hodge Laplacian in Lipschitz subdomains of Riemannian manifolds. Ann. Inst. Fourier (Grenoble) **61**(4), 1323–1349 (2012)
11. Hytönen, T., McIntosh, A.: Stability in p of the H^∞-calculus of first-order systems in L^p. Proc. Centre Math. Appl. Austral. Nat. Univ. **44**, 167–181 (2010)
12. Kunstmann, P.C., Weis, L.: Maximal L_p regularity for parabolic problems, Fourier multiplier theorems and H^∞-functional calculus. Lecture Notes in Mathematics, vol. 1855. Springer (2004)
13. McIntosh, A.: Operators which have an H^∞ functional calculus. Proc. Centre Math. Appl. Austral. Nat. Univ. **14**, 210–231 (1986)
14. McIntosh, A., Monniaux, S.: Hodge-Dirac, Hodge-Laplacian and Hodge-Stokes operators in L^p spaces on Lipschitz domains (in preparation)

15. Mitrea, M.: Sharp Hodge decompositions, Maxwell's equations, and vector Poisson problems on nonsmooth, three-dimensional Riemannian manifolds. Duke Math. J. **125**(3), 467–547 (2004)
16. Mitrea, D., Mitrea, M., Monniaux, S.: The Poisson problem for the exterior derivative operator with Dirichlet boundary condition on nonsmooth domains. Commun. Pure Appl. Anal. **7**, 1295–1333 (2008)
17. Mitrea, M., Monniaux, S.: On the analyticity of the semigroup generated by the Stokes operator with Neumann-type boundary conditions on Lipschitz subdomains of Riemannian manifolds. Trans. Am. Math. Soc. **361**(6), 3125–3157 (2009)
18. Schwarz, G.: Hodge decomposition-a method for solving boundary value problems. Lecture Notes in Mathematics, vol. 1607. Springer, Berlin (1995)

The Study of Complex Shapes of Fluid Membranes, the Helfrich Functional and New Applications

Zhong-Can Ou-Yang and Zhan-Chun Tu

Abstract The theoretical study of complex configurations of fluid membranes is reported on the basis of the Helfrich functional. Series of analytical results on the governing equations of closed lipid vesicles and open lipid vesicles with holes are surveyed. The concepts of stress tensor and moment tensor in fluid membranes are investigated from two different viewpoints: the balance of forces (moments) and the generalized variational principle of free energy. Several examples on new applications of the Helfrich functional in understanding the growth mechanism of some mesoscopic structures are illustrated.

Keywords Helfrich functional · Configuration · Membrane

Mathematics Subject Classification (2010). Primary 49Q10 · Secondary 53Z05

1 Introduction: From Soap Films to Red Blood Cells

There exist many structures whose one dimension is much smaller than the other two in our world. This kind of structures are usually called membranes, which may be thought of as 2-dimensional (2D) smooth surfaces in 3-dimensional (3D) Euclidean space. The identities formed by membranes display a variety of configurations. For example, soap bubbles at rest are always spherical; Human red blood cells are of biconcave discoid under the normal physiological condition.

The issue of equilibrium configurations of membranes has attracted much attention of mathematicians and physicists. As early as in 1803, Plateau investigated a soap film attaching to a metallic ring when the ring passed through soap water [1].

Z.-C. Ou-Yang
Institute of Theoretical Physics, Chinese Academy of Sciences, Beijing 100080, China
e-mail: oy@itp.ac.cn

Z.-C. Tu (✉)
Department of Physics, Beijing Normal University, Beijing 100875, China
e-mail: tuzc@bnu.edu.cn

T. Qian and L.G. Rodino (eds.), *Mathematical Analysis, Probability and Applications – Plenary Lectures*, Springer Proceedings in Mathematics & Statistics 177, DOI 10.1007/978-3-319-41945-9_4

He proposed that equilibrium configuration of the soap film corresponds to the minimum surface area of the film, which is mathematically equivalent to minimizing the functional

$$F = \int_M \mathrm{d}A, \tag{1.1}$$

where M and $\mathrm{d}A$ represent the membrane surface and the area element of the surface, respectively. The first-order variation of the Plateau functional (1.1) leads to a minimal surface with vanishing mean curvature $H = 0$. From 1805 to 1806, Young [2] and Laplace [3] studied soap bubbles. They proposed that the equilibrium configuration of a soap bubble corresponds to minimizing the surface area of the bubble for given volume enclosed in the bubble, which is mathematically equivalent to minimizing the functional

$$F = \lambda \int_M \mathrm{d}A + p \int \mathrm{d}V, \tag{1.2}$$

where λ and p represent the surface tension of the membrane and the osmotic pressure (pressure difference between the outer and the inner sides) of a soap bubble, respectively. $\mathrm{d}V$ represents the element of volume enclosed by the bubble. The first-order variation of the Young–Laplace functional (1.2) leads to a surface with constant mean curvature $H = p/2\lambda$. The reason that we can merely observe spherical bubbles is ascribed to the Alexandrov theorem—an embedded compact surface with constant mean curvature in 3D Euclidian space must be a spherical surface [4].

In 1812, Poisson [5] considered a solid shell and put forward an energy functional

$$F = \int_M H^2 \mathrm{d}A. \tag{1.3}$$

This functional was deeply investigated by Willmore [6, 7], thus, it is now called the Willmore functional in mathematics. Since the Willmore functional (1.3) is an invariant under conformal transformations, any configuration and its images under conformal transformations correspond to the same energy. The first-order variation of the Willmore functional (1.3) leads to

$$\nabla^2 H + 2H(H^2 - K) = 0, \tag{1.4}$$

a equation satisfied by the Willmore surfaces. The symbol K represents the Gauss curvature of the surface. The symbol ∇^2 represents the Laplace operator of the first kind defined on a 2D surface. Willmore showed that round spheres (as well as their images under conformal transformations) correspond to the least minimum of the Willmore functional (1.3) among all compact surfaces in 3D Euclidian space. In other words, all compact surfaces in 3D Euclidian space make the Willmore functional (1.3) to take values no less than 4π. Willmore further conjectured that all compact surfaces of genus one in 3D Euclidian space make the Willmore functional (1.3) to take values no less than $2\pi^2$, where the least minimum corresponds to the Willmore tori (as well as their images under conformal transformations), which are special tori with the ratio of their two generation radii being $\sqrt{2}$ [8]. Recently, the Willmore conjecture has been proved by Marques and Neves via min-max theory [9].

Human red blood cells are unique since there are no internal cellular organelles inside the cells. They may be regarded as closed vesicles enclosed by cell membranes. In the energy scale of physiological condition, cell membranes are almost inextensible, and the volumes of red blood cells are hardly compressed. To explain the biconcave discoidal shape of red blood cells, Canham argued that the biconcave configuration might minimize the bending energy of membranes under the constraints of fixed area of membranes and fixed volume of the cells [10].

The cell membrane consists of lipid molecules and proteins, where lipid molecules form a bilayer while proteins are mosaicked in the bilayer [11]. In 1973, Helfrich recognized that the lipid bilayer is in the liquid crystal state at the physiological temperature. According to the elastic theory of liquid crystals, he proposed that the bending energy of the bilayer could be expressed as a functional

$$F_{\mathrm{H}} = \int_M [(k_c/2)(2H + c_0)^2 + \bar{k}K]\mathrm{d}A, \tag{1.5}$$

where $k_c > 0$ and $\bar{k}$ are two bending moduli of the bilayer [12]. The parameter c_0 represents the spontaneous curvature of the lipid bilayer, which reflects the asymmetric factors in the two leaflets of the bilayer. The numerical results implied that the biconcave configuration indeed minimizes the bending energy of the membrane under the constraints of fixed area of the membrane and fixed volume of the vesicle [13]. Henceforth, the elastic theory of lipid membranes based on the Helfrich functional (1.5) began to flourish [14–16]. In this review, we will survey several key theoretical results during the development of the elastic theory of lipid membranes according to our personal preferences. In Sect. 2, we will introduce a mathematical preliminary—calculus of variation in a deformable surface. In Sect. 3, we will present some theoretical results on configurations of closed lipid vesicles. In Sect. 4, we will present some theoretical results on configurations of open lipid vesicles with holes. In Sect. 5, we will discuss the concepts of stress tensor and moment tensor in fluid membranes. In Sect. 6, we will probe into new applications of the Helfrich functional and understand the growth mechanism of some mesoscopic structures. In the last section, we will give a brief summary and propose some perspectives.

2 Calculus of Variation in a Deformable Surface

In this section, we introduce the theory of surfaces and the variation problem in a deformable surface, which are based on the method of moving frames.

2.1 Theory of Surfaces Based on the Method of Moving Frames

Consider a 2D surface in 3D Euclidean space. Any point on the surface may be represented by a position vector $\mathbf{r}$. At point $\mathbf{r}$ we may construct a right-handed orthonormal frame $\{\mathbf{e}_1, \mathbf{e}_2, \mathbf{e}_3\}$ with $\mathbf{e}_3$ being the normal vector at that point. The set $\{\mathbf{r}; \mathbf{e}_1, \mathbf{e}_2, \mathbf{e}_3\}$ is called a moving frame.

The differentiation of the frame may be defined as [17]:

$$\mathrm{d}\mathbf{r} = \omega_1 \mathbf{e}_1 + \omega_2 \mathbf{e}_2, \tag{2.1}$$

and

$$\mathrm{d}\mathbf{e}_i = \omega_{ij} \mathbf{e}_j, \quad (i = 1, 2, 3) \tag{2.2}$$

where ω_1, ω_2, and $\omega_{ij} = \omega_{ji}$ $(i, j = 1, 2, 3)$ are 1-forms, and 'd' is the exterior differential operator. The repeated subscripts in this paper abide by the Einstein summation convention.

The area element can be expressed as [17]:

$$\mathrm{d}A \equiv \omega_1 \wedge \omega_2. \tag{2.3}$$

The structure equations of the surface can be expressed as [17]:

$$\begin{cases} \mathrm{d}\omega_1 = \omega_{12} \wedge \omega_2, \\ \mathrm{d}\omega_2 = \omega_{21} \wedge \omega_1, \\ \mathrm{d}\omega_{ij} = \omega_{ik} \wedge \omega_{kj} \quad (i, j = 1, 2, 3), \end{cases} \tag{2.4}$$

and

$$\begin{pmatrix} \omega_{13} \\ \omega_{23} \end{pmatrix} = \begin{pmatrix} a & b \\ b & c \end{pmatrix} \begin{pmatrix} \omega_1 \\ \omega_2 \end{pmatrix}. \tag{2.5}$$

Then we can define a curvature tensor as

$$\mathcal{C} = a\mathbf{e}_1\mathbf{e}_1 + b\mathbf{e}_1\mathbf{e}_2 + b\mathbf{e}_2\mathbf{e}_1 + c\mathbf{e}_2\mathbf{e}_2, \tag{2.6}$$

where $\mathbf{e}_i\mathbf{e}_j$ $(i, j = 1, 2)$ represents the dyad of $\mathbf{e}_i$ and $\mathbf{e}_j$. The mean curvature and the Gauss curvature are respectively defined as

$$H = \mathrm{tr}(\mathcal{C})/2 = (a + c)/2, \tag{2.7}$$

and

$$K = \det(\mathcal{C}) = ac - b^2. \tag{2.8}$$

For a curve on the surface, at each point in the curve we can construct the tangent vector **t**. The normal curvature, the geodesic curvature, and the geodesic torsion of the curve may be expressed as

$$\kappa_n = a\cos^2\phi + 2b\cos\phi\sin\phi + c\sin^2\phi, \tag{2.9}$$

$$\tau_g = b\cos 2\phi + (c-a)\cos\phi\sin\phi \tag{2.10}$$

$$\kappa_g = (\mathrm{d}\phi + \omega_{12})/\mathrm{d}s \tag{2.11}$$

respectively, where ϕ represents the angle between **t** and $\mathbf{e}_1$.

2.2 *Calculus of Variations Based on the Method of Moving Frames*

Calculus of variation based on the method of moving frames was developed in the previous work by the present authors [18–20]. The main ideas are sketched as follows.

Any infinitesimal deformation of a surface can be achieved by a displacement vector

$$\delta\mathbf{r} \equiv \mathbf{\Omega} = \Omega_i\mathbf{e}_i \tag{2.12}$$

at each point on the surface, where δ can be understood as a variational operator. The frame is also changed due to the deformation of the surface. Its variation is denoted as

$$\delta\mathbf{e}_i = \Omega_{ij}\mathbf{e}_j \quad (i = 1, 2, 3), \tag{2.13}$$

where $\Omega_{ij} = -\Omega_{ji}$ $(i, j = 1, 2, 3)$. Ω_{23}, Ω_{31}, and Ω_{12} correspond to the infinitesimal rotation of the frame around direction $\mathbf{e}_1$, $\mathbf{e}_2$, and $\mathbf{e}_3$, respectively.

From $\delta\mathrm{d}\mathbf{r} = \mathrm{d}\delta\mathbf{r}$, $\delta\mathrm{d}\mathbf{e}_j = \mathrm{d}\delta\mathbf{e}_j$, we can derive:

$$\delta\omega_1 + \omega_2\Omega_{21} = \mathrm{d}\mathbf{\Omega}\cdot\mathbf{e}_1 = \mathrm{d}\Omega_1 + \Omega_2\omega_{21} + \Omega_3\omega_{31}, \tag{2.14}$$

$$\delta\omega_2 + \omega_1\Omega_{12} = \mathrm{d}\mathbf{\Omega}\cdot\mathbf{e}_2 = \mathrm{d}\Omega_2 + \Omega_1\omega_{12} + \Omega_3\omega_{32}, \tag{2.15}$$

$$\Omega_{13}\omega_1 + \Omega_{23}\omega_2 = \mathrm{d}\mathbf{\Omega}\cdot\mathbf{e}_3 = \mathrm{d}\Omega_3 + \Omega_1\omega_{13} + \Omega_2\omega_{23}, \tag{2.16}$$

$$\delta\omega_{ij} = \mathrm{d}\Omega_{ij} + \Omega_{il}\omega_{lj} - \omega_{il}\Omega_{lj}. \tag{2.17}$$

These equations are the essential equations of the variational method based on the moving frames.

With essential Eqs. (2.14)–(2.17), we may derive

$$\delta\mathrm{d}A = (\nabla\cdot\mathbf{\Omega} - 2H\Omega_3)\mathrm{d}A, \tag{2.18}$$

$$\delta(2H) = [\nabla^2 + (4H^2 - 2K)]\Omega_3 + \nabla(2H)\cdot\mathbf{\Omega}, \tag{2.19}$$

$$\delta K = \nabla \cdot \tilde{\nabla}\Omega_3 + 2KH\Omega_3 + \nabla K \cdot \mathbf{\Omega}. \tag{2.20}$$

In the above equations, the gradient operators and the Laplace operators are defined according to the differential operator, the Hodge star, and their generalizations as follows.

The 2D Hodge star operator ($*$) satisfies $*\omega_1 = \omega_2$ and $*\omega_2 = -\omega_1$ [21]. The generalized Hodge star operator ($\tilde{*}$) satisfies $\tilde{*}\omega_{13} = \omega_{23}$ and $\tilde{*}\omega_{23} = -\omega_{13}$ [19]. The generalized differential operator ($\tilde{\mathrm{d}}$) satisfies $\tilde{\mathrm{d}} f = f_1\omega_{13} + f_2\omega_{23}$ if $\mathrm{d} f = f_1\omega_1 + f_2\omega_2$ [19]. Then, we may define the gradient operator (of the first kind) and the gradient operator of the second kind as [19]:

$$\nabla f \cdot \mathrm{d}\mathbf{r} = \mathrm{d} f, \tag{2.21}$$

and

$$\tilde{\nabla} f \cdot *\mathrm{d}\mathbf{r} = \tilde{*}\tilde{\mathrm{d}} f, \tag{2.22}$$

respectively. Simultaneously, we may define the Laplace operator (of the first kind) and the Laplace operator of the second kind as [19]:

$$(\nabla^2 f)\,\mathrm{d}A = \mathrm{d} * \mathrm{d} f, \tag{2.23}$$

and

$$(\nabla \cdot \tilde{\nabla} f)\,\mathrm{d}A = \mathrm{d}\tilde{*}\tilde{\mathrm{d}} f, \tag{2.24}$$

respectively.

Let us consider, a functional which depends on the mean curvature and the Gauss curvature of a surface. In general, the functional may be expressed as the following form:

$$F_G = \int_M G(2H, K)\mathrm{d}A, \tag{2.25}$$

where $G = G(2H, K)$ is a function of $2H$ and K. It is not hard to calculate the first-order variation of functional (2.25) by using Eqs. (2.18)–(2.20). From tedious calculations, we obtain

$$\begin{aligned} \delta F_G = & \int_M [\nabla^2 G_{2H} + \nabla \cdot \tilde{\nabla} G_K + (4H^2 - 2K)G_{2H} + 2HKG_K - 2HG]\Omega_3 \mathrm{d}A \\ & + \oint_{\partial M} (G_{2H} * \mathrm{d}\Omega_3 - \Omega_3 * \mathrm{d}G_{2H} + G_K \tilde{*}\tilde{\mathrm{d}}\Omega_3 - \Omega_3 \tilde{*}\tilde{\mathrm{d}} G_K + G * \mathbf{\Omega} \cdot \mathrm{d}\mathbf{r}). \end{aligned} \tag{2.26}$$

where G_{2H} and G_K represent the partial derivatives of G with respect to $2H$ and K, respectively. $\oint_{\partial M}$ represents the integration along the boundary of surface M, which is vanishing for a closed surface.

3 Configurations of Closed Lipid Vesicles

As a model system, we will investigate configurations of a closed vesicle formed by a lipid bilayer. First, we will introduce the general shape equation for closed vesicles. Second, we will discuss the shape equation for axisymmetrical vesicles and its first integral. Finally, we will present several special solutions to the shape equation.

3.1 *Energy Functional and the Corresponding Euler–Lagrange Equation*

The bending energy of a closed vesicle may be described by the Helfrich functional (1.5). Since the area of lipid bilayer is almost inextensible and the volume of the closed vesicle is hardly compressed, we may introduce two Lagrange multipliers λ and p to replace these constraints. The extended energy functional of the closed vesicle may be expressed as

$$F = \int_M [(k_c/2)(2H + c_0)^2 + \bar{k}K + \lambda]\mathrm{d}A + pV, \tag{3.1}$$

where V represents the total volume enclosed in the vesicle. The Lagrange multiplier λ can be physically interpreted as the surface tension of the lipid bilayer. The Lagrange multiplier p can be regarded as the osmotic pressure of the vesicle, i.e., the pressure difference between the outer side and the inner side of the vesicle.

To derive the Euler–Lagrange equation corresponding to functional (3.1), we assume $G = (k_c/2)(2H + c_0)^2 + \bar{k}K + \lambda$. Substituting it into (2.26) and considering $\delta V = \int_M \Omega_3 \mathrm{d}A$, one can obtain

$$\delta F = \int_M [p - 2\lambda H + k_c(2H + c_0)(2H^2 - c_0 H - 2K) + 2k_c\nabla^2 H]\Omega_3. \tag{3.2}$$

The equilibrium configurations satisfy $\delta F = 0$, which leads to

$$p - 2\lambda H + k_c(2H + c_0)(2H^2 - c_0 H - 2K) + 2k_c\nabla^2 H = 0. \tag{3.3}$$

This equation was first derived by Ou-Yang and Helfrich [22, 23]. Now it is called the shape equation of lipid vesicles. Obviously, if $k_c = 0$, the above equation degenerates into the Young–Laplace equation $p - 2\lambda H = 0$. If $p = 0$ and $\lambda = 0$, the above equation degenerates into the Willmore Eq. (1.4).

3.2 Axisymmetrical Closed Vesicles

An axisymmetrical vesicle may be generated by its outline which is represented by $z = z(\rho)$ with ρ being the revolution radius. Take ϕ as the rotation angle and ψ as the tangent angle of the outline. The axisymmetrical vesicle may be parameterized as

$$x = \rho \cos \phi, \; y = \rho \sin \phi, \; z = \int \tan \psi(\rho) \mathrm{d}\rho. \tag{3.4}$$

According to Sect. 2, we can derive the mean curvature

$$H = -(\rho \sin \psi)'/2\rho, \tag{3.5}$$

the Gauss curvature

$$K = (\sin^2 \psi)'/2\rho, \tag{3.6}$$

and the Laplace operator

$$\nabla^2 = \frac{1}{\rho^2}\frac{\partial^2}{\partial \phi^2} + \frac{\cos\psi}{\rho}\frac{\partial}{\partial \rho}\left(\rho \cos\psi \frac{\partial}{\partial \rho}\right). \tag{3.7}$$

Substituting the above three equations into the general shape Eq. (3.3), one can derive the shape equation for axisymmetrical vesicles:

$$-\frac{\cos\psi}{\rho}\left\{\rho\cos\psi\left[\frac{(\rho\sin\psi)'}{\rho}\right]'\right\}' - \frac{1}{2}\left[\frac{(\rho\sin\psi)'}{\rho}\right]^3 + \frac{(\rho\sin\psi)'\left(\sin^2\psi\right)'}{\rho^2} - \frac{c_0\left(\sin^2\psi\right)'}{\rho} + \frac{\tilde{\lambda}\left(\rho\sin\psi\right)'}{\rho} + \tilde{p} = 0, \tag{3.8}$$

where $\tilde{\lambda} \equiv \lambda/k_c + c_0^2/2$ and $\tilde{p} \equiv p/k_c$. In addition, the prime represents the derivative with respect to radius ρ. The above equation is a third-order ordinary differential equation, which was first derived by Hu and Ou-Yang [24]. It is found that the above equation is integrable [25]. This equation may be further transformed into a second-order ordinary differential equation:

$$\frac{\Psi^3 - \Psi(\rho\Psi')^2}{2\rho} - \rho(1-\Psi^2)\left[\frac{(\rho\Psi)'}{\rho}\right]' - c_0\Psi^2 + \tilde{\lambda}\rho\Psi + \frac{\tilde{p}\rho^2}{2} = \eta_0, \tag{3.9}$$

where $\Psi \equiv \sin\psi$ and η_0 being the first integral.

3.3 *Analytical Special Solutions*

The shape Eq. (3.3) and its axisymmetrical counterparts (3.8) and (3.9) are nonlinear differential equations, to which one cannot achieve general solutions. Till now, researchers have known some special solutions to these equations, such as minimal surfaces (including catenoid, helicoid, etc.), surfaces with constant mean curvature (including sphere, cylinder, unduloid [26, 27], etc.), Willmore surfaces (including Clifford torus [28], Dupin Cyclide [29], inverted catenoid [30], etc.), cylinder-like surfaces [31–33], and circular biconcave discoid [34, 35]. Among these solutions, only sphere, Clifford torus, Dupin cyclide, and circular biconcave discoid correspond to closed vesicles without self-intersections.

3.3.1 Sphere

The mean curvature and the Gauss curvature of a spherical surface with radius R are $H = -1/R$ and $K = 1/R^2$, respectively. Substituting them into the shape Eq. (3.3), one can derive

$$\tilde{p}R^2 + 2\tilde{\lambda}R - 2c_0 = 0. \tag{3.10}$$

This equation gives the relation between the radius R, the spontaneous curvature c_0, the reduced osmotic pressure $\tilde{p} \equiv p/k_c$, and the reduced surface tension $\tilde{\lambda} \equiv \lambda/k_c + c_0^2/2$. Obviously, if $\tilde{\lambda}^2 + 2c_0\tilde{p} < 0$, there is no spherical vesicle satisfying the shape equation. If $\tilde{\lambda}^2 + 2c_0\tilde{p} = 0$, there merely exists one spherical vesicle satisfying the shape equation. If $\tilde{\lambda}^2 + 2c_0\tilde{p} > 0$, there are two spherical vesicles satisfying the shape equation, which might correspond to the exocytosis or endocytosis of cells.

3.3.2 Clifford Torus

The Clifford torus is a revolution surface generated by a circle with radius r which rotates around an axis in the same plane of the circle. The revolution radius R should be larger than r. The torus may be parameterized as $\{(R + r\cos\varphi)\cos\phi, (R + r\cos\varphi)\sin\phi, r\sin\varphi\}$. The mean curvature and the Gauss curvature are $H = -(R + 2r\cos\varphi)/2r(R + r\cos\varphi)$ and $K = \cos\varphi/r(R + r\cos\varphi)$, respectively. Substituting them into the shape Eq. (3.3), one can derive $\tilde{\lambda} = 2c_0/r$, $\tilde{p} = -2c_0/r^2$, and

$$R/r = \sqrt{2}. \tag{3.11}$$

That is, there exists a lipid torus with the ratio of its two generation radii being $\sqrt{2}$ [28], which was confirmed in the experiment [36]. This kind of Clifford torus is called the Willmore torus [8] in mathematics. It is also found that nonaxisymmetric tori [37] constructed from conformal transformations of the Willmore torus also satisfy the shape equation. In addition, it is not hard to check that $\eta_0 = -2c_0 - 1/r$ from equation (3.9) when $\tilde{\lambda} = 2c_0/r$, $\tilde{p} = -2c_0/r^2$, and $R/r = \sqrt{2}$.

3.3.3 Dupin Cyclide

The Dupin cyclide may be expressed as

$$(x^2 + y^2 + z^2 + a^2 - c^2 - \mu^2)^2 = 4(ax - c\mu)^2 + 4(a^2 - c^2)y^2, \qquad (3.12)$$

where $a > \mu > c$ are three real parameters. Ou-Yang [29] found that the Dupin cyclide could satisfy the shape Eq. (3.3) when $p = 0$, $\lambda = 0$, $c_0 = 0$ and $\mu^2 = (a^2 + c^2)/2$. This kind of lipid vesicles were also observed in the experiment by Fourcade and his coworkers [38]. The Dupin cyclide and conformal transformations of the Willmore torus mentioned above are two classes of the few known asymmetric solutions to the shape Eq. (3.3) up to now.

3.3.4 Circular Biconcave Discoid

Naito et al. [34, 35] found that the parametric equation

$$\begin{cases} \Psi \equiv \sin\psi = -c_0\rho\ln(\rho/\rho_B) \\ z = z_0 + \int_0^\rho \tan\psi \mathrm{d}\rho \end{cases} \qquad (3.13)$$

corresponds to the contour line of a circular biconcave discoid when $0 < |c_0\rho_B| < \mathrm{e}$. Substituting it into Eq. (3.9), one obtains $\tilde{p} = 0$, $\tilde{\lambda} = c_0^2/2$, and $\eta_0 = -2c_0 \neq 0$. Fitting the experimental results by Evans and Fung [39], Naito et al. obtained $c_0R_0 = -1.618$ where R_0 is the reduced radius of a red blood cell [35], i.e., $4\pi R_0^2$ corresponds to the surface area of the red blood cell. It is quite interesting that $c_0R_0 = -1.618 = -1/0.618$ happens to correspond the golden ratio.

4 Configurations of Open Lipid Vesicles with Holes

Open bilayer configurations can be stabilized by some proteins [40]. This experimental fact gave rise to investigating the configurations of lipid membranes with free exposed edges based on the Helfrich functional [18, 41–45].

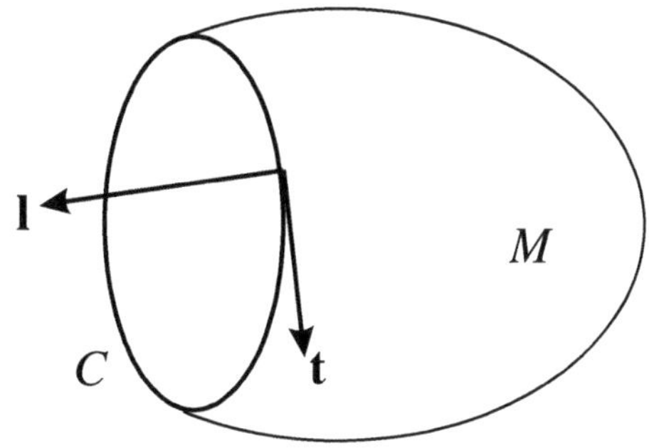

Fig. 1 An open smooth surface (M) with a boundary curve (C)

4.1 Energy Functional and the Corresponding Euler-Lagrange Equation

A lipid vesicle with a hole (i.e., a free edge) may be expressed as an open smooth surface M with a boundary curve $C = \partial M$ shown in Fig. 1. $\mathbf{t}$ represents the unit tangent vector of curve C. $\mathbf{l}$ is a unit vector which is perpendicular to $\mathbf{t}$ and the normal vector of the surface.

Based on the Helfrich functional, the energy functional for a lipid bilayer with a free edge may be expressed as

$$F = \int_M [(k_c/2)(2H + c_0)^2 + \bar{k}K + \lambda]\mathrm{d}A + \gamma \oint_C \mathrm{d}s, \tag{4.1}$$

where γ represents the line tension due to energy cost of the exposed edge. ds is the arc length element of curve C. According to the variational method in Sect. 2, from $\delta F = 0$ we can obtain the shape equation

$$2k_c\nabla^2 H + k_c(2H + c_0)(2H^2 - c_0 H - 2K) - 2\lambda H = 0, \tag{4.2}$$

and three boundary conditions [18, 41]

$$\left[k_c(2H + c_0) + \bar{k}\kappa_n\right]_C = 0, \tag{4.3}$$

$$\left[2k_c\partial H/\partial\mathbf{l} + \bar{k}\mathrm{d}\tau_g/\mathrm{d}s + \gamma\kappa_n\right]_C = 0, \tag{4.4}$$

$$\left[(k_c/2)(2H + c_0)^2 + \bar{k}K + \lambda + \gamma\kappa_g\right]_C = 0, \tag{4.5}$$

where κ_n, κ_g, and τ_g are the normal curvature, geodesic curvature, and geodesic torsion of the boundary curve, respectively. The above boundary conditions represent the force balance and the moment balance at each point in boundary curve C. They are also available for vesicles with more than one hole.

4.2 Axisymmetrical Situation

Consider an axisymmetric surface generated by a planar curve $z = z(\rho)$, which may be expressed as a vector form $\mathbf{r} = \{\rho\cos\phi, \rho\sin\phi, z(\rho)\}$ where ρ and ϕ are the rotation radius and azimuth angle, respectively. Under the axisymmetrical situation, the shape Eq. (4.2) is just the same as (3.8) with vanishing p. This equation is integrable and can be further transformed into

$$\frac{\Psi^3 - \Psi(\rho\Psi')^2}{2\rho} - \rho(1-\Psi^2)\left[\frac{(\rho\Psi)'}{\rho}\right]' - c_0\Psi^2 + \tilde{\lambda}\rho\Psi = \eta_0, \tag{4.6}$$

which is just the same as Eq. (3.9) with vanishing $\tilde{p}$.

For the boundary point C, we define a sign function $\sigma = \mathbf{t}\cdot\partial\mathbf{r}/\partial\phi$. The above boundary conditions (4.3)–(4.5) may be transformed into [18, 42]:

$$\Psi'|_C = c_0 - (1+\tilde{k})(\Psi/\rho)|_C, \tag{4.7}$$

$$\Psi''|_C = \left[\frac{\tilde{\gamma}\Psi}{\rho\sigma\cos\psi} + (2+\tilde{k})\frac{\Psi}{\rho^2} - \frac{c_0}{\rho}\right]_C, \tag{4.8}$$

$$\left[c_0\tilde{k}\left(\frac{\Psi}{\rho}\right) - \tilde{k}\left(1+\frac{\tilde{k}}{2}\right)\left(\frac{\Psi}{\rho}\right)^2 - \sigma\tilde{\gamma}\frac{\cos\psi}{\rho}\right]_C = \frac{c_0^2}{2} - \tilde{\lambda}, \tag{4.9}$$

where $\tilde{k} \equiv \bar{k}/k_c$, $\tilde{\gamma} \equiv \gamma/k_c$, $\Psi \equiv \sin\psi$, and $\tilde{\lambda} \equiv \lambda/k_c + c_0^2/2$. Since the boundary point is also in the surface, Eq. (4.6) should still hold for the boundary point C. From Eqs. (4.8) and (4.9) we can eliminate $\tilde{\gamma}$ and obtain the expression of $\Psi''|_C$. Substituting it and Eq. (4.7) into (4.6), we obtain a compatibility condition between the shape equation and boundary conditions for axisymmetrical open lipid vesicles:

$$\eta_0 = 0. \tag{4.10}$$

Under this condition, the above boundary conditions are not independent of each other. We may keep Eqs. (4.7) and (4.9) as boundary conditions. The shape equation may be expressed as (4.6) with vanishing η_0.

4.3 Analytical Special Solutions

Since the shape equation and boundary conditions are nonlinear, one may take the following procedure to find analytical special solutions: (i) finding a surface satisfying the shape equation; (ii) finding a curve C on that surface such that the boundary conditions are satisfied; (iii) the domain enclosed by boundary curve C on that surface being the solution. However, for a given surface satisfying the shape equation, we may not always find a curve C on that surface such that the boundary conditions are satisfied. On what kind of surface satisfying the shape equation can we find a curve C such that the boundary conditions are satisfied? This issue was named as compatibility between the shape equation and boundary conditions [42]. For example, the compatibility condition for axisymmetrical solutions is just Eq. (4.10), i.e., the vanishing first integral.

In general case without axisymmetry, we may obtain the compatibility condition [42]

$$2\int_M (c_0 H + \tilde{\lambda})\mathrm{d}A + \tilde{\gamma}\oint_C \mathrm{d}s = 0. \tag{4.11}$$

through scaling analysis. Here, we do not exclude the possibility to achieve the other compatibility conditions through specific method. Using the compatibility conditions (4.10) and (4.11), we can verify a theorem of nonexistence [42, 45]: For finite line tension, there does NOT exist an open membrane being a part of surfaces with nonvanishing constant mean curvature (including sphere, cylinder, and unduloid etc.), Willmore surfaces (including Willmore torus, Dupin cyclide, and inverted catenoid etc.), and circular biconcave discoid.

The above theorem of nonexistence merely leaves a small window for the surfaces simultaneously satisfying the shape equation and the boundary conditions that we have known till now. When c_0 is vanishing, the shape equation holds for minimal surfaces. Three boundary conditions (4.3)–(4.5) are degenerated to

$$\kappa_n = 0, \ \kappa_g = -\lambda/\gamma = \text{constant}, \tag{4.12}$$

which implies that the boundary should be an asymptotic curve with constant geodesic curvature. A domain in a minimal surface with a smooth boundary being an asymptotic curve with constant geodesic curvature is called a minimal geodesic disk. Obviously, a planar circular disk is a trivially minimal geodesic disk since a plane is a special minimal surface with vanishing Gauss curvature. We have conjecture that a planar disk is the unique minimal geodesic disk [46, 47]. This conjecture is probably true. Recently, we have noted that, following the work on flat points of minimal surfaces by Koch and Fischer [48], Giomi and Mahadevan argued that there does not exist a simple domain bounded by a smooth asymptotic curve in a minimal surface with nonvanishing Gauss curvature [49]. If their argument is true, then our conjecture is straightforward since a circle in a plane is the unique planar curve with constant geodesic curvature.

5 Stress Tensor and Moment Tensor in Fluid Membranes

The concepts of stress tensor and moment tensor in fluid membranes were mainly developed by Guven and his coworkers [50–53]. These concepts may be used to derive the boundary conditions of an open lipid vesicle with a hole [41] and the linking conditions of a lipid vesicle with two-phase domains [54].

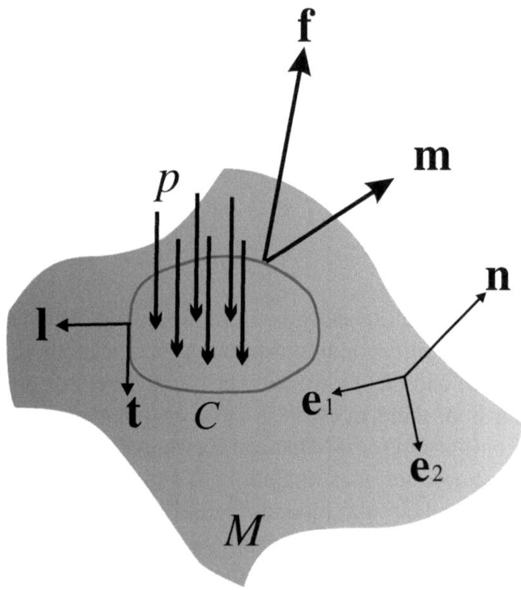

Fig. 2 Force and moment loaded on a domain cut from a fluid membrane

5.1 Balances of Local Forces and Moments

The concepts of stress tensor and moment tensor come from the force balance and the moment balance for any domain in a lipid membrane. As shown in Fig. 2, we cut a domain D bounded by a curve C from the lipid membrane. $\{\mathbf{e}_1, \mathbf{e}_2, \mathbf{n}\}$ is a right-handed orthogonal frame with $\mathbf{n} \equiv \mathbf{e}_3$ being the unit normal vector of the surface. A pressure p is loaded on the surface against the normal direction. The notations of $\mathbf{t}$ and $\mathbf{l}$ are the same as those in the above section. Vectors $\mathbf{f}$ and $\mathbf{m}$, respectively represent the density of force and the density of moment loaded on curve C by the lipids outside the domain.

According to Newtonian mechanics, the force balance and the moment balance may be expressed as

$$\oint_C \mathbf{f} \mathrm{d}s - \int p\mathbf{n} \mathrm{d}A = 0, \tag{5.1}$$

$$\oint_C \mathbf{m} \mathrm{d}s + \oint_C \mathbf{r} \times \mathbf{f} \mathrm{d}s - \int \mathbf{r} \times p\mathbf{n} \mathrm{d}A = 0. \tag{5.2}$$

If defining two second-order tensors $\mathcal{S}$ and $\mathcal{M}$ such that

$$\mathcal{S} \cdot \mathbf{l} = \mathbf{f}, \text{ and } \mathcal{M} \cdot \mathbf{l} = \mathbf{m}, \tag{5.3}$$

one can derive the equilibrium equations [20]:

$$\mathrm{div}\,\mathcal{S} = p\mathbf{n}, \tag{5.4}$$

$$\mathrm{div}\,\mathcal{M} = \mathcal{S}_1 \times \mathbf{e}_1 + \mathcal{S}_2 \times \mathbf{e}_2, \tag{5.5}$$

with $\mathcal{S}_1 \equiv \mathcal{S} \cdot \mathbf{e}_1$ and $\mathcal{S}_2 \equiv \mathcal{S} \cdot \mathbf{e}_2$ from the Stokes theorem. The tensors $\mathcal{S}$ and $\mathcal{M}$ are called the stress tensor and the moment tensor, respectively.

5.2 Explicit Expressions of Stress Tensor and Moment Tensor

One may understand the equilibrium configuration of the cut lipid domain D in Fig. 2 from the point of energy view. That is, the equilibrium configuration abides by the following generalized variational principle [54]:

$$\begin{aligned}
&\delta \int_D [(k_c/2)(2H + c_0)^2 + \bar{k}K + \lambda]\mathrm{d}A \\
&+ \int_D p\mathbf{n} \cdot \boldsymbol{\Omega}\mathrm{d}A - \oint_C \mathbf{f} \cdot \boldsymbol{\Omega}\mathrm{d}s - \oint_C \mathbf{m} \cdot \boldsymbol{\Theta}\mathrm{d}s \\
&+ \oint \mu[\Theta_1\omega_2 - \Theta_2\omega_1 - \Omega_1\omega_{13} - \Omega_2\omega_{23} - \mathrm{d}\Omega_3] = 0
\end{aligned} \tag{5.6}$$

The first line of the above equation represents the variation of bending energy of the lipid bilayer. The second line of the above equation reflects the potential energy increment due to the external loads. In the third line of the above equation, μ is a Lagrange multiplier due to the geometric constraint (2.16). The angular vector is defined as $\boldsymbol{\Theta} \equiv \Theta_i\mathbf{e}_i \equiv \Omega_{23}\mathbf{e}_1 + \Omega_{31}\mathbf{e}_2 + \Omega_{12}\mathbf{e}_3$.

Using the variational method mentioned in Sect. 2 and considering the definition (5.3), one can derive the explicit expressions of stress tensor and moment tensor as follows [54]:

$$\mathcal{S} = [(k_c/2)(2H + c_0)^2 + \lambda]\mathcal{I} - k_c(2H + c_0)\mathcal{C} - 2k_c\mathbf{n}\nabla H - (\mu\mathcal{C} - \mathbf{n}\nabla\mu) \times \mathbf{n}, \tag{5.7}$$

and

$$\mathcal{M} = \mu\mathcal{I} - [k_c(2H + c_0)\mathcal{I} + \bar{k}\mathcal{C}] \times \mathbf{n}, \tag{5.8}$$

where $\mathcal{I} \equiv \mathbf{e}_1\mathbf{e}_1 + \mathbf{e}_2\mathbf{e}_2$ represents the unit tensor, and $\mathcal{C}$ is the curvature tensor (2.6). It is not hard to verify that (5.5) automatically holds from the above two equations while Eq. (5.4) is equivalent to the shape Eq. (3.3). Substituting Eqs. (5.7) and (5.8) into (5.3), one may obtain the force and moment on the boundary C [54]:

$$\begin{aligned}
\mathbf{f} = {} & [k_c(2H + c_0)\tau_g - \mu\kappa_n]\mathbf{t} + [\nabla\mu \cdot \mathbf{t} - 2k_c\nabla H \cdot \mathbf{l}]\mathbf{n} \\
& + [k_c(2H + c_0)(c_0/2 - H + \kappa_n) + \lambda + \mu\tau_g]\mathbf{l},
\end{aligned} \tag{5.9}$$

and

$$\mathbf{m} = -[k_c(2H + c_0) + \bar{k}\kappa_n]\mathbf{t} + (\mu + \bar{k}\tau_g)\mathbf{l}, \tag{5.10}$$

where κ_n and τ_g represents the normal curvature and the geodesic torsion of the boundary, respectively.

Here two remarks should be mentioned. First, we only present the expressions of $\mathcal{S}$, $\mathcal{M}$, $\mathbf{f}$, and $\mathbf{m}$ based on the Helfrich functional. Their general forms can be found in Ref. [54]. Second, there is a Lagrange multiplier μ in the above expressions, which comes from the geometric constraint (2.16). Its physical meaning is still unknown. The terms related to μ in the expressions of $\mathcal{S}$, $\mathcal{M}$, $\mathbf{f}$, and $\mathbf{m}$ have not been included in the previous researches [20, 47, 50–53].

5.3 *Simple Applications of Stress Tensor and Moment Tensor*

Here, we will survey two applications of the concepts of stress tensor and moment tensor. One is the derivation of the boundary conditions of an open lipid vesicle with a hole [41]; another is the derivation of the linking conditions of a lipid vesicle with two-phase domains [54]. The basic ideas are as follows.

Consider a string loaded by a force density $\mathbf{f}$ and a moment density $\mathbf{m}$. The line tension γ induces a stretching force along the tangent vector of the string. From the force balance and the moment balance, one can easily derive two equilibrium equations [54]:

$$\gamma\kappa(s)\mathbf{N} + \mathbf{f}(s) = 0, \tag{5.11}$$

$$\mathbf{m}(s) = 0, \tag{5.12}$$

where s is the arc length parameter of the string. $\kappa(s)$ is the curvature of the string at s.

Now cut a very thin ribbon along the edge from the membrane as shown in Fig. 3. $\mathbf{t}$ and $\mathbf{t}_i$ represent the tangent vector of the boundary curve and that of the cutting line, respectively. $\mathbf{l}$ is perpendicular to the normal vector of membrane surface and the tangent vector of the boundary curve. $\mathbf{l}_i$ is perpendicular to the normal vector of membrane surface and the tangent vector of the cutting line. $\mathbf{f}$ and $\mathbf{m}$ represent the force density and the moment density induced by the membrane, respectively. Since $\mathbf{t}_i = -\mathbf{t}$ and $\mathbf{l}_i = -\mathbf{l}$, according Eqs. (5.9) and (5.10), we have

$$\begin{aligned}\mathbf{f} = &-[k_c(2H + c_0)\tau_g - \mu\kappa_n]\mathbf{t} - [\nabla\mu\cdot\mathbf{t} - 2k_c\nabla H\cdot\mathbf{l}]\mathbf{n}\\ &- [k_c(2H + c_0)(c_0/2 - H + \kappa_n) + \lambda + \mu\tau_g]\mathbf{l},\end{aligned} \tag{5.13}$$

and

$$\mathbf{m} = [k_c(2H + c_0) + \bar{k}\kappa_n]\mathbf{t} - (\mu + \bar{k}\tau_g)\mathbf{l}. \tag{5.14}$$

Fig. 3 Thin ribbon cut from the membrane along the edge

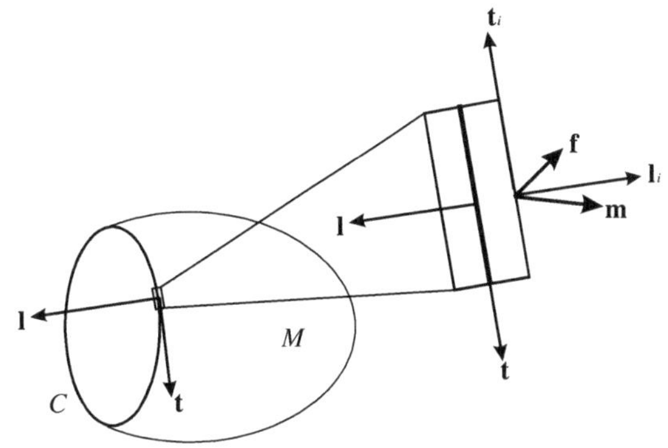

Substituting Eq. (5.14) into (5.12), we can obtain

$$\mu = -\bar{k}\tau_g, \tag{5.15}$$

and the boundary condition (4.3). If we substituting Eqs. (4.3) and (5.15) into (5.13), the force density is transformed into

$$\mathbf{f} = -[(k_c/2)(2H + c_0)^2 + \bar{k}K + \lambda]\mathbf{l} + [\bar{k}\mathrm{d}\tau_g/\mathrm{d}s + 2k_c\partial H/\partial\mathbf{l}]\mathbf{n}. \tag{5.16}$$

In the above derivation, we have used $\mathrm{d}\tau_g/\mathrm{d}s = \nabla\tau_g \cdot \mathbf{t}$, $\partial H/\partial\mathbf{l} = \nabla H \cdot \mathbf{l}$, and $(2H - \kappa_n)\kappa_n - \tau_g^2 = K$. Substituting Eq. (5.16) into (5.11) and considering $\kappa_n = \kappa\mathbf{n} \cdot \mathbf{N}$ and $\kappa_g = -\kappa\mathbf{l} \cdot \mathbf{N}$, we can obtain the boundary conditions (4.4) and (4.5).

Similar procedure is also available to derive the linking conditions of a lipid vesicle with two-phase domains as shown in Fig. 4. The separation line between domain I (D^{I}) and domain II (D^{II}) is denoted as curve C. $\mathbf{t}$ is the tangent vector of curve C. $\mathbf{l}^{\mathrm{I}}$ is perpendicular to $\mathbf{t}$ and the normal vector of surface. Different from Fig. 1, $\mathbf{l}^{\mathrm{I}}$ points to the side of domain I. The definition of $\mathbf{l}^{\mathrm{II}}$ is similar. The subtle difference is that $\mathbf{l}^{\mathrm{II}}$ points to the side of domain II. Assume that the membrane surface is so smooth that $\mathbf{l}^{\mathrm{II}} = -\mathbf{l}^{\mathrm{I}}$.

An axisymmetrical vesicle with two-phase domains was investigated by Jülicher and Lipowsky [55] many years ago. The general cases without the axisymmetrical

Fig. 4 A lipid vesicle with two-phase domains

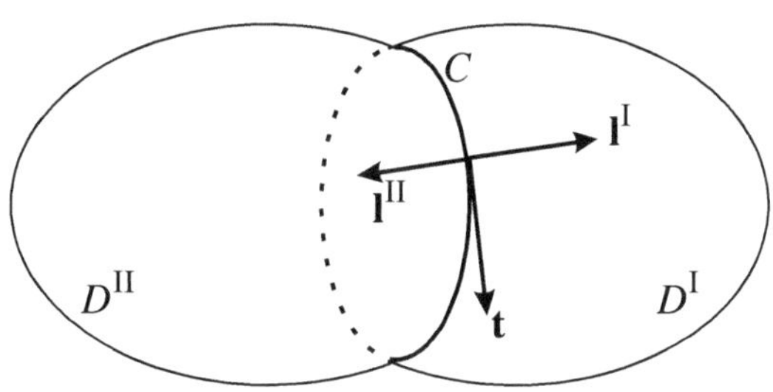

precondition have also been discussed by several researchers in recent years [19, 54, 56]. Following the Helfrich functional, the free energy of a lipid vesicle with two-phase domains may be expressed as [55]

$$F = \sum_{i=\mathrm{I}}^{\mathrm{II}} \int_{D^i} [(k_c^i/2)(2H^i + c_0^i)^2 + \bar{k}^i K^i + \lambda^i] \mathrm{d}A^i + pV + \gamma \oint_C \mathrm{d}s, \quad (5.17)$$

where superscript i labels the quantities of domain i ($i = \mathrm{I}, \mathrm{II}$). The linking conditions of separation curve C may also be derived from the concepts of stress tensor and moment tensor, which are listed as follows [54]:

$$k_c^{\mathrm{I}}(2H^{\mathrm{I}} + c_0^{\mathrm{I}}) + \bar{k}^{\mathrm{I}} \kappa_n = k_c^{\mathrm{II}}(2H^{\mathrm{II}} + c_0^{\mathrm{II}}) + \bar{k}^{\mathrm{II}} \kappa_n, \quad (5.18)$$

$$\frac{\partial \left[k_c^{\mathrm{I}}(2H^{\mathrm{I}} + c_0^{\mathrm{I}})\right]}{\partial \mathbf{l}^{\mathrm{I}}} + \frac{\partial \left[k_c^{\mathrm{II}}(2H^{\mathrm{II}} + c_0^{\mathrm{II}})\right]}{\partial \mathbf{l}^{\mathrm{II}}} = (\bar{k}^{\mathrm{I}} - \bar{k}^{\mathrm{II}}) \frac{\mathrm{d}\tau_g}{\mathrm{d}s} + \gamma \kappa_n, \quad (5.19)$$

and

$$\frac{k_c^{\mathrm{I}}}{2}(4H^{\mathrm{I}2} - c_0^{\mathrm{I}2}) - \frac{k_c^{\mathrm{II}}}{2}(4H^{\mathrm{II}2} - c_0^{\mathrm{II}2}) + (\bar{k}^{\mathrm{I}} - \bar{k}^{\mathrm{II}}) \left(\kappa_n^2 + \tau_g^2\right) = \lambda^{\mathrm{I}} - \lambda^{\mathrm{II}} + \gamma \kappa_g. \quad (5.20)$$

It should be noted that the mean curvature could be discontinuous across the separation curve. Using the above linking conditions, the Jülicher-Lipowsky conjecture on the general neck condition for the limit shape of budding vesicles was verified [54].

6 Understanding the Growth Mechanism of Some Mesoscopic Structures Based on the Helfrich Functional

The growth of mesoscopic structures is different from that of macroscopic structures. Macroscopic structures usually correspond to the least minimal free energy. But in the mesoscopic scale, there is no enough time for the structures to release energy as heat. Thus most mesoscopic structures exist in a metastable state where the different kinds of energies are balanced each other. With the consideration of the Helfrich functional, this idea has been used to explain the formation of focal conic domain in smectic-A liquid crystals [57, 58], the pitch angle of helices of multi-walled carbon nanotubes [59], and the reversible transition between peptide nanotubes and spherical vesicles [60].

6.1 Focal Conic Domain in Smectic-A Liquid Crystals

In Smectic-A (SmA) liquid crystals [61], the molecules are stacked layer-by-layer. In each layer, the orientations of all molecules are aligned to the normal of the layer. Generally, the flat configuration is energetically favorable. However, Dupin cyclides are usually formed when liquid crystals cool from the isotropic (Iso) phase to the SmA phase. Series of Dupin cyclides constitute a focal conic domain. Bragg argued that "There must be a reason why the cyclides are preferred, and it must be based on energy considerations" [62]. Natio et al. proposed that the relieved energy of the difference in the Gibbs free energy of Iso-SmA transition must be balanced by the curvature elastic energy of the smectic layers [58].

The formation energy of a focal conic domain includes three kinds of contributions. First is the volume free energy change due to the Iso-SmA transition [57]:

$$F_V = -g_0 \oint (D - D^2 H + D^3 K/3)\mathrm{d}A, \tag{6.1}$$

where $g_0 > 0$ is the difference in the Gibbs free energy density between SmA and Iso phases. H and K represent the mean and Gauss curvatures of the inner surface, respectively. Second is the surface energy of inner and outer SmA-Iso interfaces [58]:

$$F_A = \lambda \oint (1 + |1 - 2DH + D^2 K|)\mathrm{d}A, \tag{6.2}$$

where λ is the surface energy per area. Third is the curvature elastic energy, which is the sum of the energy (in the Helfrich form) of each layers. In the continuum limit, the curvature elastic energy may be expressed as [58]:

$$F_c = k_c \oint \sqrt{H^2 - K} \ln\left(\frac{1 - DH + D\sqrt{H^2 - K}}{1 - DH - D\sqrt{H^2 - K}}\right)\mathrm{d}A + \bar{k}D \oint K\mathrm{d}A. \tag{6.3}$$

Then the total formation energy may be expressed as $F = F_V + F_A + F_c \oint \Phi(H, K, D)\mathrm{d}A$ with

$$\Phi(H, K, D) \equiv k_c\sqrt{H^2 - K} \ln\left(\frac{1 - DH + D\sqrt{H^2 - K}}{1 - DH - D\sqrt{H^2 - K}}\right) + \bar{k}DK$$
$$+\lambda(1 + |1 - 2DH + D^2 K|) - g_0(D - D^2 H + D^3 K/3). \tag{6.4}$$

From $\delta F = 0$, Natio et al. obtained

$$(\nabla^2/2)\Phi_H + \nabla \cdot \tilde{\nabla}\Phi_K + (2H^2 - K)\Phi_H + 2HK\Phi_K - 2H\Phi = 0 \tag{6.5}$$

and

$$\oint \Phi_D \mathrm{d}A = 0, \tag{6.6}$$

where $\Phi_H \equiv \partial\Phi/\partial H$, $\Phi_K \equiv \partial\Phi/\partial K$, and $\Phi_D \equiv \partial\Phi/\partial D$. ∇^2 and $\nabla \cdot \tilde{\nabla}$ are the Laplace operators mentioned in Sect. 2. Natio et al. showed that the growth of the focal conic domain in SmA liquid crystals could be well explained by using the above two equations [58].

6.2 Helices of Multi-walled Carbon Nanotubes

The formation mechanism of a multi-walled carbon nanotube is similar to that of focal conic domain in SmA liquid crystals mentioned above. The formation energy of the multi-walled carbon nanotube also consists of three terms: (i) the volume term which may be expressed in the same form of (6.1); (ii) the surface term which may be expressed as the same form of (6.2); (iii) the curvature energy which may be expressed as the same form of (6.3) since the bending energy of a single layer of graphene was proven to have the Helfrich form [59]. If considering that the radius of the carbon nanotube is much smaller than the curvature radius of the central axis of the carbon nanotube, the total formation energy may be transformed into

$$F = m \int \mathrm{d}s + \alpha \int \kappa^2 \mathrm{d}s, \tag{6.7}$$

where m and α are two elastic constants. κ and s represent the curvature and the arc length of the central axis of the carbon nanotube, respectively. The first-order variation $\delta F = 0$ yields the equilibrium-shape equations of a string [63]:

$$2\mathrm{d}^2\kappa/\mathrm{d}s^2 + \kappa^3 - 2\kappa\tau^2 - (m/\alpha)\kappa = 0, \tag{6.8}$$

$$\kappa^2\tau = \text{constant}. \tag{6.9}$$

One solution to the above shape equations is a straight multi-walled carbon nanotube with vanishing κ and τ. The other solution to the above shape equations is a helix with pitch angle θ. From Eqs. (6.8) and (6.7), one may calculate the total formation energy for the helix

$$F = ml[1 + 1/(1 - 2\tan^2\theta)], \tag{6.10}$$

where l represents the total length of the central axis of the helix. The threshold condition for formation of helix is $F = 0$, which requires $\theta = \pi/4$. This value is in a good agreement with the pitch angle observed in the experiment [64].

6.3 Reversible Transition Between Peptide Nanotubes and Spherical Vesicles

In recent work, Yan et al. have observed the reversible transition between peptide nanotubes and vesicle-like structures [60]. It was found that the dilution of a peptide-nanotube dispersion solution results in the formation of vesicle-like structures, which can be reassembled into the nanotubes by concentrating the solution [60]. The mechanism underlying these phenomena is the same as the formation of the focal conic domain in SmA liquid crystals mentioned above.

As shown in reference [57], the outward growth of a layer with small thickness h on the top of the outermost equilibrium dipeptide aggregate (the nanotube or vesicle-like structure) leads to three kinds of free energy accumulations. First is the increment of the volume free energy:

$$F_V = -g_0 \oint (h - h^2 H + h^3 K/3)\mathrm{d}A, \tag{6.11}$$

where H and K are the mean curvature and the Gauss curvature of the outer surface of the dipeptide aggregate, respectively. g_0 is the difference in the Gibbs free energy density between the solution phase and the aggregate phase. Its value could be estimated with the ideal gas model, which reads

$$g_0 = C_A k_B T \ln(C_A/C_S), \tag{6.12}$$

where C_A and C_S are the concentrations of dipeptide in the aggregate phase and the solution phase, respectively [60]. k_B and T are the Boltzman constant and the temperature of the solution. Second is the extra interfacial free energy:

$$F_A = \lambda \oint (-2hH + h^2 K)\mathrm{d}A, \tag{6.13}$$

where λ is the surface energy per area of the solution/aggregate interface. Third is the extra curvature elastic energy, which can be expressed as the Helfrich form [57]:

$$F_c = \frac{k_1 h}{2} \oint (2H)^2 \mathrm{d}A + k_5 h \oint K \mathrm{d}A, \tag{6.14}$$

where k_1 and k_5 are related to the elastic constants of liquid crystals.

The equilibrium shape of the aggregate should satisfy $\partial F/\partial h = 0$, which leads to the Weingarten equation

$$2k_1 H^2 + k_5 K - g_0 - 2\lambda H = 0. \tag{6.15}$$

It is easy to verify that a sphere of radius r_0 and a cylinder of radius ρ_0 are two solutions to Eq. (6.15) provided that

$$r_0 = \frac{2k_1 + k_5}{\sqrt{\lambda^2 + g_0(2k_1 + k_5)} - \lambda} \approx \frac{2\lambda}{g_0}, \tag{6.16}$$

and

$$\rho_0 = \frac{k_1}{\sqrt{\lambda^2 + 2g_0 k_1} - \lambda} \approx \frac{\lambda}{g_0}. \tag{6.17}$$

The approximations in the above two equations have been done according to the experiment conditions [60]. From these equations, one can calculate the formation energy of sphere and tube, respectively. The results are $F_{\text{shpere}} = -(g_0^3 h^3/12\lambda^2 + g_0^2 h^2/4\lambda)$ and $F_{\text{tube}} = -g_0^2 h^2/2\lambda$. Thus, the condition for transition from a tube-to-a spherical structure is $F_{\text{tube}} > F_{\text{shpere}}$, that is, $g_0 h > 3\lambda$. Substituting $g_0 h = 3\lambda$ into Eq. (6.12) one can obtain the critical concentration for tube-to-vesicle-transition [60]

$$C_* = C_A \exp(-3\lambda/C_A h k_B T). \tag{6.18}$$

When $C_S < C_*$, peptide nanotubes will transform into spherical vesicle-like structures.

7 Conclusion

In the above discussions, we have presented several theoretical investigations based on the Helfrich functional (1.5). The configurations of closed lipid vesicles and open lipid vesicles with holes, and the concepts of stress tensor and moment tensor in fluid membranes were surveyed in detail. It was shown that the Helfrich functional could be extended to understand the growth mechanism of some mesoscopic structures.

The study of the Willmore functional (1.3) enters the epilog stage as the Willmore conjecture has been proved [9]. We believe that the times of studying the Helfrich functional is coming soon. Although the aforementioned theoretical achievements based on the Helfrich functional have been made in recent years, the substantial researches on the Helfrich functional are still in their infancy. If $c_0 > 0$, it is not hard to verify that among all compact embedded surfaces of genus 0, the round sphere with radius $R = 2/c_0$ corresponds to the least minimum of the Helfrich functional (1.5) from the Alexandrov theorem [4]. In other words, all compact embedded surfaces of genus 0 have energies no less than $4\pi\bar{k}$ for positive c_0. What will happen for $c_0 < 0$ or for embedded surfaces of nonvanishing genus? This is still an open question.

We are lack of good enough mathematical tools to deal with the Helfrich functional since the Helfrich functional, different from the Willmore functional, is not an invariant under conformal transformations. However, every coin has two sides. The breaking of conformal invariance also brings benefit to us. The critical configuration corresponding to the minimal value of the Helfrich functional should have the specific size. Introduce a scaling transformation $\mathbf{r} \to \Lambda\mathbf{r}$, where $\mathbf{r}$ represents the position vector of point on the critical configuration. Under the scaling transformation, the Helfrich functional is transformed into

$$\begin{aligned} F_H(\Lambda) = & \int_M [(k_c/2)(2H)^2 + \bar{k}K]\mathrm{d}A \\ & + 2k_c c_0 \Lambda \int_M H \mathrm{d}A + (k_c c_0^2/2)\Lambda^2 \int_M \mathrm{d}A. \end{aligned} \tag{7.1}$$

The critical configuration corresponds to $\Lambda = 1$, which implies that $F_H(\Lambda)$ takes minimal value when $\Lambda = 1$. If $c_0 \neq 0$, we may derive the necessary condition of the critical configuration:

$$\bar{H} \equiv \frac{\int_M H \mathrm{d}A}{\int_M \mathrm{d}A} = -\frac{c_0}{2}. \tag{7.2}$$

This necessary condition is quite similar to the known Minkowski formula [65]. In addition, the critical configuration of the Helfrich functional should also satisfy the shape Eq. (3.3) with vanishing p and λ. Integrating this equation and considering the Stokes theorem, we can obtain

$$\int_M H(4H^2 - 4K - c_0^2)\mathrm{d}A = 4\pi c_0 \chi(M), \tag{7.3}$$

where $\chi(M)$ is the characteristic number of surface M. The above Eqs. (7.2) and (7.3) might be helpful to the further study of the Helfrich functional.

Acknowledgments The authors are grateful to the financial support from the National Natural Science Foundation of China (Grant Nos. 11274046 and 10704009).

References

1. Plateau, J.: Statique Expérimentale et Théorique des Liquides Soumis aux Seules Forces Moléculaires. Gauthier-Villars, Paris (1873)
2. Young, T.: An essay on the cohesion of fluids. Philos. Trans. R. Soc. Lond. **95**, 65–87 (1805)
3. Laplace, P.: Traité de Mécanique Céleste. Gauthier-Villars, Paris (1839)
4. Alexandrov, A.: Uniqueness theorems for surfaces in the large. Amer. Math. Soc. transl. **21**, 341–416 (1962)
5. Poisson, S.: Traité de Mécanique. Bachelier, Paris (1833)
6. Willmore, T.: Total Curvature in Riemannian Geometry. Wiley, New York (1982)
7. Marques, F., Neves, A.: The Willmore conjecture. Jahresber. Dtsch. Math-Ver. **116**, 201–222 (2014)
8. Willmore, T.: Note on embedded surfaces. An. Ştiinţ. Univ. 'Al.I. Cuza' Iaşi, Mat. (N.S.) B **11**, 493–496 (1965)
9. Marques, F., Neves, A.: Min-max theory and the Willmore conjecture. Ann. Math. **179**, 683–782 (2014)

10. Canham, P.: The minimum energy of bending as a possible explanation of the biconcave shape of the human red blood cell. J. Theor. Biol. **26**, 61–81 (1970)
11. Singer, S., Nicolson, G.: The fluid mosaic model of cell membranes. Science **175**, 720–731 (1972)
12. Helfrich, W.: Elastic properties of lipid bilayers-theory and possible experiments. Z. Naturforsch. C **28**, 693–703 (1973)
13. Deuling, H., Helfrich, W.: Red blood cell shapes as explained on the basis of curvature elasticity. Biophys. J. **16**, 861–868 (1976)
14. Lipowsky, R.: The conformation of membranes. Nature **349**, 475–481 (1991)
15. Ou-Yang, Z., Liu, J., Xie, Y.: Geometric Methods in the Elastic Theory of Membranes in Liquid Crystal Phases. World Scientific, Singapore (1999)
16. Seifert, U.: Configurations of fluid membranes and vesicles. Adv. Phys. **46**, 13–137 (1997)
17. Chern, S., Chen, W.: Lecture on Differential Geometry. Beijing University Press, Beijing (1983)
18. Tu, Z., Ou-Yang, Z.: Lipid membranes with free edges. Phys. Rev. E **68**, 061915 (2003)
19. Tu, Z., Ou-Yang, Z.: A geometric theory on the elasticity of bio-membranes. J. Phys. A Math. Gen. **37**, 11407–11429 (2004)
20. Tu, Z., Ou-Yang, Z.: Elastic theory of low-dimensional continua and its applications in bio- and nano-structures. J. Comput. Theor. Nanosci. **5**, 422–448 (2008)
21. Westenholz, C.: Differential Forms in Mathematical Physics. North-Holland, Amsterdam (1981)
22. Ou-Yang, Z., Helfrich, W.: Instability and deformation of a spherical vesicle by pressure. Phys. Rev. Lett. **59**, 2486–2488 (1987)
23. Ou-Yang, Z., Helfrich, W.: Bending energy of vesicle membranes: general expressions for the first, second, and third variation of the shape energy and applications to spheres and cylinders. Phys. Rev. A **39**, 5280–5288 (1989)
24. Hu, J., Ou-Yang, Z.: Shape equations of the axisymmetric vesicles. Phys. Rev. E **47**, 461–467 (1993)
25. Zheng, W., Liu, J.: Helfrich shape equation for axisymmetric vesicles as a first integral. Phys. Rev. E **48**, 2856–2860 (1993)
26. Naito, H., Okuda, M., Ou-Yang, Z.: New solutions to the helfrich variation problem for the shapes of lipid bilayer vesicles: beyond delaunay's surfaces. Phys. Rev. Lett. **74**, 4345–4348 (1995)
27. Mladenov, I.: New solutions of the shape equation. Eur. Phys. J. B **29**, 327–330 (2002)
28. Ou-Yang, Z.: Anchor ring-vesicle membranes. Phys. Rev. A **41**, 4517–4520 (1990)
29. Ou-Yang, Z.: Selection of toroidal shape of partially polymerized membranes. Phys. Rev. E **47**, 747–749 (1993)
30. Castro-Villarreal, P., Guven, J.: Inverted catenoid as a fluid membrane with two points pulled together. Phys. Rev. E **76**, 011922 (2007)
31. Zhang, S., Ou-Yang, Z.: Periodic cylindrical surface solution for fluid bilayer membranes. Phys. Rev. E **53**, 4206–4208 (1996)
32. Vassilev, V., Djondjorov, P., Mladenov, I.: Cylindrical equilibrium shapes of fluid membranes. J. Phys. A Math. Theor. **41**, 435201 (2008)
33. Zhou, X.: Periodic-cylinder vesicle with minimal energy. Chin. Phys. B **19**, 058702 (2010)
34. Naito, H., Okuda, M., Ou-Yang, Z.: Counterexample to some shape equations for axisymmetric vesicles. Phys. Rev. E **48**, 2304–2307 (1993)
35. Naito, H., Okuda, M., Ou-Yang, Z.: Polygonal shape transformation of a circular biconcave vesicle induced by osmotic pressure. Phys. Rev. E **54**, 2816–2826 (1996)
36. Mutz, M., Bensimon, D.: Observation of toroidal vesicles. Phys. Rev. A **43**, 4525–4527 (1991)
37. Seifert, U.: Vesicles of toroidal topology. Phys. Rev. Lett. **66**, 2404–2407 (1991)
38. Fourcade, B., Mutz, M., Bensimon, D.: Experimental and theoretical study of toroidal vesicles. Phys. Rev. Lett. **68**, 2551–2554 (1992)
39. Evans, E., Fung, Y.: Improved measurements of the erythrocyte geometry. Microvasc. Res. **4**, 335–347 (1972)

40. Saitoh, A., Takiguchi, K., Tanaka, Y., Hotani, H.: Opening-up of liposomal membranes by Talin. Proc. Natl. Acad. Sci. **95**, 1026–1031 (1998)
41. Capovilla, R., Guven, J., Santiago, J.: Lipid membranes with an edge. Phys. Rev. E **66**, 021607 (2002)
42. Tu, Z.: Compatibility between shape equation and boundary conditions of lipid membranes with free edges. J. Chem. Phys. **132**, 084111 (2010)
43. Umeda, T., Suezaki, Y., Takiguchi, K., Hotani, H.: Theoretical analysis of opening-up vesicles with single and two holes. Phys. Rev. E **71**, 011913 (2005)
44. Wang, X., Du, Q.: Modelling and simulations of multi-component lipid membranes and open membranes via diffuse interface approaches. J. Math. Biol. **56**, 347–371 (2008)
45. Tu, Z.: Geometry of membranes. J. Geom. Symmetry Phys. **24**, 45–75 (2011)
46. Tu, Z.: Challenges in theoretical investigations of configurations of lipid membranes. Chin. Phys. B **22**, 028701 (2013)
47. Tu, Z., Ou-Yang, Z.: Recent theoretical advances in elasticity of membranes following Helfrich's spontaneous curvature model. Adv. Colloid Interface Sci. **208**, 66–75 (2014)
48. Koch, E., Fischer, W.: Flat points of minimal balance surfaces. Acta Cryst. A **46**, 33–40 (1990)
49. Giomi, L., Mahadevan, L.: Minimal surfaces bounded by elastic lines. Proc. R. Soc. A **468**, 1851–1864 (2012)
50. Capovilla, R., Guven, J.: Stresses in lipid membranes. J. Phys. A: Math. Gen. **35**, 6233–6247 (2002)
51. Müller, M., Deserno, M., Guven, J.: Interface-mediated interactions between particles: a geometrical approach. Phys. Rev. E **72**, 061407 (2005)
52. Müller, M., Deserno, M., Guven, J.: Balancing torques in membrane-mediated interactions: exact results and numerical illustrations. Phys. Rev. E **76**, 011921 (2007)
53. Deserno, M.: Fluid lipid membranes: from differential geometry to curvature stresses. Chem. Phys. Lipids **185**, 11–45 (2015)
54. Yang, P., Tu, Z.: General neck condition for the limit shape of budding vesicles. arXiv:1508.02151
55. Jülicher, F., Lipowsky, R.: Shape transformations of vesicles with intramembrane domains. Phys. Rev. E **53**, 2670–2683 (1996)
56. Du, Q., Guven, J., Tu, Z., Vázquez-Montejo, P.: Fluid membranes bounded by semi-flexible polymers (in preparation)
57. Naito, H., Okuda, M., Ou-Yang, Z.: Equilibrium shapes of smectic-A phase grown from isotropic phase. Phys. Rev. Lett. **70**, 2912–2915 (1993)
58. Naito, H., Okuda, M., Ou-Yang, Z.: Preferred equilibrium structures of a smectic-A phase grown from an isotropic phase: origin of focal conic domains. Phys. Rev. E **52**, 2095–2098 (1995)
59. Ou-Yang, Z., Su, Z., Wang, C.: Coil formation in multishell carbon nanotubes: competition between curvature elasticity and interlayer adhesion. Phys. Rev. Lett. **78**, 4055–4058 (1997)
60. Yan, X., Cui, Y., He, Q., Wang, K., Li, J., Mu, W., Wang, B., Ou-Yang, Z.: Reversible transitions between peptide nanotubes and vesicle-like structures including theoretical modeling studies. Chem. Eur. J. **14**, 5974–5980 (2008)
61. Friedel, G.: Les états mésomorphes de la matiére. Ann. Phys. **18**, 273–474 (1922)
62. Bragg, W.: Liquid crystals. Nature **133**, 445–456 (1934)
63. Langer, J., Singer, D.: The total squared curvature of closed curves. J. Differ. Geom. **20**, 1–22 (1984)
64. Zhang, X., Zhang, X., Bernaerts, D., Vantendeloo, G., Amelinckx, S., Vanlanduyt, J., Ivanov, V., Nagy, J., Lambin, P., Lucas, A.: The texture of catalytically grown coil-shaped carbon nanotubules. Europhys. Lett. **27**, 141–146 (1994)
65. Sabitov, I.: Some integral formulas for compact surfaces. TWMS J. Pure Appl. Math. **1**, 123–131 (2010)

Multiplication and Composition in Weighted Modulation Spaces

Maximilian Reich and Winfried Sickel

Abstract We study the existence of the product of two weighted modulation spaces. For this purpose, we discuss two different strategies. The more simple one allows transparent proofs in various situations. However, our second method allows a closer look onto associated norm inequalities under restrictions in the Fourier image. This will give us the opportunity to treat the boundedness of composition operators.

Keywords Weighted modulation spaces · Short-time Fourier transform · Frequency-uniform decomposition · Multiplication of distributions · Multiplication algebras · Composition of functions

Mathematics Subject Classification (2010). 46E35 · 47B38 · 47H30

1 Introduction

Since modulation spaces have been introduced by Feichtinger [7] they have become canonical for both time-frequency and phase-space analysis. However, in recent time modulation spaces have been found useful also in connection with linear and nonlinear partial differential equations, see e.g., Wang et al. [35–38], Ruzhansky et al. [26], or Bourdaud et al. [5]. Investigations of partial differential equations require partly different tools than used in time-frequency and phase-space analysis. In particular, Fourier multipliers, pointwise multiplication and composition of functions need to be studied. In our contribution, we will concentrate on pointwise multiplication and composition of functions. Already Feichtinger [7] was aware of the importance of pointwise multiplication in modulation spaces. In the meanwhile

M. Reich
Institut für Angewandte Analysis, TU Bergakademie Freiberg, 09596 Freiberg, Germany
e-mail: maximilian.reich@math.tu-freiberg.de

W. Sickel (✉)
Friedrich Schiller University, Ernst-Abbe-Platz 2, 07737 Jena, Germany
e-mail: winfried.sickel@uni-jena.de

T. Qian and L.G. Rodino (eds.), *Mathematical Analysis, Probability and Applications – Plenary Lectures*, Springer Proceedings in Mathematics & Statistics 177, DOI 10.1007/978-3-319-41945-9_5

several authors have studied this problem, we refer, e.g., to [6, 13, 29, 30, 32]. In Sect. 3, we will give a survey about the known results. Therefore, we will discuss two different proof strategies. The more simple one, due to Toft [30, 32] and Sugimoto et al. [29], allows transparent proofs in various situations, in particular one can deal with those situations where the modulation spaces form algebras with respect to pointwise multiplication. As a consequence, Sugimoto et al. [29] are able to deal with composition operators on modulation spaces induced by analytic functions. Our second method, much more complicated, allows a closer look onto associated norm inequalities under restrictions in the Fourier image. This will give us the possibility to discuss the boundedness of composition operators on weighted modulation spaces based on a technique which goes back to Bourdaud [3], see also Bourdaud et al. [5] and Reich et al. [23]. Our approach will allow to deal with the boundedness of nonlinear operators $T_f : g \mapsto f \circ g$ without assuming f to be analytic. However, as the case of $M^s_{2,2}$ shows, our sufficient conditions are not very close to the necessary conditions. There is still a certain gap.

The paper is organized as follows. In Sect. 2, we collect what is needed about the weighted modulation spaces we are interested in. The next section is devoted to the study of pointwise multiplication. In particular, we are interested in embeddings of the type

$$M^{s_1}_{p,q} \cdot M^{s_2}_{p,q} \hookrightarrow M^{s_0}_{p,q} ,$$

where s_1, s_2, p and q are given and we are asking for an optimal s_0. These results will be applied to problems around the regularity of composition of functions in Sect. 4. For convenience of the reader we also recall what is known in the more general situation

$$M^{s_1}_{p_1,q_1} \cdot M^{s_2}_{p_2,q_2} \hookrightarrow M^{s_0}_{p,q} .$$

Special attention will be paid to the algebra property. Here, the known sufficient conditions are supplemented by necessary conditions, see Theorem 3.5. Also only partly new is our main result in Sect. 3 stated in Theorem 3.22. Here we investigate multiplication of distributions (possibly singular) with regular functions (which are not assumed to be C^∞). Partly we have found necessary and sufficient conditions also in this more general situation. Finally, Sect. 4 deals with composition operators. As direct consequences of the obtained results for pointwise multiplication we can deal with the mappings $g \mapsto g^\ell$, $\ell \geq 2$, see Sect. 4.1. In Sect. 4.4, we shall investigate $g \mapsto f \circ g$, where f is not assumed to be analytic. Sufficient conditions, either in terms of a decay for $\mathcal{F}f$ or in terms of regularity of f, are given.

Notation

We introduce some basic notation. As usual, $\mathbb{N}$ denotes the natural numbers, $\mathbb{N}_0 := \mathbb{N} \cup \{0\}$, $\mathbb{Z}$ the integers and $\mathbb{R}$ the real numbers, $\mathbb{C}$ refers to the complex numbers. For a real number a, we put $a_+ := \max(a, 0)$. For $x \in \mathbb{R}^n$ we use $\|x\|_\infty := \max_{j=1,\dots,n} |x_j|$. Many times we shall use the abbreviation $\langle \xi \rangle := (1 + |\xi|^2)^{\frac{1}{2}}$, $\xi \in \mathbb{R}^n$.

The symbols $c, c_1, c_2, \ldots, C, C_1, C_2, \ldots$ denote positive constants which are independent of the main parameters involved but whose values may differ from line to line. The notation $a \lesssim b$ is equivalent to $a \leq Cb$ with a positive constant C. Moreover, by writing $a \asymp b$ we mean $a \lesssim b \lesssim a$.

Let X and Y be two Banach spaces. Then the symbol $X \hookrightarrow Y$ indicates that the embedding is continuous. By $\mathcal{L}(X, Y)$ we denote the collection of all linear and continuous operators which map X into Y. By $C_0^\infty(\mathbb{R}^n)$ the set of compactly supported infinitely differentiable functions $f : \mathbb{R}^n \to \mathbb{C}$ is denoted. Let $\mathcal{S}(\mathbb{R}^n)$ be the Schwartz space of all complex-valued rapidly decreasing infinitely differentiable functions on $\mathbb{R}^n$. The topological dual, the class of tempered distributions, is denoted by $\mathcal{S}'(\mathbb{R}^n)$ (equipped with the weak topology). The Fourier transform on $\mathcal{S}(\mathbb{R}^n)$ is given by

$$\mathcal{F}\varphi(\xi) = (2\pi)^{-n/2} \int_{\mathbb{R}^n} e^{ix\cdot\xi} \varphi(x)\, dx\,, \qquad \xi \in \mathbb{R}^n\,.$$

The inverse transformation is denoted by $\mathcal{F}^{-1}$. We use both notations also for the transformations defined on $\mathcal{S}'(\mathbb{R}^n)$.

Convention. If not otherwise stated all functions will be considered on the Euclidean n-space $\mathbb{R}^n$. Therefore $\mathbb{R}^n$ will be omitted in notation.

2 Basics on Modulation Spaces

2.1 Definitions

A general reference for definition and properties of weighted modulation spaces is Gröchenig's monograph [10, Chap. 11].

Definition 2.1 Let $\phi \in \mathcal{S}$ be nontrivial. Then the short-time Fourier transform of a function f with respect to ϕ is defined as

$$V_\phi f(x, \xi) = (2\pi)^{-\frac{n}{2}} \int_{\mathbb{R}^n} f(s)\overline{\phi(s-x)}e^{-is\cdot\xi}\, ds \qquad (x, \xi \in \mathbb{R}^n).$$

The function ϕ is usually called the window function. For $f \in \mathcal{S}'$ the short-time Fourier transform $V_\phi f$ is a continuous function of at most polynomial growth on $\mathbb{R}^{2n}$, see [10, Theorem 11.2.3].

Definition 2.2 Let $1 \leq p, q \leq \infty$. Let $\phi \in \mathcal{S}$ be a fixed window and assume $s \in \mathbb{R}$. Then the weighted modulation space $M^s_{p,q}$ is the collection of all $f \in \mathcal{S}'$ such that

$$\|f\|_{M^s_{p,q}} = \Big(\int_{\mathbb{R}^n} \Big(\int_{\mathbb{R}^n} |V_\phi f(x, \xi)\, \langle\xi\rangle^s|^p dx\Big)^{\frac{q}{p}} d\xi\Big)^{\frac{1}{q}} < \infty$$

(with obvious modifications if $p = \infty$ and/or $q = \infty$).

Formally these spaces $M^s_{p,q}$ depend on the window ϕ. However, for different windows ϕ_1, ϕ_2 the resulting spaces coincide as sets and the norms are equivalent, see [10, Proposition 11.3.2]. For that reason we do not indicate the window in the notation (we do not distinguish spaces which differ only by an equivalent norm).

Remark 2.3 (i) General references with respect to weighted modulation spaces are Feichtinger [7], Gröchenig [10, Chap. 11], Gol'dman [9], Guo et al. [11], Toft [30–32], Triebel [34] and Wang et al. [38] to mention only a few.

(ii) There is an important special case. In case of $p = q = 2$ we obtain $M^s_{2,2} = H^s$ in the sense of equivalent norms, see Feichtinger [7], Gröchenig [10, Proposition 11.3.1]. Here H^s is nothing but the standard Sobolev space built on L_2, at least for $s \in \mathbb{N}$. In general H^s is the collection of all $f \in \mathcal{S}'$ such that

$$\|f\|_{H^s} := \Big(\int_{\mathbb{R}^n} (1 + |\xi|^2)^s \, |\mathcal{F} f(\xi)|^2 \, d\xi \Big)^{1/2} < \infty .$$

For us of grcat use will be another alternative approach to the spaces $M^s_{p,q}$. This will be more close to the standard techniques used in connection with Besov spaces. We shall use the so-called frequency-uniform decomposition, see e.g., Wang [37]. Therefore, let $\rho : \mathbb{R}^n \mapsto [0,1]$ be a Schwartz function which is compactly supported in the cube

$$Q_0 := \{\xi \in \mathbb{R}^n : -1 \le \xi_i \le 1, \; i = 1, \dots, n\} .$$

Moreover, we assume

$$\rho(\xi) = 1 \qquad \text{if} \quad |\xi_i| \le \frac{1}{2}, \qquad i = 1, 2, \dots, n.$$

With $\rho_k(\xi) := \rho(\xi - k)$, $\xi \in \mathbb{R}^n$, $k \in \mathbb{Z}^n$, it follows

$$\sum_{k \in \mathbb{Z}^n} \rho_k(\xi) \ge 1 \qquad \text{for all} \quad \xi \in \mathbb{R}^n .$$

Finally, we define

$$\sigma_k(\xi) := \rho_k(\xi) \Big(\sum_{k \in \mathbb{Z}^n} \rho_k(\xi) \Big)^{-1}, \qquad \xi \in \mathbb{R}^n , \quad k \in \mathbb{Z}^n .$$

The following properties are obvious:

- $0 \le \sigma_k(\xi) \le 1$ for all $\xi \in \mathbb{R}^n$;
- $\operatorname{supp} \sigma_k \subset Q_k := \{\xi \in \mathbb{R}^n : -1 \le \xi_i - k_i \le 1, \; i = 1, \dots, n\}$;
- $\sum_{k \in \mathbb{Z}^n} \sigma_k(\xi) \equiv 1$ for all $\xi \in \mathbb{R}^n$;
- There exists a constant $C > 0$ such that $\sigma_k(\xi) \ge C$ if $\max_{i=1,\dots,n} |\xi_i - k_i| \le \frac{1}{2}$;
- For all $m \in \mathbb{N}_0$ there exist positive constants C_m such that for $|\alpha| \le m$

$$\sup_{k\in\mathbb{Z}^n}\sup_{\xi\in\mathbb{R}^n} |D^\alpha \sigma_k(\xi)| \le C_m \, .$$

We shall call the mapping

$$\Box_k f := \mathcal{F}^{-1}\left[\sigma_k(\xi)\,\mathcal{F}f(\xi)\right](\cdot), \qquad k\in\mathbb{Z}^n, \quad f\in\mathcal{S}',$$

frequency-uniform decomposition operator.

As it is well-known there is an equivalent description of the modulation spaces by means of the frequency-uniform decomposition operators.

Proposition 2.4 *Let* $1\le p,q\le\infty$ *and assume* $s\in\mathbb{R}$. *Then the weighted modulation space* $M^s_{p,q}$ *consists of all tempered distributions* $f\in\mathcal{S}'$ *such that*

$$\|f\|^*_{M^s_{p,q}} = \Big(\sum_{k\in\mathbb{Z}^n}\langle k\rangle^{sq}\,\|\Box_k f\|^q_{L^p}\Big)^{\frac{1}{q}} < \infty\, .$$

Furthermore, the norms $\|f\|_{M^s_{p,q}}$ *and* $\|f\|^*_{M^s_{p,q}}$ *are equivalent.*

We refer to Feichtinger [7] or Wang and Hudzik [37]. In what follows, we shall work with both characterizations. In general, we shall use the same notation $\|\cdot\|_{M^s_{p,q}}$ for both norms.

Lemma 2.5 *(i) The modulation space* $M^s_{p,q}$ *is a Banach space.*
(ii) $M^s_{p,q}$ *is independent of the choice of the window* $\rho\in C_0^\infty$ *in the sense of equivalent norms.*
(iii) $M^s_{p,q}$ *is continuously embedded into* $\mathcal{S}'$.
(iv) $M^s_{p,q}$ *has the Fatou property, i.e., if* $(f_m)_{m=1}^\infty\subset M^s_{p,q}$ *is a sequence such that* $f_m \rightharpoonup f$ *(weak convergence in* $\mathcal{S}'$*) and*

$$\sup_{m\in\mathbb{N}} \|\, f_m\,\|_{M^s_{p,q}} < \infty\, ,$$

then $f\in M^s_{p,q}$ *follows and*

$$\|\, f\,\|_{M^s_{p,q}} \le \sup_{m\in\mathbb{N}}\|\, f_m\,\|_{M^s_{p,q}} < \infty\, .$$

Proof For (i), (ii), (iii) we refer to [10].

We comment on a proof of (iv). Therefore, we follow [8] and work with the norm $\|\cdot\|^*_{M^s_{p,q}}$. From the assumption, we obtain that for all $k\in\mathbb{Z}^n$ and $x\in\mathbb{R}^n$,

$$\mathcal{F}^{-1}[\sigma_k\,\mathcal{F}f_m](x) = (2\pi)^{-n/2}\, f_m(x-\cdot)(\sigma_k) \to f(x-\cdot)(\sigma_k) = \mathcal{F}^{-1}[\sigma_k\,\mathcal{F}f](x)$$

as $m \to \infty$. Fatou's lemma yields

$$\sum_{|k|\le N} \Big(\int_{\mathbb{R}^n} |\mathcal{F}^{-1}[\sigma_k \mathcal{F} f](x)|^p dx \Big)^{\frac{q}{p}} \le \liminf_{m\to\infty} \sum_{|k|\le N} \Big(\int_{\mathbb{R}^n} |\mathcal{F}^{-1}[\sigma_k \mathcal{F} f_m](x)|^p dx \Big)^{\frac{q}{p}}.$$

An obvious monotonicity argument completes the proof. ∎

2.2 *Embeddings*

Obviously the spaces $M^s_{p,q}$ are monotone in s and q. But they are also monotone with respect to p. To show this we recall Nikol'skij's inequality, see e.g., Nikol'skij [21, 3.4] or Triebel [33, 1.3.2].

Lemma 2.6 *Let $1 \le p \le q \le \infty$ and f be an integrable function with* $\operatorname{supp} \mathcal{F} f \subset B(y,r)$*, i.e., the support of the Fourier transform of f is contained in a ball with radius $r > 0$ and center in $y \in \mathbb{R}^n$. Then it holds*

$$\|f\|_{L_q} \le C r^{n(\frac{1}{p}-\frac{1}{q})} \|f\|_{L_p}$$

with a constant $C > 0$ independent of r and y.

This implies $\|\Box_k f\|_{L_q} \le c\, \|\Box_k f\|_{L_p}$ if $p \le q$ with c independent of k and f which results in the following corollary (by using the norm $\|\cdot\|^*_{M^s_{p,q}}$).

Corollary 2.7 *Let $s_0 > s$, $p_0 < p$ and $q_0 < q$. Then the following embeddings hold and are continuous:*

$$M^{s_0}_{p,q} \hookrightarrow M^s_{p,q}, \qquad M^s_{p_0,q} \hookrightarrow M^s_{p,q}$$

and

$$M^s_{p,q_0} \hookrightarrow M^s_{p,q};$$

i.e., for all p, q, $1 \le p, q \le \infty$, we have

$$M^s_{1,1} \hookrightarrow M^s_{p,q} \hookrightarrow M^s_{\infty,\infty}.$$

Of some importance are embeddings with respect to different metrics. To find sufficient conditions is not difficult when working with $\|\cdot\|^*_{M^s_{p,q}}$. A bit more tricky are the necessity parts. We refer to the recent paper by Guo et al. [11].

Proposition 2.8 *Let $s_0, s_1 \in \mathbb{R}$ and $1 \le p_0, p_1 \le \infty$. Then*

$$M^{s_0}_{p_0,q_0} \hookrightarrow M^{s_1}_{p_1,q_1}$$

holds if and only if either

- $p_0 \le p_1$ *and* $s_0 - s_1 > n\Big(\frac{1}{q_1} - \frac{1}{q_0}\Big)$
- *or* $p_0 \le p_1$, $s_0 = s_1$ *and* $q_0 = q_1$.

Remark 2.9 Embeddings of modulation spaces are treated at various places, we refer to Feichtinger [7], Wang and Hudzik [37], Cordero and Nicola [6], Iwabuchi [13] and Guo et al. [11].

The weighted modulation spaces $M^s_{p,q}$ cannot distinguish between boundedness and continuity (as Besov spaces). Let C_{ub} denote the class of all uniformly continuous and bounded functions $f : \mathbb{R}^n \to \mathbb{C}$ equipped with the supremum norm. If $f \in M^s_{p,q}$ is a regular distribution it is determined (as a function) almost everywhere. We shall say that f is a continuous function if there is one continuous function g which equals f almost everywhere.

Corollary 2.10 *Let $s \in \mathbb{R}$ and $1 \le p, q \le \infty$. Then the following assertions are equivalent:*

- $M^s_{p,q} \hookrightarrow L_\infty$;
- $M^s_{p,q} \hookrightarrow C_{ub}$;
- $M^s_{p,q} \hookrightarrow M^0_{\infty,1}$;
- *either* $s \ge 0$ *and* $q = 1$ *or* $s > n/q'$.

Proof We shall work with $\| \cdot \|^*_{M^s_{p,q}}$.
Step 1. Sufficiency. By Proposition 2.8 it will be enough to show $M^0_{\infty,1} \hookrightarrow C_{ub}$. From the definition of $M^0_{\infty,1}$ it follows that

$$\sum_{k \in \mathbb{Z}^n} \Box_k f(x)$$

is pointwise convergent (for all $x \in \mathbb{R}^n$). Furthermore, since $\Box_k f \in C^\infty$, there is a continuous representative in the equivalence class f, given by $\sum_{k \in \mathbb{Z}^n} \Box_k f(x)$. In what follows, we shall work with this representative. Boundedness of $f \in M^0_{\infty,1}$ is obvious, we have

$$|f(x)| = |\sum_{k \in \mathbb{Z}^n} \Box_k f(x)| \le \| f \|_{M^0_{\infty,1}} .$$

It remains to prove uniform continuity. For fixed $\varepsilon > 0$ we choose N such that

$$\sum_{|k|>N} \| \Box_k f \|_{L_\infty} < \varepsilon/2 .$$

In case $|k| \le N$ we observe that

$$|\square_k f(x) - \square_k f(y)| \le \| \nabla(\square_k f) \|_{L_\infty} |x-y| \, .$$

It follows from [33, Theorem 1.3.1] that

$$\| \nabla(\square_k f) \|_{L_\infty} \le c_1 \, \| (M \square_k f) \|_{L_\infty}$$

with a constant c_1 independent of f and k. Here M denotes the Hardy–Littlewood maximal function. In the quoted reference, the assumption $\square_k f \in \mathcal{S}$ is used. A closer look at the proof shows that $\square_k f \in L_1^{\ell oc}$ satisfying

$$\int_{Q_k} |\square_k f(x)| \, dx \le c_2 \, (1 + |k|)^N \, , \qquad k \in \mathbb{Z}^n \, ,$$

for some $N \in \mathbb{N}$ is sufficient. Since $\square_k f \in L_\infty$ this is obvious. Consequently we obtain

$$\begin{aligned} |\square_k f(x) - \square_k f(y)| &\le c_1 \, \| (M \square_k f) \|_{L_\infty} |x-y| \le c_1 \, \| \square_k f \|_{L_\infty} |x-y| \\ &\le c_2 \, \| f \|_{L_\infty} |x-y| \, , \end{aligned}$$

where in the last step we used the standard convolution inequality $\| g * h \|_{L_\infty} \le \| g \|_{L_1} \| f \|_{L_\infty}$. This implies uniform continuity of $\square_k f$ and therefore of $\sum_{|k| \le N} \square_k f$. In particular, we find

$$\begin{aligned} |f(x) - f(y)| &= \Big| \sum_{k \in \mathbb{Z}^n} (\square_k f(x) - \square_k f(y)) \Big| \\ &\le \sum_{|k|>N} (|\square_k f(x)| + |\square_k f(y)|) + c_2 \, \| f \|_{L_\infty} |x-y| \sum_{|k| \le N} 1 \\ &\le \varepsilon + c_2 \, \| f \|_{L_\infty} |x-y| \, (2N+1)^n \, . \end{aligned}$$

Choosing $\delta = (c_2 \, \| f \|_{L_\infty} (2N+1)^n)^{-1} \, \varepsilon$ we arrive at

$$|f(x) - f(y)| < 2\,\varepsilon \qquad \text{if} \qquad |x-y| < \delta \, .$$

Step 2. Necessity. Let $\psi \in \mathcal{S}$ be a real-valued function such that $\psi(0) = 1$ and

$$\operatorname{supp} \mathcal{F}\psi \subset \{\xi : \max_{j=1,\ldots,n} |\xi_j| < \varepsilon\} \qquad \text{with} \quad \varepsilon < 1/2.$$

We define f by

$$\mathcal{F} f(\xi) := \sum_{k \in \mathbb{Z}^n} a_k \, \mathcal{F}\psi(\xi - k) \, .$$

Clearly,

$$\square_k f(x) = a_k\, e^{ikx}\, \psi(x)\,, \qquad k \in \mathbb{Z}^n\,.$$

Substep 2.1. Let $s = 0$ and $1 \le p \le \infty$. The above arguments imply $f \in M^0_{p,q}$ if and only if $(a_k)_k \in \ell_q$. On the other hand,

$$f(x) = \psi(x) \sum_{k\in\mathbb{Z}^n} a_k\, e^{ikx} \tag{2.1}$$

which implies that f is unbounded in 0 if $\sum_{k\in\mathbb{Z}^n} a_k = \infty$. Choosing

$$a_k := \begin{cases} (k_1 \log(2+k_1))^{-1} & \text{if } \; k_1 \in \mathbb{N}\,, \quad k = (k_1, 0, \ldots 0)\,; \\ 0 & \text{otherwise;} \end{cases}$$

then $f \in M^0_{p,q} \setminus L_\infty$, $q > 1$, follows.

Substep 2.2. Let $1 \le p \le \infty$ and $q = \infty$. Then we choose $a_k := \langle k\rangle^{-n}$. It follows $f \in M^n_{p,\infty}$ but $f(0) = +\infty$.

Substep 2.3. Let $1 \le p \le \infty$, $1 < q < \infty$ and $s = n/q'$. Then, with $\delta > 0$, we choose

$$a_k := \begin{cases} \langle k\rangle^{-n}\, (\log\langle k\rangle)^{-(1+\delta)/q} & \text{if } \; |k| > 0\,; \\ 0 & \text{otherwise.} \end{cases}$$

It follows

$$\begin{aligned} \| f \|_{M^{n/q'}_{p,q}} &= \|\psi\|_{L_p} \left(\sum_{|k|>0} \langle k\rangle^{-nq+nq/q'}\, (\log\langle k\rangle)^{-(1+\delta)} \right)^{\frac{1}{q}} \\ &= \|\psi\|_{L_p} \left(\sum_{|k|>0} \langle k\rangle^{-n}\, (\log\langle k\rangle)^{-(1+\delta)} \right)^{\frac{1}{q}} < \infty\,. \end{aligned}$$

On the other hand, we have

$$f(0) = \sum_{|k|>0} \langle k\rangle^{-n}\, (\log\langle k\rangle)^{-(1+\delta)/q} = \infty$$

if $(1+\delta)/q \le 1$. Hence, for choosing $\delta = q - 1$ the claim follows. ■

Remark 2.11 Sufficient conditions for embeddings of modulation spaces into spaces of continuous functions can be found at several places, in particular in Feichtinger's original paper [7]. We did not find references for the necessity.

3 Pointwise Multiplication in Modulation Spaces

We are interested in embeddings of the type

$$M^{s_1}_{p,q} \cdot M^{s_2}_{p,q} \hookrightarrow M^{s_0}_{p,q} \, ,$$

where s_1, s_2, p and q are given and we are asking for an optimal s_0. These results will be applied in connection with our investigations on the regularity of compositions of functions in Sect. 4. However, several times we shall deal with the slightly more general problem

$$M^{s_1}_{p_1,q} \cdot M^{s_2}_{p_2,q} \hookrightarrow M^{s_0}_{p,q} \, , \qquad \frac{1}{p} = \frac{1}{p_1} + \frac{1}{p_2} \, .$$

In view of Corollary 2.7 this always yields

$$M^{s_1}_{p_1,q} \cdot M^{s_2}_{p_2,q} \hookrightarrow M^{s_0}_{p,q} \, , \qquad \frac{1}{p} \le \frac{1}{p_1} + \frac{1}{p_2} \, .$$

For convenience of the reader we also recall what is known in the more general situation

$$M^{s_1}_{p_1,q_1} \cdot M^{s_2}_{p_2,q_2} \hookrightarrow M^{s_0}_{p,q} \, .$$

At first we shall deal with the algebra property. Afterwards we turn to the existence of the product in more general situations.

3.1 On the Algebra Property

The main aim consists in giving necessary and sufficient conditions for the embedding $M^s_{p,q} \cdot M^s_{p,q} \hookrightarrow M^s_{p,q}$. To prepare this we recall a nice identity due to Toft [30], see also Sugimoto et al. [29].

Lemma 3.1 *Let $\varphi_1, \varphi_2 \in \mathcal{S}$ be nontrivial. Let $f, g \in L_2^{\ell oc}$ such that there exist $c > 0$ and $M > 0$ with*

$$\int_{Q_k} |f(x)|^2 + |g(x)|^2 \, dx \le c \, (1 + |k|)^M \, , \qquad k \in \mathbb{Z}^n \, .$$

For all $x, \xi \in \mathbb{R}^n$ the following identity takes place

$$V_{\varphi_1 \cdot \varphi_2}(fg)(x, \xi) = (2\pi)^{-n/2} \int V_{\varphi_1}(f)(x, \xi - \eta) \, V_{\varphi_2}(g)(x, \eta) \, d\eta \, . \qquad (3.1)$$

Proof The main tool will be the Plancherel identity. Observe, that for any fixed $x \in \mathbb{R}^n$ the functions $f(t)\,\overline{\varphi_1(t-x)}, \overline{g(t)}\,\varphi_2(t-x)$ belong to L_2 and therefore their Fourier transforms as well. For brevity we put

$$I := \int V_{\varphi_1}(f)(x,\xi-\eta)\, V_{\varphi_2}(g)(x,\eta)\, d\eta\,.$$

Applying the Plancherel identity, we conclude

$$\begin{aligned} I &= \int \mathcal{F}(f(t)\,\overline{\varphi_1(t-x)}\, e^{-i\xi t})(-\eta)\, \overline{\mathcal{F}(\overline{g(t)}\,\varphi_2(t-x))(-\eta)}\, d\eta \\ &= \int f(t)\,\overline{\varphi_1(t-x)}\, e^{-i\xi t}\, \overline{\overline{g(t)}\,\varphi_2(t-x))}\, dt \\ &= \int f(t)\, g(t)\, \overline{\varphi_1(t-x)\,\varphi_2(t-x)}\, e^{-i\xi t}\, dt \\ &= (2\pi)^{n/2}\, V_{\varphi_1\cdot\varphi_2}(fg)(x,\xi)\,. \end{aligned}$$

The proof is complete. ∎

Remark 3.2 It is clear that the assertion does not extend very much. For example, if $f, g \in L_p^{\ell oc}$ for some $p < 2$ then the above claim is not true. We may take

$$f(x) = g(x) = \psi(x)\, |x|^{-n/2}\,, \qquad x \in \mathbb{R}^n\,,$$

where ψ is a smooth and compactly supported cut-off function s.t. $\psi(0) = 1$. Then $f \cdot g$ is not longer a distribution, i.e., the integral

$$V_{\varphi_1\cdot\varphi_2}(fg)(x,\xi) = (2\pi)^{-\frac{n}{2}} \int_{\mathbb{R}^n} f(s)\, g(s)\, \overline{\varphi_1(s-x)\,\varphi_2(s-x)}\, e^{-is\cdot\xi}\, ds$$

does not make sense in general.

In [29, 30], the identity (3.1) is applied either in case $f, g \in \mathcal{S}$ or $f, g \in L_\infty$. Here, we shall apply it in the wider context of Lemma 3.1.

Lemma 3.3 *Let $1 \le p, q \le \infty$ and assume $M^s_{p,q} \hookrightarrow M^0_{\infty,1}$. Then there exists a constant c such that*

$$\| f \cdot g \|_{M^s_{p,q}} \le c \Big(\| f \|_{M^0_{\infty,1}}\, \| g \|_{M^s_{p,q}} + \| f \|_{M^s_{p,q}}\, \| g \|_{M^0_{\infty,1}} \Big)$$

holds for all $f, g \in M^s_{p,q}$.

Proof The main idea in the proof consists in the fact that the modulation space can be characterized by different window functions. Since $M^0_{\infty,1} \hookrightarrow L_\infty$ we know that f, g satisfy the conditions in Lemma 3.1. Hence

$$\| f \cdot g \|_{M^s_{p,q}} = \Big\{ \int \Big[\int \Big| V_{\varphi^2}(f \cdot g)(x,\xi) \Big|^p dx \Big]^{q/p} \langle \xi \rangle^{sq} d\xi \Big\}^{1/q} \tag{3.2}$$
$$\leq \Big\{ \int \Big[\int \Big| \int V_\varphi f(x,\xi-\eta) V_\varphi g(x,\eta) d\eta \Big|^p dx \Big]^{q/p} \langle \xi \rangle^{sq} d\xi \Big\}^{1/q} .$$

We split the integration with respect to η into two parts

$$\Omega_\xi := \{\eta : \ |\xi - \eta| \geq |\eta|\} \quad \text{and} \quad \Gamma_\xi := \{\eta : \ |\xi-\eta| < |\eta|\}, \quad \xi \in \mathbb{R}^n . \tag{3.3}$$

It follows

$$\| f \cdot g \|_{M^s_{p,q}} \leq 2^s \, (A + B) ,$$

where

$$A := \Big\{ \int \Big[\int \Big| \int_{\Omega_\xi} V_\varphi f(x,\xi-\eta) \, (1+|\xi-\eta|^2)^{s/2} V_\varphi g(x,\eta) \, d\eta \Big|^p dx \Big]^{q/p} d\xi \Big\}^{1/q}$$

and

$$B := \Big\{ \int \Big[\int \Big| \int_{\Gamma_\xi} V_\varphi f(x,\xi-\eta) \, V_\varphi g(x,\eta) \, (1+|\eta|^2)^{s/2} d\eta \Big|^p dx \Big]^{q/p} d\xi \Big\}^{1/q} .$$

We continue by applying the generalized Minkowski inequality, see [18, Theorem 2.4]. This yields

$$A \leq \int \Big\{ \int \Big[\int |V_\varphi f(x,\xi-\eta) \, \langle \xi-\eta \rangle^s \, V_\varphi g(x,\eta) \, |^p \, dx \Big]^{q/p} d\xi \Big\}^{1/q} d\eta$$
$$\leq \int \Big(\sup_{x \in \mathbb{R}^n} |V_\varphi g(x,\eta)| \Big) \Big\{ \int \Big[\int |V_\varphi f(x,\xi-\eta) \, \langle \xi - \eta \rangle^s \, |^p \, dx \Big]^{q/p} d\xi \Big\}^{1/q} d\eta$$
$$= \| g \|_{M^0_{\infty,1}} \, \| f \|_{M^s_{p,q}} .$$

Analogously one can prove

$$B \leq \| f \|_{M^0_{\infty,1}} \, \| g \|_{M^s_{p,q}} .$$

The proof is complete. ∎

Remark 3.4 (i) We proved a bit more than stated. In fact, we have shown

$$\| f \cdot g \|_{M^s_{p,q}} \leq 2^s \Big(\| f \|_{M^0_{\infty,1}} \, \| g \|_{M^s_{p,q}} + \| f \|_{M^s_{p,q}} \, \| g \|_{M^0_{\infty,1}} \Big)$$

But here one has to notice that the norm on the left-hand side is generated by the window φ^2, whereas the norms on the right-hand side are generated by the window φ.

(ii) Lemma 3.3 has been proved by Sugimoto et al. [29]. For partial results with a different proof we refer to Feichtinger [7].

Next, we turn to necessary and sufficient conditions for the algebra property.

Theorem 3.5 *Let $1 \le p, q \le \infty$ and $s \in \mathbb{R}$. Then $M^s_{p,q}$ is an algebra with respect to pointwise multiplication if and only if either $s \ge 0$ and $q = 1$ or $s > n/q'$.*

Remark 3.6 (i) By Corollary 2.10 the Theorem 3.5 can be reformulated as

$$M^s_{p,q} \quad \text{is an algebra} \qquad \Longleftrightarrow \qquad M^s_{p,q} \hookrightarrow L_\infty\,.$$

This is in some sense natural because otherwise one could increase local singularities by pointwise multiplication.
(ii) Theorem 3.5 has a partial counterpart for Besov spaces. Here one knows that $B^s_{p,q}$ is an algebra if and only if $B^s_{p,q} \hookrightarrow L_\infty$ and $s > 0$. We refer to Peetre [22, Theorem 11, p. 147], Triebel [33, Theorem 2.8.3] (sufficiency) and to [25, Theorem 4.6.4/1] (necessity).

To prepare the proof, we need the following lemma which is of interest for its own.

Lemma 3.7 *Let $1 \le p, q < \infty$ and $s \in \mathbb{R}$. Let $f \in \mathcal{S}'$ and let there exists a constant $c > 0$ such that*

$$\|f \cdot g\|_{M^s_{p,q}} \le c\, \|g\|_{M^s_{p,q}}$$

holds for all $g \in \mathcal{S}$. Then $f \in L_\infty$ follows.

Proof Let $T_f(g) := f \cdot g$, $g \in \mathcal{S}$. Let $\mathring{M}^s_{p,q}$ denote the closure of $\mathcal{S}$ in $M^s_{p,q}$. Hence, there is a unique extension of T_f to a continuous operator belonging to $\mathcal{L}(\mathring{M}^s_{p,q}, M^s_{p,q})$. Next we employ duality. We fix p, q and s $(1 \le p, q \le \infty, s \in \mathbb{R})$. Let (g, h) denote the standard dual pairing on $\mathcal{S}' \times \mathcal{S}$. Then

$$\|g\| := \sup\Big\{|(g,h)| \;:\quad h \in \mathcal{S},\; \|h\|_{M^{-s}_{p',q'}} \le 1\Big\}$$

is an equivalent norm on $M^s_{p,q}$, see Feichtinger [7] or Toft [30]. In view of this equivalent norm our assumption on T_f implies $\mathcal{L}(M^{-s}_{p',q'}, M^{-s}_{p',q'})$. Next we continue by complex interpolation. Let $0 < \Theta < 1$. It is known that

$$M^s_{p,q} = [M^{s_1}_{p_1,q_1}, M^{s_2}_{p_2,q_2}]_\Theta$$

if $1 \le p_1, q_1 < \infty$, $1 \le p_2, q_2 \le \infty$, $s_1, s_2 \in \mathbb{R}$ and

$$s := (1-\Theta)s_1 + \Theta\, s_2\,, \quad \frac{1}{p} := \frac{1-\Theta}{p_1} + \frac{\Theta}{p_2}\,, \quad \frac{1}{q} := \frac{1-\Theta}{q_1} + \frac{\Theta}{q_2}\,,$$

see Feichtinger [7]. Thanks to the interpolation property of the complex method we conclude

$$T_f \in \mathcal{L}\Big([\mathring{M}^s_{p,q}, M^{-s}_{p',q'}]_{1/2}, [\mathring{M}^s_{p,q}, M^{-s}_{p',q'}]_{1/2}\Big).$$

Because of $\mathring{M}^s_{p,q} = M^s_{p,q}$ if $\max(p,q) < \infty$ we find

$$T_f \in \mathcal{L}\Big(M^0_{2,2}, M^0_{2,2}\Big) = \mathcal{L}\Big(L_2, L_2\Big).$$

But this implies $f \in L_\infty$. ■

Proof of Theorem 3.5.
Step 1. Sufficiency is covered by Lemma 3.3.
Step 2. Necessity in case $1 \le p, q < \infty$ and $s \in \mathbb{R}$. In view of Lemma 3.7 the embedding $M^s_{p,q} \cdot M^s_{p,q} \hookrightarrow M^s_{p,q}$ implies $M^s_{p,q} \subset L_\infty$.
Step 3. To treat the remaining cases $\max(p,q) = \infty$ we argue by using explicit counterexamples.

Substep 3.1. Let $1 \le p \le \infty$, $s = 0$ and $1 < q \le \infty$. We assume that $M^0_{p,q}$ is an algebra. This implies the existence of a constant $c > 0$ such that

$$\| f \cdot g \|_{M^0_{p,q}} \le c \, \| f \|_{M^0_{p,q}} \, \| g \|_{M^0_{p,q}} \tag{3.4}$$

holds for all $f, g \in M^0_{p,q}$. Let

$$f(x) = \psi(x) \sum_{k=1}^{\infty} a_k \, e^{ikx_1}, \qquad x = (x_1, \ldots, x_n) \in \mathbb{R}^n,$$

be as in (2.1). Then, as shown above,

$$\| f \|_{M^0_{p,q}} = \|\psi\|_{L_p} \, \|(a_k)_k\|_{\ell_q}$$

follows. Let

$$f_N(x) := \psi(x) \sum_{k=1}^{N} a_k \, e^{ikx_1}, \qquad x = (x_1, \ldots, x_n) \in \mathbb{R}^n, \quad N \in \mathbb{N}.$$

Obviously $f_N \in \mathcal{S}$. We assume that

$$\operatorname{supp} \mathcal{F}\psi \subset \{\xi : \max_{j=1,\ldots,n} |\xi_j| < \varepsilon\} \qquad \text{with } \ \varepsilon < 1/4.$$

Then, because of

$$f_N(x) \cdot f_N(x) = \psi^2(x) \sum_{m=2}^{2N} \Big(\sum_{k=1}^{m-1} a_k\, a_{m-k} \Big) e^{imx} ,$$

we conclude

$$\| f_N \cdot f_N \|_{M^0_{p,q}} = \|\psi^2\|_{L_p} \Big(\sum_{m=2}^{2N} \Big| \sum_{k=1}^{m-1} a_k\, a_{m-k} \Big|^q \Big)^{1/q} .$$

Inequality (3.4) implies

$$\Big(\sum_{m=2}^{2N} \Big| \sum_{k=1}^{m-1} a_k\, a_{m-k} \Big|^q \Big)^{1/q} \le c\, \frac{\|\psi\|^2_{L_p}}{\|\psi^2\|_{L_p}} \Big(\sum_{k=1}^{N} |a_k|^q \Big)^{2/q} .$$

Clearly, in case $q > 1$ this is impossible in this generality. Explicit counterexamples are given by

$$a_k := k^{-1/q} \qquad \text{if} \quad 1 < q < \infty$$

and

$$a_k = 1 \qquad \text{if} \quad q = \infty .$$

In case $1 < q < \infty$ (3.4) yields

$$\| f_N \cdot f_N \|_{M^0_{p,q}} \asymp N^{1-1/q} \qquad \text{and} \qquad \| f_N \|^2_{M^0_{p,q}} \asymp (\log N)^{2/q} .$$

For $q = \infty$ we obtain

$$\| f_N \cdot f_N \|_{M^0_{p,\infty}} \asymp N \qquad \text{and} \qquad \| f_N \|^2_{M^0_{p,\infty}} \asymp 1 .$$

For $N \to \infty$ we find a contradiction in both situations.

Substep 3.2. Let $1 \le p \le \infty$, $q = \infty$ and $0 < s \le n$. We argue as in Substep 3.1 and assume $M^s_{p,\infty}$ is an algebra with respect to pointwise multiplication. This leads to the existence of a constant $c > 0$ such that

$$\| f \cdot g \|_{M^s_{p,\infty}} \le c\, \| f \|_{M^s_{p,\infty}} \| g \|_{M^s_{p,\infty}}$$

holds for all $f, g \in M^s_{p,\infty}$. We choose

$$f(x) = g(x) = f_N(x) := \psi(x) \sum_{\|k\|_\infty \le N} a_k\, e^{ikx} , \qquad x \in \mathbb{R}^n ,$$

and obtain

$$\| f_N \|_{M^s_{p,q}} = \|\psi\|_{L_p} \Big(\sum_{\|k\|_\infty \le N} |a_k \langle k\rangle^s|^q \Big)^{1/q} ,$$

$$\| f_N \cdot f_N \|_{M^s_{p,q}} = \|\psi^2\|_{L_p} \Big(\sum_{\|m\|_\infty \le 2N} \langle m\rangle^{sq} \Big| \sum_{\substack{k:\ \|k\|_\infty \le N \\ \|m-k\|_\infty \le N}} a_k\, a_{m-k} \Big|^q \Big)^{1/q} .$$

In case $s < n$ we choose $a_k := 1$ for all k and obtain

$$\| f_N \cdot f_N \|_{M^s_{p,\infty}} \asymp N^{n+s} \qquad \text{and} \qquad \| f_N \|^2_{M^s_{p,\infty}} \asymp N^{2s} .$$

This yields a contradiction if $s < n$. For $s = n$ we consider $a_k := \langle k\rangle^{-n}$ for all k. This yields

$$\log N \lesssim \| f_N \cdot f_N \|_{M^n_{p,\infty}} \qquad \text{and} \qquad \| f_N \|^2_{M^n_{p,\infty}} \asymp 1 ,$$

yielding a contradiction as well.

Substep 3.3. Let $s < 0$ and $1 \le p, q \le \infty$. We choose $a_k := \langle k\rangle^{2|s|}$ for all k and obtain

$$N^{3|s|+n+n/q} \lesssim \| f_N \cdot f_N \|_{M^s_{p,q}} \qquad \text{and} \qquad \| f_N \|^2_{M^s_{p,q}} \asymp N^{2|s|+2n/q} .$$

For $N \to \infty$ this implies $|s| + n \le n/q$. Since $|s| > 0$ this is impossible. The proof is complete. ■

Corollary 3.8 *Let $1 \le p, q \le \infty$ and $s \ge 0$. Then $M^s_{p,q} \cap M^0_{\infty,1}$ is an algebra with respect to pointwise multiplication and there exist a constant c such that*

$$\| f \cdot g \|_{M^s_{p,q}} \le c \Big(\| f \|_{M^0_{\infty,1}} \| g \|_{M^s_{p,q}} + \| f \|_{M^s_{p,q}} \| g \|_{M^0_{\infty,1}} \Big)$$

holds for all $f, g \in M^s_{p,q} \cap M^0_{\infty,1}$.

Proof The same arguments as in Lemma 3.3 apply. ■

Remark 3.9 Corollary 3.8 has a counterpart for Besov spaces. Here, one knows that $B^s_{p,q} \cap L_\infty$ is an algebra if $1 \le p, q \le \infty$ and $s > 0$. We refer to Peetre [22, Theorem 11, p. 147] and to [25, Theorem 4.6.4/2].

3.2 More General Products of Functions

Here, we consider the problem

$$M^{s_1}_{p_1,q_1} \cdot M^{s_2}_{p_2,q} \hookrightarrow M^s_{p,q} \,.$$

As a first result, we mention a generalization of Lemma 3.3.

Lemma 3.10 *Let $1 \le p_1, p_2, q \le \infty$ and $s \ge 0$. We put $1/p := (1/p_1) + (1/p_2)$. If $p \in [1, \infty]$, then there exists a constant c such that*

$$\| f \cdot g \|_{M^s_{p,q}} \le c \Big(\| f \|_{M^0_{p_1,1}} \| g \|_{M^s_{p_2,q}} + \| f \|_{M^s_{p_1,q}} \| g \|_{M^0_{p_2,1}} \Big)$$

holds for all $f \in M^s_{p_1,q} \cap M^0_{p_1,1}$ and all $g \in M^s_{p_2,q} \cap M^0_{p_2,1}$.

Proof We argue similar as above but using Hölder's inequality with respect to p before applying the generalized Minkowski inequality. ■

Remark 3.11 Observe that $M^0_{p_1,1}, M^0_{p_2,1} \hookrightarrow M^0_{\infty,1} \hookrightarrow L_\infty$.

Lemma 3.12 *Let $1 \le p_1, p_2, q \le \infty$ and $s \le 0$. We put $1/p := (1/p_1) + (1/p_2)$. If $p \in [1, \infty]$, then there exists a constant c such that*

$$\| f \cdot g \|_{M^s_{p,q}} \le c \, \| f \|_{M^{|s|}_{p_1,1}} \| g \|_{M^s_{p_2,q}} \tag{3.5}$$

holds for all $f \in M^{|s|}_{p_1,1}$, $g \in M^s_{p_2,q}$ such that g satisfies $g \in L_2^{\ell oc}$ and

$$\int_{Q_k} |g(x)|^2 \, dx \le C \, (1 + |k|)^M \,, \tag{3.6}$$

for some $C > 0$ and $M > 0$ independent of $k \in \mathbb{Z}^n$.

Proof Point of departure is the formula (3.2). Instead of the splitting in (3.3) we use now the elementary inequality

$$1 + |\eta|^2 \le 2 \, (1 + |\xi|^2) \, (1 + |\xi - \eta|^2)$$

which implies

$$(1 + |\xi|^2)^{s/2} \le 2^{|s|/2} \, (1 + |\xi - \eta|^2)^{|s|/2} \, (1 + |\eta|^2)^{s/2} \,.$$

This leads to the estimate

$$\begin{aligned} &\| f \cdot g \|_{M^s_{p,q}} \\ &\lesssim \Big\{ \int \Big[\int \Big| \int V_\varphi f(x, \xi - \eta) \, \langle \xi - \eta \rangle^{|s|} \, V_\varphi g(x, \eta) \, \langle \eta \rangle^s \, d\eta \, \Big|^p dx \Big]^{q/p} d\xi \Big\}^{1/q} \\ &= \Big\{ \int \Big[\int \Big| \int V_\varphi f(x, \tau) \, \langle \tau \rangle^{|s|} \, V_\varphi g(x, \xi - \tau) \, \langle \xi - \tau \rangle^s \, d\tau \, \Big|^p dx \Big]^{q/p} d\xi \Big\}^{1/q} . \end{aligned}$$

We continue by applying the generalized Minkowski inequality and Hölder's inequality (with respect to p) and obtain

$$\begin{aligned}
&\| f \cdot g \|_{M^s_{p,q}} \\
&\lesssim \int \Big\{ \int \Big[\| V_\varphi f(x,\tau) \langle \tau \rangle^{|s|} \|_{L_{p_1}} \, \| V_\varphi g(x,\xi-\tau) \langle \xi-\tau \rangle^s \|_{L_{p_2}} \Big]^q d\xi \Big\}^{1/q} d\tau \\
&\lesssim \int \| V_\varphi f(x,\tau) \langle \tau \rangle^{|s|} \|_{L_{p_1}} d\tau \, \| g \|_{M^s_{p_2,q}} \\
&\lesssim \| f \|_{M^{|s|}_{p_1,1}} \, \| g \|_{M^s_{p_2,q}} \, .
\end{aligned}$$

■

Remark 3.13 Observe that $M^{|s|}_{p_1,1} \hookrightarrow M^{|s|}_{\infty,1} \hookrightarrow L_\infty$. In addition we would like to mention that the constant c in (3.5) does not depend on the constant C in (3.6).

We recall a final result of Cordero and Nicola [6] concentrating on $s = 0$. These authors study $M^0_{p_1,q_1} \cdot M^0_{p_2,q_2} \hookrightarrow M^0_{p,q}$.

Proposition 3.14 *Let* $1 \le p_1, p_2, q_1, q_2 \le \infty$. *Then* $M^0_{p_1,q_1} \cdot M^0_{p_2,q_2} \hookrightarrow M^0_{p,q}$ *holds if and only if*

$$\frac{1}{p} \le \frac{1}{p_1} + \frac{1}{p_2} \qquad \textit{and} \qquad 1 + \frac{1}{q} \le \frac{1}{q_1} + \frac{1}{q_2} \, .$$

Remark 3.15 (i) Proposition 3.14 shows that in case $s = 0$ in Lemma 3.12 we proved an optimal estimate.
(ii) Necessity of the restrictions in Proposition 3.14 is shown by studying products of Gaussian functions. For extensions of Proposition 3.14 to the case of products with more than two factors we refer to Guo et al. [11] and Toft [30].

3.3 *Products of a Distribution with a Function*

Up to now, we considered only products of either L_∞-functions or $L_2^{\ell oc}$-functions with L_∞-functions. But now we turn to the product of a distribution with a function which is not assumed to be C^∞. This requires a definition.

The Definition of the Product in $\mathcal{S}'$

Let $\psi \in \mathcal{S}$ be a function in C_0^∞ such that $\psi(\xi) = 1$ in a neighbourhood of the origin. We define

$$S^j f(x) = \mathcal{F}^{-1}[\psi(2^{-j}\xi)\, \mathcal{F} f(\xi)](x), \quad j = 0, 1, \ldots .$$

The Paley-Wiener theorem tells us that $S^j f$ is anentire analytic function of exponential type. Hence, if $f,\ g \in \mathcal{S}'$ the products $S^j f \cdot S^j g$ makes sense for any j. Further,

$$\lim_{j\to\infty} \mathcal{F}^{-1}[\psi(2^{-j}\xi)\,\mathcal{F}f(\xi)](\cdot) = f \qquad \text{(convergence in } \mathcal{S}'\text{)}$$

for any $f \in \mathcal{S}'$.

Definition 3.16 Let $f,\ g \in \mathcal{S}'$. We define

$$f \cdot g = \lim_{j\to\infty} S^j f \cdot S^j g$$

whenever the limit on the right-hand side exists in $\mathcal{S}'$. We call $f \cdot g$ the product of f and g.

Remark 3.17 In defining the product we followed a usual practice, see e.g., [22], [33, 2.8], [14, 15] and [25, 4.2]. For basic properties of this notion, we refer to [14, 15] and [25, 4.2].

Theorem 3.18 *Let $1 \le p_1, p_2, q \le \infty$ and $s \le 0$. We put $1/p := (1/p_1) + (1/p_2)$. If $p \in [1, \infty]$, then there exists a constant c such that*

$$\| f \cdot g \|_{M^s_{p,q}} \le c \, \| f \|_{M^{|s|}_{p_1,1}} \, \| g \|_{M^s_{p_2,q}})$$

holds for all $f \in M^{|s|}_{p_1,1}$ and $g \in M^s_{p_2,q}$.

Proof We have to show that the limit of $(S^j f \cdot S^j g)_j$ exists in $\mathcal{S}'$. The remaining assertions, $\lim_{j\to\infty} S^j f \cdot S^j g \in M^s_{p,q}$ and the norm estimates will follow by employing the Fatou property, see Lemmas 2.5 and 3.12.
Step 1. Let $1 \le q < \infty$. We have

$$\lim_{j\to\infty} \|S^j f - f\|_{M^s_{p,q}} = 0 \qquad \text{for all} \quad f \in M^s_{p,q}\,.$$

In addition it is easily seen that

$$\sup_{j\in\mathbb{N}_0} \|S^j f\|_{M^s_{p,q}} \le \|\mathcal{F}^{-1}\psi\|_{L_1} \, \| f \|_{M^s_{p,q}} \tag{3.7}$$

holds for all $f \in M^s_{p,q}$. Hence, we conclude by means of Lemma 3.12

$$\begin{aligned}
&\|S^k f\, S^k g - S^j f\, S^j g\|_{M^s_{p,q}} \\
&\quad \le \|(S^k f - S^j f)\, S^k g\|_{M^s_{p,q}} + \|S^j f\, (S^k g - S^j g)\|_{M^s_{p,q}} \\
&\quad \le c\,(\| S^k g \|_{M^s_{p_2,q}} \| S^k f - S^j f \|_{M^{|s|}_{p_1,1}} + \| S^k g - S^j g \|_{M^s_{p_2,q}} \| S^j f \|_{M^{|s|}_{p_1,1}})
\end{aligned}$$

the convergence of $(S^k f \cdot S^k g)_k$ in $M^s_{p,q}$ and therefore in $\mathcal{S}'$, see Lemma 2.5.
Step 2. Let $q = \infty$ and suppose $p = 1$. Let $\psi, \psi^* \in C_0^\infty$ be functions such that $\psi(\xi) = 1$, $|\xi| \le 1$, $\psi(\xi) = 0$ if $|\xi| > 3/2$ and $\psi^*(\xi) = 1$, $|\xi| \le 6$. Then checking the Fourier support of the product $S^k f\, S^k g$ and using linearity of $\mathcal{F}$ we conclude

$$\Big\langle S^k f\, S^k g - S^j f\, S^j g, \varphi \Big\rangle$$
$$= \Big\langle S^k f\, S^k g - S^j f\, S^j g, \mathcal{F}^{-1}[(\psi^*(2^k\xi) - \psi^*(2^j\xi))\mathcal{F}\varphi(\xi)](\,\cdot\,)\Big\rangle .$$

For brevity we put

$$h_1 := S^k f\, S^k g - S^j f\, S^j g \quad \text{and} \quad h_2 := \mathcal{F}^{-1}[(\psi^*(2^k\xi) - \psi^*(2^j\xi))\mathcal{F}\varphi(\xi)](\,\cdot\,)\,.$$

h_1, h_2 are smooth functions with compactly supported Fourier transform. Hence,

$$h_1 = \sum_{k\in I_1} \Box_k h_1 \quad \text{and} \quad h_2 = \sum_{k\in I_2} \Box_k h_2\,,$$

where I_1, I_2 are finite subsets of $\mathbb{Z}^n$. This allows us to rewrite $\Big\langle S^k f\, S^k g - S^j f\, S^j g, \varphi\Big\rangle$ as follows

$$\Big\langle S^k f\, S^k g - S^j f\, S^j g, \varphi\Big\rangle = \sum_{k\in I_1}\sum_{\ell\in I_2} \int \Box_k h_1(x)\, \Box_\ell h_2(x)\, dx$$
$$= \sum_{\ell\in I_2}\ \sum_{k\in I_1:\ Q_k\cap Q_\ell\neq\emptyset} \int \Box_k h_1(x)\, \Box_\ell h_2(x)\, dx\,.$$

Application of Hölder's inequality yields

$$\Big|\Big\langle S^k f\, S^k g - S^j f\, S^j g, \varphi\Big\rangle\Big| \le 2^n \sup_{k\in\mathbb{Z}^n} \langle k\rangle^s \|\, \Box_k h_1 \,\|_{L_{p_1}} \Big(\sum_{\ell\in\mathbb{Z}^n} \langle \ell\rangle^{-s} \|\, \Box_\ell h_2 \,\|_{L_{p_2}}\Big)$$
$$\le 2^n \,\|\, h_1 \,\|_{M^s_{p_1,\infty}} \,\|\, h_2 \,\|_{M^{-s}_{p_2,1}}\,. \tag{3.8}$$

By means of Lemma 3.12 and (3.7) we know that

$$\|\, h_1 \,\|_{M^s_{p_1,\infty}} = \|\, S^k f\, S^k g - S^j f\, S^j g \,\|_{M^s_{p_1,\infty}}$$
$$\le c_1 \sup_{j\in\mathbb{N}_0} \|\, S^j g \,\|_{M^s_{p_2,q}} \,\|\, S^j f \,\|_{M^{|s|}_{p_1,1}}$$
$$\le c_2 \,\|\, g \,\|_{M^s_{p_2,q}} \,\|\, f \,\|_{M^{|s|}_{p_1,1}}\,.$$

On the other hand, if $j \le k$, a standard Fourier multiplier argument yields

$$\|\, h_2 \,\|_{M^{-s}_{p_2,1}} = \|\, \mathcal{F}^{-1}[(\psi^*(2^k\xi) - \psi^*(2^j\xi))\mathcal{F}\varphi(\xi)](\,\cdot\,) \,\|_{M^{-s}_{p_2,1}}$$
$$\le C \sum_{A\,2^j \le |\ell| \le B\,2^k} \langle \ell\rangle^{-s} \,\|\, \Box_\ell \varphi \,\|_{L_{p_2}}$$

for appropriate positive constants A, B, C independent of j, k and φ. Since $\varphi \in \mathcal{S} \subset M^{-s}_{p_2,1}$ we conclude that the right-hand side tends to 0 if $j \to \infty$. This finally proves

$$\left|\left\langle S^k f\, S^k g - S^j f\, S^j g, \varphi\right\rangle\right| < \varepsilon \qquad \text{if} \quad j, k \geq j_0(\varepsilon)\,.$$

Hence $(S^k f\, S^k g)_k$ is weakly convergent in $\mathcal{S}'$. Now, Lemma 3.12 yields the claim also for $q = \infty$.

Step 3. Let $q = \infty$ and suppose $1 < p \leq \infty$. We employ (3.8) with $p_1 = \infty$ and $p_2 = 1$ and afterwards Proposition 2.8. It follows

$$\begin{aligned}\left|\left\langle S^k f\, S^k g - S^j f\, S^j g, \varphi\right\rangle\right| &\leq 2^n \,\|\, h_1 \,\|_{M^s_{\infty,\infty}} \,\|\, h_2 \,\|_{M^{-s}_{1,1}} \\ &\leq c_1 \,\|\, h_1 \,\|_{M^s_{p,q}} \,\|\, h_2 \,\|_{M^{-s}_{1,1}}\,.\end{aligned}$$

Now we can argue as in Step 2. ■

Remark 3.19 For a partial result concerning Theorem 3.18 we refer to Feichtinger [7].

3.4 One Example

We consider the Dirac δ distribution. Since

$$\mathcal{F}\delta(\xi) = (2\pi)^{-n/2}\,, \qquad \xi \in \mathbb{R}^n\,,$$

it is easily seen that $\delta \in M^0_{p,\infty}$ for all p. Also not difficult to see is that $M^0_{1,\infty}$ is the smallest space of type $M^s_{p,q}$ to which δ belongs to. Theorem 3.18 yields

$$\|\, f \cdot \delta \,\|_{M^0_{p,\infty}} \leq c \,\|\, \delta \,\|_{M^0_{p,\infty}} \,\|\, f \,\|_{M^0_{\infty,1}}$$

with some c independent of $f \in M^0_{\infty,1}$. With other words, we can multiply δ with a modulation space $M^s_{p,q}$ if this space is embedded into C_{ub}, see Corollary 2.10. This looks reasonable.

3.5 The Second Method

Finally, we would like to investigate also the cases $\min(s_1, s_2) \leq n/q'$. For dealing with this special situation we turn to a different method which will allow a better localization in the Fourier image. Therefore we shall work with the frequency-uniform decomposition $(\sigma_k)_k$. Recall that $\operatorname{supp}\sigma_k \subset Q_k := \{\xi \in \mathbb{R}^n : -1 \le \xi_i - k_i \le 1,\ i = 1, \ldots, n\}$. For brevity we put

$$f_k(x) := \mathcal{F}^{-1}[\sigma_k(\xi)\mathcal{F}f(\xi)](x)\,, \qquad x \in \mathbb{R}^n\,, \quad k \in \mathbb{Z}^n\,.$$

Then, at least formally, we have the following representation of the product $f \cdot g$ as

$$f \cdot g = \sum_{k,l \in \mathbb{Z}^n} f_k \cdot g_l.$$

In what follows we shall study bounds for related partial sums.

Lemma 3.20 *Let $1 \le p_1, p_2 \le \infty$, $1 < q \le \infty$ and $s_0, s_1, s_2 \in \mathbb{R}$. Define p by $\frac{1}{p} := \frac{1}{p_1} + \frac{1}{p_2}$. If $p \in [1, \infty]$, $0 \le s_0 \le \min(s_1, s_2)$ and $s_2 + s_1 - s_0 > n/q'$, then there exists a constant c such that*

$$\| \sum_{k,l \in \mathbb{Z}^n} f_k \cdot g_l \|_{M^{s_0}_{p,q}} \le c\, \|f\|_{M^{s_1}_{p_1,q}}\, \|g\|_{M^{s_2}_{p_2,q}}$$

holds for all $f, g \in \mathcal{S}'$ such that $\operatorname{supp}\mathcal{F}f$ and $\operatorname{supp}\mathcal{F}g$ are compact. The constant c is independent from $\operatorname{supp}\mathcal{F}f$ and $\operatorname{supp}\mathcal{F}g$, respectively.

Proof Later on, we shall use the same strategy of proof as below in slightly different situations. For this reason and later use we shall take care of all constants showing up in our estimates below.
Step 1. Preparations. Determining the Fourier support of $f_j \cdot g_l$ we see that

$$\begin{aligned}\operatorname{supp}\mathcal{F}(f_j \cdot g_l) &= \operatorname{supp}(\mathcal{F}f_j * \mathcal{F}g_l)\\ &\subset \{\xi \in \mathbb{R}^n : j_i + l_i - 2 \le \xi_i \le j_i + l_i + 2,\ i = 1, \ldots, n\}.\end{aligned}$$

Hence, the term $\mathcal{F}^{-1}(\sigma_k\mathcal{F}(f_j \cdot g_l))$ vanishes if $\|k - (j + l)\|_\infty \ge 3$. In addition, since $\operatorname{supp}\mathcal{F}f$ and $\operatorname{supp}\mathcal{F}g$ are compact, the sum $\sum_{j,l\in\mathbb{Z}^n} f_j \cdot g_l$ is a finite sum. We obtain

$$\begin{aligned}\sigma_k\mathcal{F}(f \cdot g) = \sigma_k\mathcal{F}\Big(\sum_{j,l\in\mathbb{Z}^n} f_j \cdot g_l\Big) &= \sigma_k\mathcal{F}\Big(\sum_{\substack{j,l\in\mathbb{Z}^n,\\ k_i-3<j_i+l_i<k_i+3,\\ i=1,\ldots,n}} f_j \cdot g_l\Big)\\ &\overset{[r=j+l]}{=} \sum_{\substack{r\in\mathbb{Z}^n,\\ k_i-3<r_i<k_i+3,\\ i=1,\ldots,n}} \sum_{l\in\mathbb{Z}^n} \sigma_k\mathcal{F}\big(f_{r-l} \cdot g_l\big)\,.\end{aligned}$$

Consequently

$$\begin{aligned}\big\|\mathcal{F}^{-1}\big(\sigma_k\mathcal{F}(f\cdot g)\big)\big\|_{L_p} &\le \sum_{\substack{r\in\mathbb{Z}^n,\\ k_i-3<r_i<k_i+3,\\ i=1,\dots,n}} \sum_{l\in\mathbb{Z}^n} \|\mathcal{F}^{-1}\big(\sigma_k\mathcal{F}(f_{r-l}\cdot g_l)\big)\|_{L_p}\\ &\overset{[t=r-k]}{=} \sum_{\substack{t\in\mathbb{Z}^n,\\ -3<t_i<3,\\ i=1,\dots,n}} \sum_{l\in\mathbb{Z}^n} \|\mathcal{F}^{-1}\big(\sigma_k\mathcal{F}(f_{t-(l-k)}\cdot g_l)\big)\|_{L_p}.\end{aligned}$$

Step 2. Norm estimates. These preparations yield the following estimates

$$\begin{aligned}&\Big(\sum_{k\in\mathbb{Z}^n}\langle k\rangle^{s_0q}\|\mathcal{F}^{-1}\big(\sigma_k\mathcal{F}(f\cdot g)\big)\|_{L_p}^q\Big)^{\frac{1}{q}}\\ &\quad\le\Bigg(\sum_{k\in\mathbb{Z}^n}\langle k\rangle^{s_0q}\Big[\sum_{\substack{t\in\mathbb{Z}^n,\\ -3<t_i<3,\\ i=1,\dots,n}}\sum_{l\in\mathbb{Z}^n}\|\mathcal{F}^{-1}\big(\sigma_k\mathcal{F}(f_{t-(l-k)}\cdot g_l)\big)\|_{L_p}\Big]^q\Bigg)^{\frac{1}{q}}\\ &\quad\le\sum_{\substack{t\in\mathbb{Z}^n,\\ -3<t_i<3,\\ i=1,\dots,n}}\Bigg(\sum_{k\in\mathbb{Z}^n}\langle k\rangle^{s_0q}\Big[\sum_{l\in\mathbb{Z}^n}\|\mathcal{F}^{-1}\big(\sigma_k\mathcal{F}(f_{t-(l-k)}\cdot g_l)\big)\|_{L_p}\Big]^q\Bigg)^{\frac{1}{q}}.\end{aligned}$$

Observe

$$\begin{aligned}\|\mathcal{F}^{-1}\big(\sigma_k\mathcal{F}(f_{t-(l-k)}\cdot g_l)\big)\|_{L^p} &= (2\pi)^{-n/2}\|(\mathcal{F}^{-1}\sigma_k)*(f_{t-(l-k)}\cdot g_l)\|_{L_p}\\ &\le(2\pi)^{-n/2}\|\mathcal{F}^{-1}\sigma_k\|_{L^1}\|f_{t-(l-k)}\cdot g_l\|_{L_p}\\ &=(2\pi)^{-n/2}\|\mathcal{F}^{-1}\sigma_0\|_{L^1}\|f_{t-(l-k)}\cdot g_l\|_{L_p},\end{aligned}$$

where we used Young's inequality. We put $c_1 := (2\pi)^{-n/2}\|\mathcal{F}^{-1}\sigma_0\|_{L_1}$. This implies

$$\begin{aligned}&\Big(\sum_{k\in\mathbb{Z}^n}\langle k\rangle^{s_0q}\|\mathcal{F}^{-1}\big(\sigma_k\mathcal{F}(f\cdot g)\big)\|_{L_p}^q\Big)^{\frac{1}{q}}\\ &\qquad\le c_1\sum_{\substack{t\in\mathbb{Z}^n,\\ -3<t_i<3,\\ i=1,\dots,n}}\Bigg(\sum_{k\in\mathbb{Z}^n}\langle k\rangle^{s_0q}\Big[\sum_{l\in\mathbb{Z}^n}\|f_{t-(l-k)}\cdot g_l\|_{L_p}\Big]^q\Bigg)^{\frac{1}{q}}.\end{aligned}$$

We continue by using Hölder's inequality to get

$$\Big(\sum_{k\in\mathbb{Z}^n} \langle k\rangle^{s_0 q} \|\mathcal{F}^{-1}\big(\sigma_k \mathcal{F}(f\cdot g)\big)\|_{L_p}^q\Big)^{\frac{1}{q}}$$

$$\le c_2 \max_{\substack{t\in\mathbb{Z}^n,\\ -3<t_i<3,\\ i=1,\dots,n}} \left(\sum_{k\in\mathbb{Z}^n} \langle k\rangle^{s_0 q}\left[\sum_{l\in\mathbb{Z}^n} \|f_{t-(l-k)}\|_{L_{p_1}} \|g_l\|_{L_{p_2}}\right]^q\right)^{\frac{1}{q}}$$

with $c_2 := c_1\, 5^n$. Since $s_0 \ge 0$ elementary calculations yield

$$\langle k\rangle^{s_0}\Big[\sum_{l\in\mathbb{Z}^n} \|f_{t-(l-k)}\|_{L_{p_1}} \|g_l\|_{L_{p_2}}\Big]$$

$$\le 2^{s_0} \sum_{\substack{l\in\mathbb{Z}^n,\\ |l|\le|l-k|}} \langle k-l\rangle^{s_0} \|f_{t-(l-k)}\|_{L_{p_1}} \|g_l\|_{L_{p_2}}$$

$$+2^{s_0} \sum_{\substack{l\in\mathbb{Z}^n,\\ |l-k|\le|l|}} \|f_{t-(l-k)}\|_{L_{p_1}} \langle l\rangle^{s_0} \|g_l\|_{L_{p_2}} .$$

Both parts of this right-hand side will be estimated separately. We put

$$S_{1,t,k} := \sum_{\substack{l\in\mathbb{Z}^n,\\ |l|\le|l-k|}} \langle k-l\rangle^{s_1} \|f_{t-(l-k)}\|_{L_{p_1}} \langle l\rangle^{s_2} \|g_l\|_{L_{p_2}} \langle k-l\rangle^{s_0-s_1} \langle l\rangle^{-s_2} ;$$

$$S_{2,t,k} := \sum_{\substack{l\in\mathbb{Z}^n,\\ |l-k|\le|l|}} \langle k-l\rangle^{s_1} \|f_{t-(l-k)}\|_{L_{p_1}} \langle l\rangle^{s_2} \|g_l\|_{L_{p_2}} \langle k-l\rangle^{-s_1} \langle l\rangle^{s_0-s_2} .$$

With $\frac{1}{q}+\frac{1}{q'}=1$ we find

$$S_{1,t,k} \overset{[j=l-k]}{=} \sum_{\substack{j\in\mathbb{Z}^n,\\ |j+k|\le|j|}} \langle j\rangle^{s_0} \|f_{t-j}\|_{L_{p_1}} \|g_{j+k}\|_{L_{p_2}}$$

$$\le \Big(\sum_{\substack{j\in\mathbb{Z}^n,\\ |j+k|\le|j|}} (\langle j\rangle^{s_1} \|f_{t-j}\|_{L_{p_1}} \langle j+k\rangle^{s_2} \|g_{j+k}\|_{L_{p_2}})^q\Big)^{1/q}$$

$$\times \Big(\sum_{\substack{j\in\mathbb{Z}^n,\\ |j+k|\le|j|}} (\langle j\rangle^{s_0-s_1} \langle j+k\rangle^{-s_2})^{q'}\Big)^{\frac{1}{q'}} .$$

Substep 2.1. Our assumptions $s_0 \le s_1, s_2 \ge 0$ and $s_1 + s_2 - s_0 > n/q'$ imply

$$\Big(\sum_{\substack{j\in\mathbb{Z}^n,\\ |j+k|\le |j|}} \Big| \langle j\rangle^{s_0-s_1} \langle j+k\rangle^{-s_2}\Big|^{q'}\Big)^{\frac{1}{q'}} \le \Big(\sum_{m\in\mathbb{Z}^n} \langle m\rangle^{(s_0-s_1-s_2)q'}\Big)^{\frac{1}{q'}} =: c_3 < \infty\,.$$

Inserting this in our previous estimate we obtain

$$\Big(\sum_{k\in\mathbb{Z}^n} S_{1,t,k}^q\Big)^{1/q} \le c_3 \Big(\sum_{k\in\mathbb{Z}^n}\sum_{\substack{j\in\mathbb{Z}^n,\\ |j+k|\le |j|}} \langle j\rangle^{s_1 q} \|f_{t-j}\|^q_{L^{p_1}} \langle j+k\rangle^{s_2 q} \|g_{j+k}\|^q_{L^{p_2}}\Big)^{1/q}$$

$$\le c_3 \Big(\sum_{j\in\mathbb{Z}^n} \langle j\rangle^{s_1 q} \|f_{t-j}\|^q_{L^{p_1}} \sum_{k\in\mathbb{Z}^n} \langle j+k\rangle^{s_2 q} \|g_{j+k}\|^q_{L^{p_2}}\Big)^{\frac{1}{q}}.$$

Because of $1 + |j|^2 \le 1 + 8n + |j-t|^2$ we know

$$\max_{\substack{t\in\mathbb{Z}^n,\\ -3<t_i<3,\\ i=1,\ldots,n}} \sup_{j\in\mathbb{Z}^n} \frac{\langle j\rangle^{s_1}}{\langle j-t\rangle^{s_1}} \le (1+8n)^{s_1/2} =: c_4 < \infty\,.$$

This implies

$$\Big(\sum_{k\in\mathbb{Z}^n} S_{1,t,k}^q\Big)^{1/q} \le c_3\, c_4\, \|g\|_{M^{s_2}_{p_2,q}} \|f\|_{M^{s_1}_{p_1,q}}, \tag{3.9}$$

where c_3, c_4 are independent of f, g and t.

Substep 2.2. Because of $0 \le s_0 \le s_1$, $s_0 \le s_2$ and $s_1 + s_2 - s_0 > n/q'$ we conclude

$$\Big(\sum_{\substack{l\in\mathbb{Z}^n,\\ |l-k|\le |l|}} \Big| \langle k-l\rangle^{-s_1} \langle l\rangle^{s_0-s_2}\Big|^{q'}\Big)^{\frac{1}{q'}} \le \Big(\sum_{m\in\mathbb{Z}^n} \langle m\rangle^{(s_0-s_1-s_2)q'}\Big)^{\frac{1}{q'}} =: c_5 < \infty\,.$$

This leads to the estimate

$$\Big(\sum_{k\in\mathbb{Z}^n} S_{2,t,k}^q\Big)^{1/q} \le c_5 c_6\, \|g\|_{M^{s_2}_{p_2,q}} \|f\|_{M^{s_1}_{p_1,q}} \tag{3.10}$$

with some constants c_6 independent from f and g. Combining the inequalities (3.9) and (3.10) we have proved the claim. ∎

Remark 3.21 Some basic ideas of the above proof are taken over from [5], see also [23].

Of course the above method of proof works as well for $q = 1$. But all spaces $M^s_{p,1}$, $s \ge 0$, are algebras.

Theorem 3.22 *Let* $1 \le p, p_1, p_2 \le \infty$ *and* $s_0, s_1, s_2 \in \mathbb{R}$. *Let* $1/p \le (1/p_1) + (1/p_2)$, $1 < q \le \infty$, $0 \le s_0 \le \min(s_1, s_2)$ *and* $s_1 + s_2 - s_0 > n/q'$. *There exists a constant c such that*

$$\| f \cdot g \|_{M^{s_0}_{p,q}} \le c \, \| f \|_{M^{s_1}_{p_1,q}} \, \| g \|_{M^{s_2}_{p_2,q}}$$

holds for all $f \in M^{s_1}_{p_1,q}$ *and all* $g \in M^{s_2}_{p_2,q}$.

Proof We only comment on the case $1/p = (1/p_1) + (1/p_2)$, see Corollary 2.7. It will be enough to prove the weak convergence of $(S^k f \cdot S^k g)_k$ in $\mathcal{S}'$. The claimed estimate will then follow from Lemma 3.20. We employ the method and the notation used in proof of Theorem 3.18 (Steps 2 and 3). There we have proved

$$\left| \left\langle S^k f \, S^k g - S^j f \, S^j g, \varphi \right\rangle \right| \le c_1 \, \| h_1 \|_{M^{s_0}_{p,q}} \, \| h_2 \|_{M^{-s_0}_{1,1}}$$

with c_1 independent of f, g, k and j. By means of Lemma 3.20 we know the uniform boundedness of $\| h_1 \|_{M^{s_0}_{p,q}}$ in k and j. The estimate of $\| h_2 \|_{M^{-s_0}_{1,1}}$ can be done as above. It follows

$$\| h_2 \|_{M^{-s_0}_{1,1}} \le \varepsilon$$

if $j, k \ge j_0(\varepsilon)$. This guarantees the weak convergence of $(S^k f \cdot S^k g)_k$ in $\mathcal{S}'$. ∎

Our sufficient conditions are not far away from necessary conditions.

Lemma 3.23 *Let* $1 \le p_1, p_2, p, q \le \infty$ *and* $s_0, s_1, s_2 \in \mathbb{R}$. *Suppose that there exists a constant c such that*

$$\| f \cdot g \|_{M^{s_0}_{p,q}} \le c \, \| f \|_{M^{s_1}_{p_1,q}} \, \| g \|_{M^{s_2}_{p_2,q}} \tag{3.11}$$

holds for all $f, g \in \mathcal{S}$.
(i) *It follows* $s_0 \le \min(s_1, s_2)$, $s_1 + s_2 \ge 0$ *and* $s_1 + s_2 - s_0 \ge n/q'$.
(ii) *If* $1 \le p_2 = p < \infty$ *and* $1 \le q < \infty$, *then either* $q = 1$ *and* $s_1 \ge 0$ *or* $1 < q < \infty$ *and* $s_1 > n/q'$.

Proof Part (ii) is an immediate consequence of Lemma 3.7. Concerning the proof of (i) we shall work with the same test functions as used in Step 2 of the proof of Corollary 2.10, see (2.1).
Step 1. We choose $a_k := \delta_{k,\ell}$, $k \in \mathbb{Z}^n$, for a fixed given $\ell \in \mathbb{Z}^n$ and put $b_k := \delta_{k,0}$, $k \in \mathbb{Z}^n$. Then we define

$$f(x) := \psi(x) \, e^{i\ell x} \qquad \text{and} \qquad g(x) := \psi(x) \, .$$

We obtain

$$\| f \|_{M^{s_1}_{p_1,q}} \cdot \| g \|_{M^{s_2}_{p_2,q}} = \| \psi \|_{L_{p_1}} \, \| \psi \|_{L_{p_2}} \, \langle \ell \rangle^{s_1}$$

as well as

$$\| f \cdot g \|_{M^{s_0}_{p,q}} = \|\psi^2\|_{L_p} \langle \ell \rangle^{s_0} .$$

Hence, (3.11) implies $s_0 \le s_1$. Interchanging the roles of f and g leads to the conclusion $s_0 \le s_2$.

Step 2. Let $\ell \in \mathbb{Z}^n$ be fixed. We choose $a_k := \delta_{k,\ell}$, $k \in \mathbb{Z}^n$, and $b_k := \delta_{k,-\ell}$, $k \in \mathbb{Z}^n$. Then we define

$$f(x) := \psi(x)\, e^{i\ell x} \qquad \text{and} \qquad g(x) := \psi(x)\, e^{-i\ell x} .$$

It follows

$$\| f \|_{M^{s_1}_{p_1,q}} \cdot \| g \|_{M^{s_2}_{p_2,q}} = \|\psi\|_{L_{p_1}} \, \|\psi\|_{L_{p_2}} \langle \ell \rangle^{s_1+s_2}$$

as well as

$$\| f \cdot g \|_{M^{s_0}_{p,q}} = \|\psi^2\|_{L_p} .$$

Hence, (3.11) implies $s_1 + s_2 \ge 0$.

Step 3. Let $\varepsilon_1, \varepsilon_2 \ge 0$. These two numbers will be chosen such that

$$\min(s_1 + \varepsilon_1 + n/q, s_2 + \varepsilon_2 + n/q) > 0 \qquad \text{and} \qquad s_0 + \varepsilon_2 + \varepsilon_1 + n > 0 .$$

We choose $a_k := \langle k \rangle^{\varepsilon_1}$, $k \in \mathbb{Z}^n$, and $b_k := \langle k \rangle^{\varepsilon_2}$, $k \in \mathbb{Z}^n$. Then we define

$$f(x) := \psi(x) \sum_{\|k\|_\infty \le N} a_k\, e^{ikx} \qquad \text{and} \qquad g(x) := \psi(x) \sum_{\|k\|_\infty \le N} b_k\, e^{ikx} .$$

By means of the same arguments as used in Substep 3.1 of the proof of Theorem 3.5, we conclude

$$\| f \|_{M^{s_1}_{p_1,q}} \asymp N^{s_1+\varepsilon_1+n/q} \qquad \text{and} \qquad \| g \|_{M^{s_2}_{p_2,q}} \asymp N^{s_2+\varepsilon_2+n/q} .$$

In addition, we have

$$\begin{aligned}
\| f \cdot g \|_{M^{s_0}_{p,q}} &\asymp \Bigg(\sum_{\|m\|_\infty \le 2N} \langle m \rangle^{s_0 q} \Bigg| \sum_{\substack{k:\ \|k\|_\infty \le N \\ \|m-k\|_\infty \le N}} a_k\, b_{m-k} \Bigg|^q \Bigg)^{1/q} \\
&\ge \frac{1}{2n} \Bigg(\sum_{\|m\|_\infty \le N} \langle m \rangle^{(s_0+\varepsilon_2) q} \Bigg| \sum_{k:\ \|k\|_\infty \le \|m\|_\infty/2} \langle k \rangle^{\varepsilon_1} \Bigg|^q \Bigg)^{1/q} \\
&\ge C_1 \Bigg(\sum_{\|m\|_\infty \le N} \langle m \rangle^{(s_0+\varepsilon_2+\varepsilon_1+n) q} \Bigg)^{1/q} \\
&\ge C_2\, N^{s_0+\varepsilon_2+\varepsilon_1+n+n/q}
\end{aligned}$$

for some C_1, C_2 independent of N, see Substep 3.2 of the proof of Theorem 3.5. The inequality (3.11) yields

$$s_0 + \varepsilon_2 + \varepsilon_1 + n + n/q \le s_1 + \varepsilon_1 + n/q + s_2 + \varepsilon_2 + n/q$$

which proves the claim. ∎

The duality argument used in the proof of Lemma 3.7 allows to treat the case $s_0 < 0$.

Theorem 3.24 *Let* $1 \le p, p_1, p_2 \le \infty$ *and* $s_0, s_1, s_2 \in \mathbb{R}$. *Let* $1/p \le (1/p_1) + (1/p_2)$, $1 \le q < \infty$, $s_0 \le s_2 \le 0$, $0 \le s_1 + s_2$ *and* $s_1 + s_2 - s_0 > n/q$. *There exists a constant c such that*

$$\| f \cdot g \|_{M^{s_0}_{p,q}} \le c \, \| f \|_{M^{s_1}_{p_1,q'}} \, \| g \|_{M^{s_2}_{p_2,q}}$$

holds for all $f \in M^{s_1}_{p_1,q'}$ *and all* $g \in M^{s_2}_{p_2,q}$.

Remark 3.25 Theorems 3.18 and 3.24 have some overlap.

3.6 Some Further Remarks to the Literature

Here, we recall results of Iwabuchi [13] and Toft et al. [32]. As Cordero and Nicola [6] also Iwabuchi considered the more general situation $M^{s_1}_{p_1,q_1} \cdot M^{s_2}_{p_2,q_2} \hookrightarrow M^{s_0}_{p,q}$. This greater flexibility with respect to the tripel q, q_1, q_2 allows to treat cases not covered by Theorems 3.22, 3.24.

Proposition 3.26 (Iwabuchi [13])
Let $1 \le p, p_1, p_2 \le \infty$, $1 < q, q_1, q_2 < \infty$ *and* $0 < s_0 < n/q$.
(i) *If* $q \ge q_1$,

$$\frac{1}{p} \le \frac{1}{p_1} + \frac{1}{p_2} \quad \textit{and} \quad 1 + \frac{1}{q} - \Big(\frac{1}{q_1} + \frac{1}{q_2}\Big) = \frac{s_0}{n}, \tag{3.12}$$

then there exists a constant c such that

$$\| f \cdot g \|_{M^{-s_0}_{p,q}} \le c \, \| f \|_{M^{0}_{p_1,q_1}} \, \| g \|_{M^{0}_{p_2,q_2}}$$

holds for all $f \in M^{0}_{p_1,q_1}$ *and all* $g \in M^{0}_{p_2,q_2}$.
(ii) *Assume* $q \ge \max(q_1, q_2)$ *and* (3.12). *Then, there exists a constant c such that*

$$\| f \cdot g \|_{M^{s_0}_{p,q}} \le c \, \| f \|_{M^{s_0}_{p_1,q_1}} \, \| g \|_{M^{s_0}_{p_2,q_2}}$$

holds for all $f \in M^{s_0}_{p_1,q_1}$ *and all* $g \in M^{s_0}_{p_2,q_2}$.

Remark 3.27 Let us take $q = q_1 = q_2$. Then (3.12) reads as $s_0 = n/q'$. In combination with $0 < s_0 < n/q$ this yields $1 < q < 2$. Hence, (i) reads as

$$\| f \cdot g \|_{M^{-n/q'}_{p,q}} \le c \, \| f \|_{M^0_{p_1,q}} \, \| g \|_{M^0_{p_2,q}} \, ,$$

whereas (ii) gives

$$\| f \cdot g \|_{M^{n/q'}_{p,q}} \le c \, \| f \|_{M^{n/q'}_{p_1,q_1}} \, \| g \|_{M^{n/q'}_{p_2,q_2}} \, .$$

Toft et al. [32] also consider the situation $M^{s_1}_{p_1,q_1} \cdot M^{s_2}_{p_2,q_2} \hookrightarrow M^{s_0}_{p,q}$. Recall, $\mathring{M}^{s_0}_{p,q}$ denotes the closure of $\mathcal{S}$ in $M^{s_0}_{p,q}$.

Proposition 3.28 (Toft et al. [32])
Let $1 \le p, p_1, p_2, q, q_1, q_2 \le \infty$ *and* $s_0, s_1, s_2 \in \mathbb{R}$.
(i) *We suppose*

(a) $1 + \frac{1}{p} - \frac{1}{p_1} - \frac{1}{p_2} \le 1$;
(b) $0 \le 1 + \frac{1}{q} - \frac{1}{q_1} - \frac{1}{q_2} \le 1/2$;
(c) $s_0 \le \min(s_1, s_2)$;
(d) $s_1 + s_2 \ge 0$;
(e) $s_1 + s_2 - s_0 - n\left(1 + \frac{1}{q} - \frac{1}{q_1} - \frac{1}{q_2}\right) \ge 0$;
(f) $s_1 + s_2 - s_0 - n\left(1 + \frac{1}{q} - \frac{1}{q_1} - \frac{1}{q_2}\right) > 0$ *if* $1 + \frac{1}{q} - \frac{1}{q_1} - \frac{1}{q_2} > 0$ *and either* s_1 *or* s_2 *or* $-s_0$ *equals* $n\left(1 + \frac{1}{q} - \frac{1}{q_1} - \frac{1}{q_2}\right)$.

Then there exists a constant c such that

$$\| f \cdot g \|_{M^{s_0}_{p,q}} \le c \, \| f \|_{M^{s_1}_{p_1,q_1}} \, \| g \|_{M^{s_2}_{p_2,q_2}} \tag{3.13}$$

holds for all $f \in \mathring{M}^{s_1}_{p_1,q_1}$ *and all* $g \in \mathring{M}^{s_2}_{p_2,q_2}$.
(ii) *If* (3.13) *holds for all* $f, g \in \mathcal{S}$, *then (c), (d) and (e) follow.*

Remark 3.29 Again we consider the case $q = q_1 = q_2$. Then (b) implies $1 \le q \le 2$ and (e) reads as $s_1 + s_2 - s_0 - n/q' \ge 0$. Hence, if we restrict us to $1 < q \le 2$, Proposition 3.28 is slightly more general than Theorem 3.22 and Theorem 3.24. However, for our purpose, see the next section on composition of functions, Theorem 3.22 is already sufficient. Let us mention that Proposition 4.5 below, which is nothing but a modification of Lemma 3.20, is of central importance for the applications to composition operators we have in mind.

3.7 An Important Special Case

We consider $M^s_{2,2}$. A simple argument, based on the frequency-uniform decomposition yields $M^s_{2,2} = H^s$ in the sense of equivalent norms, see Remark 2.3. For these Sobolev spaces H^s almost all is known.

- H^s is an algebra with respect to pointwise multiplication if and only if $s > n/2$, see Strichartz [28], Triebel [33, 2.8] or [25, Theorem 4.6.4/1]. This coincides with Theorem 3.5.
- Let E be a Banach space of functions. By $M(E)$ we denote the set of all pointwise multipliers of E, i.e., the set of all f such that T_f, defined as $T_f(g) = f \cdot g$, maps E into E. We equip $M(E)$ with the norm $\| f \|_{M(E)} := \| T_f \|_{\mathcal{L}(E)}$. For a description of $M(H^s)$ one needs the classes $H^{s,\ell oc}$. Here $H^{s,\ell oc}$ denotes the collection of all distributions $f \in \mathcal{S}'$ such that $f \cdot \varphi \in H^s$ for all $\varphi \in C_0^\infty$. In case $s > n/2$ it holds

$$M(H^s) = \Big\{ f \in H^{s,\ell oc}: \quad \| f \|^*_{M(H^s)} := \sup_{\lambda \in \mathbb{R}^n} \| \psi(\cdot - \lambda)\, f \|_{H^s} < \infty \Big\}$$

  in the sense of equivalent norms. Here ψ is a smooth nontrivial cut-off function supported around the origin. For all this we refer to Strichartz [28].
- In case $0 \le s < n/2$ also characterizations of $M(H^s)$ are known, this time more complicated, based on capacities. For all details we refer to the monograph of Maz'ya and Shaposnikova [20, Theorem 3.2.2, pp. 86].
- Now we concentrate on the situation described in Theorem 3.22 in case $0 < s < \frac{n}{2}$. As it is well-known, there exists a constant c such that

$$\| f \cdot g \|_{H^{2s-n/2}} \le c \, \| f \|_{H^s} \, \| g \|_{H^s}$$

  holds for all $f, g \in H^s$, see e.g., [25, Theorem 4.5.2]. In Theorem 3.22 we proved that for any $\varepsilon > 0$ there exists a constant c_ε such that

$$\| f \cdot g \|_{M^{2s-n/2-\varepsilon}_{1,2}} \le c_\varepsilon \, \| f \|_{H^s} \, \| g \|_{H^s}$$

  holds for all $f, g \in H^s$. We conjecture that $M^{2s-n/2-\varepsilon}_{1,2}$ and $H^{2s-n/2}$ are incomparable.

4 Composition of Functions

There are some attempts to investigate composition of functions in the framework of modulation spaces, i.e., we consider the operator

$$T_f: \quad g \mapsto f \circ g, \qquad g \in M^s_{p,q}, \tag{4.1}$$

and ask for mapping properties. Of course, we used the symbol T_f before with a different meaning, but we hope that will not cause problems. Within Sect. 4 T_f will have the meaning as in (4.1). Based on pointwise multiplication one can treat f to be a polynomial or even the more general case of f being an entire function.

4.1 Polynomials

We consider the case

$$f(z) := \sum_{\ell=1}^{m} a_\ell \, z^\ell \,, \qquad z \in \mathbb{C}\,,$$

where $m \in \mathbb{N}$, $m \geq 2$, and $a_\ell \in \mathbb{C}$, $\ell = 1, \ldots, m$. For brevity we denote the associated composition operator by T_m. In addition we need the abbreviation

$$t_m(s) := s + (m-1)(s - n/q')\,, \qquad m = 2, 3, \ldots .$$

Theorem 4.1 *Let* $1 \leq p, q \leq \infty$ *and* $m \in \mathbb{N}$, $m \geq 2$.
(i) *Let either* $s \geq 0$ *and* $q = 1$ *or* $s > n/q'$. *Then* T_m *maps* $M^s_{p,q}$ *into itself. There exists a constant c such that*

$$\| T_m g \|_{M^s_{p,q}} \leq c \, \| g \|_{M^s_{p,q}} \sum_{\ell=1}^{m} |a_\ell| \, \| g \|^{\ell-1}_{M^0_{\infty,1}}$$

holds for all $g \in M^s_{p,q}$.
(ii) *Let* $1 < q \leq \infty$, $0 < s \leq n/q'$ *and* $t_m(s) > 0$. *If* $p \in [m, \infty]$ *and* $t < t_m(s)$, *then there exists a constant c such that*

$$\| T_m g \|_{M^t_{p/m,q}} \leq c \sum_{\ell=1}^{m} |a_\ell| \, \|g\|^{\ell}_{M^s_{p,q}}$$

holds for all $g \in M^s_{p,q}$.
(iii) *Let* $q = 1$ *and* $s \geq 0$. *If* $p \in [m, \infty]$, *then there exists a constant c such that*

$$\| T_m g \|_{M^s_{p/m,1}} \leq c \sum_{\ell=1}^{m} |a_\ell| \, \|g\|^{\ell}_{M^s_{p,1}}$$

holds for all $g \in M^s_{p,1}$.

Proof Step 1. Both parts, (i) and (ii), can be proved by induction based on Theorem 3.5 or Theorem 3.22. We concentrate on the proof of (ii). Let $m = 2$. Then by assumption $t_2(s) = 2s - n/q' > 0$. Hence, we may apply Theorem 3.22 with $p_1 = p_2 = p$ and $s_1 = s_2$ and obtain

$$\| g^2 \|_{M^t_{p/2,q}} \leq c \, \|g\|^2_{M^s_{p,q}}$$

for any $t < 2s - n/q' = t_2(s)$. Now we assume that part (ii) is correct for all natural numbers in the interval $[2, m]$. We split the product g^{m+1} into the two factors g^m

and g. By assumption $g^m \in M^t_{p/m,q}$ for any $t < t_m(s)$. We put $s_1 = t = t_m(s) - \varepsilon$, $s_2 = s$, $p_1 = p/m$ and $p_2 = p$, where we assume that $\varepsilon > 0$ is sufficiently small. This guarantees

$$s_1 + s_2 - \frac{n}{q'} = s + (m-1)\Big(s - \frac{n}{q'}\Big) - \varepsilon + s - \frac{n}{q'} = t_{m+1}(s) - \varepsilon > 0\,.$$

Hence, we may choose s_0 by

$$s_0 < \min(s_1, s_2, t_{m+1}(s) - \varepsilon) = t_{m+1}(s) - \varepsilon\,.$$

Since $\varepsilon > 0$ is arbitrary, any value $< t_{m+1}(s)$ becomes admissible for s_0. An application of Theorem 3.22 yields

$$\| g^m \cdot g \|_{M^{s_0}_{p/(m+1),q}} \le c \, \| g_m \|_{M^{t_m - \varepsilon}_{p/m,q}} \, \|g\|_{M^s_{p,q}}\,.$$

Step 2. Part (iii) is an immediate consequence of Lemma 3.10. ■

Remark 4.2 For the case $s = 0$ we refer to Cordero, Nicola [6], Toft [30] and Guo et al. [11].

4.2 Entire Functions

We consider the case of f being an entire analytic function on $\mathbb{C}$, i.e.,

$$f(z) := \sum_{\ell=0}^{\infty} a_\ell \, z^\ell\,, \qquad z \in \mathbb{C}\,,$$

where $a_\ell \in \mathbb{C}$, $\ell \in \mathbb{N}_0$. Clearly, we need to assume $f(0) = a_0 = 0$. Otherwise $T_f g$ will not have global integrability properties. Let

$$f_0(r) := \sum_{\ell=1}^{\infty} |a_\ell| \, r^\ell\,, \qquad r > 0\,.$$

Theorem 4.3 *Let $1 \le p, q \le \infty$ and let either $s \ge 0$ and $q = 1$ or $s > n/q'$. Let f be an entire function satisfying $f(0) = 0$. Then T_f maps $M^s_{p,q}$ into itself. There exist two constants a, b, independent of f, such that*

$$\| T_f g \|_{M^s_{p,q}} \le a \, f_0(b \, \| g \|_{M^s_{p,q}})$$

holds for all $g \in M^s_{p,q}$.

Proof The constant c in Theorem 4.1 (i) depends on m. To clarify the dependence on m we proceed by induction. Let c_1 be the best constant in the inequality

$$\| g_1 \cdot g_2 \|_{M^s_{p,q}} \le c_1 \left(\| g_1 \|_{M^s_{p,q}} \| g_2 \|_{M^0_{\infty,1}} + \| g_2 \|_{M^s_{p,q}} \| g_1 \|_{M^0_{\infty,1}}\right), \tag{4.2}$$

see Lemma 3.3. Further, let c_2 be the best constant in the inequality

$$\| g_1 \cdot g_2 \|_{M^0_{\infty,1}} \le c_2 \| g_1 \|_{M^0_{\infty,1}} \| g_2 \|_{M^0_{\infty,1}}, \tag{4.3}$$

see also Lemma 3.3. By c_3 we denote $\max(1, c_1, c_2)$. Our induction hypothesis consists in: the inequality

$$\| g^m \|_{M^s_{p,q}} \le c_3^{m-1}\, m \, \| g \|_{M^s_{p,q}} \| g \|^{m-1}_{M^0_{\infty,1}}$$

holds for all $g \in M^s_{p,q}$ and all $m \ge 2$. This follows easily from (4.2) and (4.3). Next we need the best constant, denoted by c_4, in the inequality

$$\| g \|_{M^0_{\infty,1}} \le c_4 \| g \|_{M^s_{p,q}}, \qquad g \in M^s_{p,q}.$$

This proves that

$$\| g^m \|_{M^s_{p,q}} \le c_3^{m-1}\, m \, c_4^{m-1} \| g \|^m_{M^s_{p,q}} \tag{4.4}$$

holds for all $g \in M^s_{p,q}$ and all $m \ge 2$. Hence

$$\begin{aligned} \| T_f g \|_{M^s_{p,q}} &\le \sum_{m=1}^{\infty} |a_m| c_3^{m-1}\, m \, c_4^{m-1} \| g \|^m_{M^s_{p,q}} \\ &= \frac{1}{c_3\, c_4} \sum_{m=1}^{\infty} |a_m|\, m \, (c_3\, c_4 \| g \|_{M^s_{p,q}})^m . \end{aligned}$$

Since

$$\sup_{m \in \mathbb{N}} m^{1/m} = 3^{1/3}$$

the claimed estimate follows. ∎

Remark 4.4 Theorem 4.3 is essentially known, see e.g., Sugimoto et al. [29] or Bhimani [1].

4.3 One Example

The following example has been considered at various places. Let $f(z) := e^z - 1$, $z \in \mathbb{C}$. For appropriate constants $a, b > 0$ it follows that

$$\| e^g - 1 \|_{M^s_{p,q}} \le a\, e^{b \| g \|_{M^s_{p,q}}} \tag{4.5}$$

holds for all $g \in M^s_{p,q}$.

It will be essential for our approach to non-analytic composition results that we can improve this estimate.

4.4 Non-analytic Superposition Operators

There is a famous classical result by Katznelson [17] (in the periodic case) and by Helson, Kahane, Katznelson, Rudin [12] (nonperiodic case) which says that only analytic functions operate on the Wiener algebra $\mathcal{A}$. More exactly, the operator $T_f : u \mapsto f(u)$ maps $\mathcal{A}$ into $\mathcal{A}$ if and only if $f(0) = 0$ and f is analytic. Here, $\mathcal{A}$ is the collection of all $u \in C$ such that $\mathcal{F}u \in L_1$. Moreover, a similar result is obtained for particular standard modulation spaces. Bhimani and Ratnakumar [2], see also Bhimani [1], proved that T_f maps $M_{1,1}$ into $M_{1,1}$ if and only if $f(0) = 0$ and f is analytic. Therefore, the existence of non-analytic superposition results for weighted modulation spaces is a priori not so clear.

We shall concentrate on the algebra case. Our first aim consists in deriving a better estimate than (4.5).

To proceed we need some preparations. An essential tool in proving our main result will be a certain subalgebra property. Therefore, we consider the following decomposition of the phase space. Let $R > 0$ and $\epsilon = (\epsilon_1, \ldots, \epsilon_n)$ be fixed with $\epsilon_j \in \{0, 1\}$, $j = 1, \ldots, n$. Then a decomposition of $\mathbb{R}^n$ into $(2^n + 1)$ parts is given by

$$P_R := \{\xi \in \mathbb{R}^n : \ |\xi_j| \le R,\ j = 1, \ldots, n\}$$

and

$$P_R(\epsilon) := \{\xi \in \mathbb{R}^n : \ \mathrm{sign}\,(\xi_j) = (-1)^{\epsilon_j},\ j = 1, \ldots, n\} \setminus P_R.$$

For given $p, q, s, \epsilon = (\epsilon_1, \ldots, \epsilon_n)$ and $R > 0$ we introduce the spaces

$$M^s_{p,q}(\epsilon, R) := \{f \in M^s_{p,q} : \ \mathrm{supp}\, \mathcal{F}f \subset P_R(\epsilon)\}\,.$$

Proposition 4.5 *Let $1 \le p_1, p_2 \le \infty$, $1 < q \le \infty$ and $s_0, s_1, s_2 \in \mathbb{R}$. Define p by $\frac{1}{p} := \frac{1}{p_1} + \frac{1}{p_2}$. Let $R > 2$. If $p \in [1, \infty]$, $s_0 \le \min(s_1, s_2)$, $s_1, s_2 \ge 0$ and $s_1 + s_2 - s_0 > n/q'$, then there exists a constant c such that*

$$\| f \cdot g \|_{M^{s_0}_{p,q}} \le c\, (R-2)^{-[(s_1+s_2-s_0)-n/q']} \, \| f \|_{M^{s_1}_{p_1,q}} \, \| g \|_{M^{s_2}_{p_2,q}}$$

holds for all $f \in M^{s_1}_{p_1,q}(\epsilon, R)$ and all $g \in M^{s_2}_{p_2,q}(\epsilon, R)$. The constant c is independent from $R > 2$ and ϵ.

Proof In order to show the subalgebra property we follow the same steps as in the proof of Lemma 3.20. We start with some almost trivial observations. Let $f \in M^s_{p_1,q}(\epsilon, R)$ and $g \in M^s_{p_2,q}(\epsilon, R)$. By

$$\mathrm{supp}\,(\mathcal{F}f * \mathcal{F}g) \subset \{\xi + \eta : \xi \in \mathrm{supp}\,\mathcal{F}f, \eta \in \mathrm{supp}\,\mathcal{F}g\}$$

we have $\mathrm{supp}\,\mathcal{F}(fg) \subset P_R(\epsilon)$. Let

$$P^*_R(\epsilon) := \Big\{k \in \mathbb{Z}^n : \quad \|k\|_\infty > R-1\,, \quad \mathrm{sign}\,(k_j) = (-1)^{\epsilon_j},\ j = 1, \ldots, n\Big\}.$$

Hence, if $\mathrm{supp}\,\sigma_k \cap P_R(\epsilon) \neq \emptyset$, then $k \in P^*_R(\epsilon)$ follows. Now we continue as in proof of Lemma 3.20, Step 2, and obtain

$$\Big(\sum_{k\in P^*_R(\epsilon)} \langle k\rangle^{s_0 q}\, \|\mathcal{F}^{-1}\big(\sigma_k \mathcal{F}(f\cdot g)\big)\|^q_{L_p}\Big)^{\frac{1}{q}}$$
$$\le \sum_{\substack{t\in\mathbb{Z}^n,\\ -3<t_i<3,\\ i=1,\ldots,n}} \Big(\sum_{k\in P^*_R(\epsilon)} \langle k\rangle^{s_0 q}\Big[\sum_{\substack{l\in\mathbb{Z}^n:\\ t-l+k,l\in P^*_R(\epsilon)}} \|\mathcal{F}^{-1}\big(\sigma_k \mathcal{F}(f_{t-(l-k)}\cdot g_l)\big)\|_{L_p}\Big]^q\Big)^{\frac{1}{q}}.$$

This implies

$$\Big(\sum_{k\in P^*_R(\epsilon)} \langle k\rangle^{s_0 q}\, \|\mathcal{F}^{-1}\big(\sigma_k \mathcal{F}(f\cdot g)\big)\|^q_{L_p}\Big)^{\frac{1}{q}}$$
$$\le c_1 \sum_{\substack{t\in\mathbb{Z}^n,\\ -3<t_i<3,\\ i=1,\ldots,n}} \Big(\sum_{k\in P^*_R(\epsilon)} \langle k\rangle^{s_0 q}\Big[\sum_{\substack{l\in\mathbb{Z}^n:\\ t-l+k,l\in P^*_R(\epsilon)}} \|f_{t-(l-k)}\cdot g_l\|_{L_p}\Big]^q\Big)^{\frac{1}{q}}$$
$$\le c_2 \max_{\substack{t\in\mathbb{Z}^n,\\ -3<t_i<3,\\ i=1,\ldots,n}} \Big(\sum_{k\in P^*_R(\epsilon)} \langle k\rangle^{s_0 q}\Big[\sum_{\substack{l\in\mathbb{Z}^n:\\ t-l+k,l\in P^*_R(\epsilon)}} \|f_{t-(l-k)}\|_{L_{p_1}}\|g_l\|_{L_{p_2}}\Big]^q\Big)^{\frac{1}{q}}$$

with c_2 and c_1 as above. We put

$$S_{1,t,k} := \sum_{\substack{l\in\mathbb{Z}^n:\ t-l+k,l\in P^*_R(\epsilon),\\ |l|\le|l-k|}} \langle k-l\rangle^{s_1}\, \|f_{t-(l-k)}\|_{L_{p_1}}\, \langle l\rangle^{s_2}\, \|g_l\|_{L_{p_2}}\, \langle k-l\rangle^{s_0-s_1}\, \langle l\rangle^{-s_2}\,;$$

$$S_{2,t,k} := \sum_{\substack{l\in\mathbb{Z}^n:\ t-l+k,l\in P^*_R(\epsilon),\\ |l-k|\le|l|}} \langle k-l\rangle^{s_1}\, \|f_{t-(l-k)}\|_{L_{p_1}}\, \langle l\rangle^{s_2}\, \|g_l\|_{L_{p_2}}\, \langle k-l\rangle^{-s_1}\, \langle l\rangle^{s_0-s_2}\,.$$

Hölder's inequality leads to

$$S_{1,t,k} \le \Big(\sum_{\substack{j\in\mathbb{Z}^n,\\ |j+k|\le|j|}} (\langle j\rangle^{s_1} \| f_{t-j}\|_{L_{p_1}} \langle j+k\rangle^{s_2} \|g_{j+k}\|_{L_{p_2}})^q\Big)^{1/q}$$

$$\times \Big(\sum_{\substack{j\in\mathbb{Z}^n: t-j, j+k\in P_R^*(\epsilon)\\ |j+k|\le|j|}} (\langle j\rangle^{s_0-s_1} \langle j+k\rangle^{-s_2})^{q'}\Big)^{\frac{1}{q'}} .$$

Our assumptions $s_0 \le s_1$, $s_2 \ge 0$ and $s_1+s_2-s_0 > n/q'$ and $j+k \in P_R^*(\epsilon)$ imply

$$\Big(\sum_{\substack{j\in\mathbb{Z}^n,\\ |j+k|\le|j|}} \Big|(\langle j\rangle^{s_0-s_1} \langle j+k\rangle^{-s_2})\Big|^{q'}\Big)^{\frac{1}{q'}} \le \Big(\sum_{m\in P_R^*(\epsilon)} \langle m\rangle^{(s_0-s_1-s_2)q'}\Big)^{\frac{1}{q'}}$$

$$\le \Big(2^{-n} \int_{\|x\|_\infty > R-2} (1+|x|^2)^{(s_0-s_1-s_2)q'/2}\, dx\Big)^{1/q'}$$

$$\le \Big(2^{-n} \int_{|x|>R-2} |x|^{(s_0-s_1-s_2)q'}\, dx\Big)^{1/q'}$$

$$\le \Big(\frac{2^{-n}}{(s_1+s_2-s_0)q'-n}\Big)^{1/q'} (R-2)^{-[(s_1+s_2-s_0)-n/q']} .$$

With $c_3 := \Big(\frac{2^{-n}}{(s_1+s_2-s_0)q'-n}\Big)^{1/q'}$ we insert this in our previous estimate and obtain

$$\Big(\sum_{k\in\mathbb{Z}^n} S_{1,t,k}^q\Big)^{1/q} \le c_3\, (R-2)^{-[(s_1+s_2-s_0)-n/q']}$$

$$\times \Big(\sum_{k\in\mathbb{Z}^n} \sum_{\substack{j\in\mathbb{Z}^n,\\ |j+k|\le|j|}} \langle j\rangle^{s_1 q} \| f_{t-j}\|_{L^{p_1}}^q \langle j+k\rangle^{s_2 q} \|g_{j+k}\|_{L^{p_2}}^q\Big)^{1/q}$$

$$\le c_3\, c_4\, (R-2)^{-[(s_1+s_2-s_0)-n/q']} \|g\|_{M^{s_2}_{p_2,q}} \|f\|_{M^{s_1}_{p_1,q}} .$$

Here, c_3, c_4 are independent of f, g, ϵ and R. For the second sum the estimate

$$\Big(\sum_{k\in\mathbb{Z}^n} S_{2,t,k}^q\Big)^{1/q} \le c_5\, (R-2)^{-[(s_1+s_2-s_0)-n/q']} \|g\|_{M^{s_2}_{p_2,q}} \|f\|_{M^{s_1}_{p_1,q}}$$

follows by analogous computations. The proof is complete. ■

Of course, the above arguments have a counterpart in case $q' = \infty$.

Proposition 4.6 *Let $1 \le p_1, p_2 \le \infty$, $q = 1$ and $s_1, s_2 \in \mathbb{R}$. Define p by $\frac{1}{p} := \frac{1}{p_1} + \frac{1}{p_2}$. Let $R > 2$. If $p \in [1, \infty]$, $s_1, s_2 \ge 0$ and $s_0 := \min(s_1, s_2)$, then there exists a constant c such that*

$$\| f \cdot g \|_{M^{s_0}_{p,1}} \le c\,(R-2)^{-(s_1+s_2-s_0)}\, \|f\|_{M^{s_1}_{p_1,1}}\, \|g\|_{M^{s_2}_{p_2,1}}$$

holds for all $f \in M^{s_1}_{p_1,1}(\epsilon, R)$ and all $g \in M^{s_2}_{p_2,1}(\epsilon, R)$. The constant c is independent from $R > 2$ and ϵ.

As a consequence of Nikol'kij's inequality, see Lemma 2.6, Proposotion 4.5 (with $s_0 = s_1 = s_2$ and $p_1 = p$, $p_2 = \infty$) and Corollary 2.7 we obtain the following.

Proposition 4.7 *Let $1 \le p \le \infty$ and $R > 2$.*
(i) *Let $1 < q \le \infty$ and $s > n/q'$. Then there exists a constant c such that*

$$\| f \cdot g \|_{M^{s}_{p,q}} \le c\,(R-2)^{-(s-n/q')}\, \|f\|_{M^{s}_{p,q}}\, \|g\|_{M^{s}_{p,q}}$$

holds for all $f, g \in M^{s}_{p,q}(\epsilon, R)$. The constant c is independent from $R > 2$ and ϵ.
(ii) *Let $q = 1$ and $s \ge 0$. Then there exists a constant c such that*

$$\| f \cdot g \|_{M^{s}_{p,1}} \le c\,(R-2)^{-s}\, \|f\|_{M^{s}_{p,1}}\, \|g\|_{M^{s}_{p,1}}$$

holds for all $f, g \in M^{s}_{p,1}(\epsilon, R)$. The constant c is independent from $R > 2$ and ϵ.

Note that in the following, we assume every function to be real-valued unless it is explicitly stated that complex-valued functions are allowed. To make this more clear we switch from $g \in M^s_{p,q}$ to $u \in M^s_{p,q}$.

Next we have to recall some assertions from harmonic analysis. The first one concerns a standard estimate of Fourier multipliers, see e.g., [33, Theorem1.5.2].

Lemma 4.8 *Let $1 \le r \le \infty$ and assume that $s > n/2$. Then there exists a constant $c > 0$ such that*

$$\|\mathcal{F}^{-1}[\phi\, \mathcal{F} g](\,\cdot\,)\|_{L_r} \le c\, \|\phi\|_{H^s}\, \|g\|_{L_r}$$

holds for all $g \in L_r$ and all $\phi \in H^s$.

The next lemma is taken from [5].

Lemma 4.9 *Let $N \in \mathbb{N}$ and suppose $a_1, a_2, \ldots, a_N$ to be complex numbers. Then, it holds*

$$a_1 \cdot a_2 \cdot \ldots \cdot a_N - 1 = \sum_{l=1}^{N} \sum_{\substack{j=(j_1,\ldots,j_l),\\ 0\le j_1<\ldots<j_l\le N}} (a_{j_1} - 1) \cdot \ldots \cdot (a_{j_l} - 1).$$

In our approach the next estimate will be fundamental.

Proposition 4.10 *Let $1 < p < \infty$, $1 \le q \le \infty$ and $s > n/q'$. Then there exists a positive constant C such that*

$$\|e^{iu} - 1\|_{M^s_{p,q}} \le C\, \|u\|_{M^s_{p,q}} \left(1 + \|u\|_{M^s_{p,q}}\right)^{(s+n/q)(1+\frac{1}{s-n/q'})}$$

holds for all real-valued $u \in M^s_{p,q}$.

Proof This proof follows ideas developed in [5], but see also [23].
Step 1. Let u be a nontrivial function in $M^s_{p,q}$ satisfying $\operatorname{supp} \mathcal{F}u \subset P_R$ for some $R \ge 2$.

First we consider the Taylor expansion

$$e^{iu} - 1 = \sum_{l=1}^{r} \frac{(iu)^l}{l!} + \sum_{l=r+1}^{\infty} \frac{(iu)^l}{l!}$$

resulting in the norm estimate

$$\|e^{iu} - 1\|_{M^s_{p,q}} \le \Big\| \sum_{l=1}^{r} \frac{(iu)^l}{l!} \Big\|_{M^s_{p,q}} + \Big\| \sum_{l=r+1}^{\infty} \frac{(iu)^l}{l!} \Big\|_{M^s_{p,q}} .$$

For brevity we put

$$S_1 := \Big\| \sum_{l=1}^{r} \frac{(iu)^l}{l!} \Big\|_{M^s_{p,q}} \qquad \text{and} \qquad S_2 := \Big\| \sum_{l=r+1}^{\infty} \frac{(iu)^l}{l!} \Big\|_{M^s_{p,q}} .$$

The natural number r will be chosen later on. Next we employ the algebra property, in particular the estimate (4.4) with $C_1 := 2\, c_3\, c_4$. We obtain

$$S_2 \le \sum_{l=r+1}^{\infty} \frac{1}{l!} \|u^l\|_{M^s_{p,q}} \le \frac{1}{C_1} \sum_{l=r+1}^{\infty} \frac{(C_1\, \|u\|_{M^s_{p,q}})^l}{l!} .$$

Now we choose r as a function of $\|u\|_{M^s_{p,q}}$ and distinguish two cases:

1. $C_1\, \|u\|_{M^s_{p,q}} > 1$. Assume that

$$3\, C_1\, \|u\|_{M^s_{p,q}} \le r \le 3\, C_1\, \|u\|_{M^s_{p,q}} + 1 \qquad (4.6)$$

and recall Stirling's formula $l! = \Gamma(l+1) \ge l^l e^{-l} \sqrt{2\pi l}$. Thus, we get

$$\sum_{l=r+1}^{\infty} \frac{(C_1 \|u\|_{M^s_{p,q}})^l}{l!} \le \sum_{l=r+1}^{\infty} \Big(\frac{r}{l}\Big)^l \Big(\frac{e}{3}\Big)^l \frac{1}{\sqrt{2\pi l}} \le \sum_{l=r+1}^{\infty} \Big(\frac{e}{3}\Big)^l \le \frac{3}{3-e}.$$

2. $C_1 \|u\|_{M^s_{p,q}} \le 1$. It follows

$$\sum_{l=r+1}^{\infty} \frac{(C_1 \|u\|_{M^s_{p,q}})^l}{l!} \le C_1 \|u\|_{M^s_{p,q}} \sum_{l=1}^{\infty} \frac{1}{l!} \le C_1\, e\, \|u\|_{M^s_{p,q}}.$$

Both together can be summarized as

$$S_2 \le C_2 \|u\|_{M^s_{p,q}}, \qquad C_2 := \max\Big(e, \frac{3}{C_1(3-e)}\Big).$$

To estimate S_1 we check the support of $\mathcal{F}u^\ell$ and find

$$\begin{aligned} S_1 = \Big\| \sum_{l=1}^{r} \frac{(iu)^l}{l!} \Big\|_{M^s_{p,q}} &= \Big(\sum_{k\in\mathbb{Z}^n} \langle k\rangle^{sq} \Big\| \square_k \Big(\sum_{l=1}^{r} \frac{(iu)^l}{l!} \Big) \Big\|_{L_p}^q \Big)^{\frac{1}{q}} \\ &= \Big(\sum_{\substack{k\in\mathbb{Z}^n,\\ -Rr-1<k_i<Rr+1,\\ i=1,\ldots,n}} \langle k\rangle^{sq} \Big\| \square_k \Big(\sum_{l=1}^{r} \frac{(iu)^l}{l!} \Big) \Big\|_{L_p}^q \Big)^{\frac{1}{q}} \\ &\le \Big(\sum_{\substack{k\in\mathbb{Z}^n,\\ -Rr-1<k_i<Rr+1,\\ i=1,\ldots,n}} \langle k\rangle^{sq} \|\square_k (e^{iu}-1)\|_{L_p}^q \Big)^{\frac{1}{q}} + S_2\,. \end{aligned}$$

Concerning S_2 we proceed as above. To estimate the first part we observe that

$$C_3 := \sup_{k\in\mathbb{Z}^n} \|\, \sigma_k \,\|_{H^t} = \|\, \sigma_0 \,\|_{H^t} < \infty\,,$$

see Lemma 4.8. Furthermore, cos, sin are Lipschitz continuous and consequently we get

$$\begin{aligned} \|\square_k (e^{iu}-1)\|_{L_p} &\le C_3 \|e^{iu}-1\|_{L_p} \\ &\le C_3 \,(\|\cos u - \cos 0\|_{L_p} + \|\sin u - \sin 0\|_{L_p}) \\ &\le 2\, C_3 \,\|u - 0\|_{L_p}\,. \end{aligned}$$

This implies

$$\Big(\sum_{\substack{k\in\mathbb{Z}^n,\\ -Rr-1<k_i<Rr+1,\\ i=1,\ldots,n}} \langle k\rangle^{sq}\, \|\Box_k(e^{iu}-1)\|^q_{L_p}\Big)^{\frac{1}{q}}$$

$$\le 2\,C_3\,\|\,u\,\|_{L_p} \Big(\sum_{\substack{k\in\mathbb{Z}^n,\\ -Rr-1<k_i<Rr+1,\\ i=1,\ldots,n}} \langle k\rangle^{sq}\Big)^{\frac{1}{q}} .$$

Clearly,

$$\begin{aligned}\sum_{\substack{k\in\mathbb{Z}^n,\\ -Rr-1<k_i<Rr+1,\\ i=1,\ldots,n}} \langle k\rangle^{sq} &\le \int_{\|\,x\,\|_\infty<Rr+1} \langle x\rangle^{sq}\,dx\\ &\le \int_{|x|<\sqrt{n}(Rr+1)} \langle x\rangle^{sq}\,dx\\ &\le 2\,\frac{\pi^{n/2}}{\Gamma(n/2)} \int_0^{\sqrt{n}(Rr+1)} (1+\tau)^{n-1+sq}\,d\tau\\ &\le 2\,\frac{\pi^{n/2}}{\Gamma(n/2)}\,\frac{1}{n+sq}\,(\sqrt{n}(Rr+2))^{n+sq} .\end{aligned}$$

To simplify notation we define

$$C_4 := \Big(2\,\frac{\pi^{n/2}}{\Gamma(n/2)}\,\frac{1}{n+sq}\,\sqrt{n}^{n+sq}\Big)^{1/q} .$$

In addition we shall use in case $1<q\le\infty$

$$\|u\|_{L^p} \le C_5\,\|u\|_{M^s_{p,q}}\,, \qquad C_5 := \Big(\sum_{k\in\mathbb{Z}^n} \langle k\rangle^{-sq'}\Big)^{1/q'}$$

which follows from Hölder's inequality and in case $q=1$

$$\|u\|_{L^p} \le \|u\|_{M^s_{p,1}}$$

as a consequence of triangle inequality. Summarizing we have found

$$\|e^{iu}-1\|_{M^s_{p,q}} \le \Big(2\,C_2 + 2\,\max(C_5,1)\,C_4\,C_3\,(Rr+2)^{s+n/q}\Big)\,\|u\|_{M^s_{p,q}} .$$

Next we apply (4.6) which results in

$$\|e^{iu} - 1\|_{M^s_{p,q}} \le C_6 \, \|u\|_{M^s_{p,q}} \Big(1 + R\, \|u\|_{M^s_{p,q}}\Big)^{s+n/q} , \qquad (4.7)$$

valid for all $u \in M^s_{p,q}$ satisfying $\operatorname{supp} \mathcal{F}u \subset P_R$ and with positive constant C_6 not depending on u and $R \ge 2$.

Step 2. This time we consider $u \in M^s_{p,q}$ without any restriction on the Fourier support. Here we need the restriction $1 < p < \infty$. For those p the characteristic functions χ of cubes are Fourier multipliers in L^p by the famous Riesz Theorem and therefore also in $M^s_{p,q}$. In addition we shall make use of the fact that the norm of the operator $f \mapsto \mathcal{F}^{-1}\chi\,\mathcal{F}f$ does not depend on the size of the cube. Below we shall denote this norm by $C_7 = C_7(p)$. We refer to Lizorkin [19] for all details. For decomposing u on the phase space we introduce functions $\chi_{R,\epsilon}$ and χ_R, that is, the characteristic functions of the sets $P_R(\epsilon)$ and P_R, respectively. By defining

$$\begin{aligned} u_\epsilon(x) &= \mathcal{F}^{-1}[\chi_{R,\epsilon}(\xi)\, \mathcal{F}u(\xi)](x), \qquad x \in \mathbb{R}^n , \\ u_0(x) &= \mathcal{F}^{-1}[\chi_R(\xi)\, \mathcal{F}u(\xi)](x) , \qquad x \in \mathbb{R}^n , \end{aligned}$$

we can rewrite u as

$$u(x) = u_0(x) + \sum_{\epsilon \in I} u_\epsilon(x), \qquad (4.8)$$

where I is the set of all $\epsilon = (\epsilon_1, \ldots , \epsilon_n)$ with $\epsilon_j \in \{0, 1\}$, $j = 1, \ldots , n$. Hence

$$\|u\|_{M^s_{p,q}} \le \|u_0\|_{M^s_{p,q}} + \sum_{\epsilon \in I} \|u_\epsilon\|_{M^s_{p,q}}$$

and

$$\max \Big(\|u_0\|_{M^s_{p,q}}, \|u_\epsilon\|_{M^s_{p,q}}\Big) \le C_7 \, \| \, u \, \|_{M^s_{p,q}} .$$

Due to the representation (4.8) and using an appropriate enumeration Lemma 4.9 leads to

$$e^{iu} - 1 = \sum_{l=1}^{2^n+1} \sum_{0 \le j_1 < \ldots < j_l \le 2^n} (e^{iu_{j_1}} - 1) \cdot \ldots \cdot (e^{iu_{j_l}} - 1) .$$

The algebra property, in particular the estimate (4.4) with $C_1 := 2\, c_3\, c_4$, yields

$$\|e^{iu} - 1\|_{M^s_{p,q}} \le \sum_{l=1}^{2^n+1} C_1^{l-1} \sum_{0 \le j_1 < \ldots < j_l \le 2^n} \|e^{iu_{j_1}} - 1\|_{M^s_{p,q}} \cdot \ldots \cdot \|e^{iu_{j_l}} - 1\|_{M^s_{p,q}} . \qquad (4.9)$$

By Proposition 4.7 and (4.7) it follows

$$\|e^{iu_{j_k}} - 1\|_{M^s_{p,q}} = \Big\| \sum_{l=1}^{\infty} \frac{(iu_{j_k})^l}{l!} \Big\|_{M^s_{p,q}} \leq \frac{R^{s-n/q'}}{c} \Big(e^{c\,\|u_{j_k}\|_{M^s_{p,q}}/R^{s-n/q'}} - 1\Big)$$
$$\leq \frac{(R-2)^{s-n/q'}}{c} \Big(e^{c\,C_7\,\|u\|_{M^s_{p,q}}/(R-2)^{s-n/q'}} - 1\Big), \tag{4.10}$$

as well as

$$\|e^{iu_0} - 1\|_{M^s_{p,q}} \leq C_6\, C_7\, \|u\|_{M^s_{p,q}} \Big(1 + R\, C_7 \|u\|_{M^s_{p,q}}\Big)^{s+n/q}, \tag{4.11}$$

where we used the Fourier multiplier assertion mentioned at the beginning of this step. The final step in our proof is to choose the number R as a function of $\|u\|_{M^s_{p,q}}$ such that (4.10) and (4.11) will be approximately of the same size.
Substep 2.1. Let $\|u\|_{M^s_{p,q}} \leq 1$. We choose $R = 3$. Then (4.9) combined with (4.10) and (4.11) results in the estimate

$$\|e^{iu} - 1\|_{M^s_{p,q}} \leq C_8\, \|u\|_{M^s_{p,q}},$$

where C_8 does not depend on u.
Substep 2.2. Let $\|u\|_{M^s_{p,q}} > 1$. We choose $R \geq 3$ such that

$$(R-2)^{s-n/q'} = \|u\|_{M^s_{p,q}}.$$

Now (4.9), combined with (4.10) and (4.11), results in

$$\|e^{iu} - 1\|_{M^s_{p,q}} \leq C_9\, \|u\|_{M^s_{p,q}} \Big(1 + \|u\|_{M^s_{p,q}}\Big)^{(s+n/q)(1+\frac{1}{s-n/q'})}, \tag{4.12}$$

with a constant C_9 independent of u. ∎

Remark 4.11 The restriction of p to the interval $(1,\infty)$ is caused by our decomposition technique, see Step 2 of the preceding proof. We do not know whether Proposition 4.10 extends to $p = 1$ and/or $p = \infty$.

Next, we need again a technical lemma.

Lemma 4.12 *Let $1 < p < \infty$, $1 \leq q \leq \infty$ and $s > n/q'$.*
(i) *The mapping $u \mapsto e^{iu} - 1$ is locally Lipschitz continuous (considered as a mapping of $M^s_{p,q}$ into $M^s_{p,q}$).*
(ii) *Assume $u \in M^s_{p,q}$ to be fixed and define a function $g : \mathbb{R} \mapsto M^s_{p,q}$ by $g(\xi) = e^{iu(x)\xi} - 1$. Then the function g is continuous.*

Proof Local Lipschitz continuity follows from the identity

$$e^{iu} - e^{iv} = (e^{iv} - 1)\,(e^{i(u-v)} - 1) + (e^{i(u-v)} - 1)\,, \tag{4.13}$$

the algebra property of $M^s_{p,q}$ and Proposition 4.10.

To prove the continuity of g we also employ the identity (4.13). The claim follows by using the algebra property and Proposition 4.10. ■

Now we are in position to prove the main result of this section.

Theorem 4.13 *Let $1 < p < \infty$, $1 \le q \le \infty$ and $s > n/q'$. Let μ be a complex measure on $\mathbb{R}$ such that*

$$L := \int_{-\infty}^{\infty} (1 + |\xi|)^{1+(s+n/q)(1+\frac{1}{s-n/q'})}\, d|\mu|(\xi) < \infty \tag{4.14}$$

and such that $\mu(\mathbb{R}) = 0$. Furthermore, assume that the function f is the inverse Fourier transform of μ. Then f is a continuous function and the composition operator $T_f : u \mapsto f \circ u$ maps $M^s_{p,q}$ into $M^s_{p,q}$.

Proof Equation (4.14) yields $\int_{\mathbb{R}^n} d|\mu|(\xi) < \infty$. Thus, μ is a finite measure and $\mu(\mathbb{R}) = 0$ makes sense. Now we define the inverse Fourier transform of μ

$$f(t) = \frac{1}{\sqrt{2\pi}} \int_{\mathbb{R}^n} e^{i\xi t}\, d\mu(\xi).$$

Moreover, since

$$(s + n/q)\Big(1 + \frac{1}{s - n/q'}\Big) > n$$

we conclude that $\int_{\mathbb{R}} |(i\xi)^j|\, d|\mu|(\xi) < \infty$, $j = 1, \ldots, n+1$, which implies $f \in C^{n+1}$. Due to $\mu(\mathbb{R}) = 0$ we can also write f as follows:

$$f(t) = \frac{1}{\sqrt{2\pi}} \int_{\mathbb{R}} (e^{i\xi t} - 1)\, d\mu(\xi).$$

Since μ is a complex measure we can split it up into real part μ_r and imaginary part μ_i, where each of them is a signed measure. Without loss of generality we proceed our computations only with the positive real measure μ_r^+. For all measurable sets E we have $\mu_r^+(E) \le |\mu|(E)$.

Let $u \in M^s_{p,q}$ and define the function $g(\xi) = e^{iu(x)\xi} - 1$ analogously to Lemma 4.12. Then g is Bochner integrable because of its continuity and taking into account that the measure μ_r^+ is finite. Therefore we obtain the Bochner integral

$$\int_{-\infty}^{\infty} \big(e^{iu(x)\xi} - 1\big)\, d\mu_r^+(\xi) = \int_{-\infty}^{\infty} g(\xi)\, d\mu_r^+(\xi)$$

with values in $M^s_{p,q}$. By applying Minkowski inequality it follows

$$\left\| \int_{-\infty}^{\infty} \left(e^{iu(\cdot)\xi} - 1\right) d\mu_r^+(\xi) \right\|_{M^s_{p,q}} \leq \int_{-\infty}^{\infty} \left\| e^{iu(\cdot)\xi} - 1 \right\|_{M^s_{p,q}} d|\mu|(\xi).$$

Using the abbreviation $\|u\| := \|u\|_{M^s_{p,q}}$, Proposition 4.10 together with equation (4.14) yields

$$\begin{aligned} &\int_{|\xi|\|u\|\geq 1} \left\| e^{iu(\cdot)\xi} - 1 \right\|_{M^s_{p,q}} d|\mu|(\xi) \\ &\qquad \leq C' \, \| \, u \, \|_{M^s_{p,q}}^{1+(s+n/q)(1+\frac{1}{s-n/q'})} \int_{|\xi|\|u\|\geq 1} |\xi|^{1+(s+n/q)(1+\frac{1}{s-n/q'})} \, d|\mu|(\xi) \\ &\qquad < \infty. \end{aligned}$$

In a similar way the remaining part $|\xi| \leq 1/\|u\|$ of the integral can be treated. The same estimates also hold for the measures μ_r^-, μ_i^+ and μ_i^-. Thus, the result is obtained by

$$\begin{aligned} &\|\sqrt{2\pi} f(u(x))\|_{M^s_{p,q}} \\ &= \left\| \int_{-\infty}^{\infty} g(\xi)\, d\mu_r^+ - \int_{-\infty}^{\infty} g(\xi)\, d\mu_r^- + i \int_{-\infty}^{\infty} g(\xi)\, d\mu_i^+ - i \int_{-\infty}^{\infty} g(\xi)\, d\mu_i^- \right\|_{M^s_{p,q}} \\ &\leq \int_{-\infty}^{\infty} \|g(\xi)\|_{M^s_{p,q}} \, d|\mu_r^+| + \int_{-\infty}^{\infty} \|g(\xi)\|_{M^s_{p,q}} \, d|\mu_r^-| \\ &\qquad + \int_{-\infty}^{\infty} \|g(\xi)\|_{M^s_{p,q}} \, d|\mu_i^+| + \int_{-\infty}^{\infty} \|g(\xi)\|_{M^s_{p,q}} \, d|\mu_i^-|, \end{aligned}$$

where every integral on the right-hand side is finite. Thus, the statement is proved. ■

A bit more transparent sufficient conditions can be obtained by using Szasz theorem, see Peetre [22, pp. 9–11] and [27, Proposition 1.7.5]. By $B^s_{p,q}(\mathbb{R})$ we denote the Besov spaces on $\mathbb{R}$, see e.g., [33] or [25] for details.

Lemma 4.14 *Let $t \geq 0$ and suppose $f \in B^{t+1/2}_{2,1}(\mathbb{R})$. Then the Fourier transform of f is a regular distribution and*

$$\int_{-\infty}^{\infty} (1 + |\xi|^2)^{t/2} |\mathcal{F} f(\xi)| \, d\xi \leq c \, \| f \|_{B^{t+1/2}_{2,1}(\mathbb{R})}$$

follows with some c independent of f.

Based on Lemma 4.14 and Theorem 4.13 one obtains the next result.

Corollary 4.15 *Let $1 < p < \infty$, $1 \leq q \leq \infty$ and $s > n/q'$. Let $f \in B^t_{2,1}(\mathbb{R})$ for some*

$$t \geq \frac{3}{2} + (s + n/q)\Big(1 + \frac{1}{s - n/q'}\Big)$$

and suppose $f(0) = 0$. *Then the composition operator* $T_f : u \mapsto f \circ u$ *maps real-valued functions in* $M^s_{p,q}$ *boundedly into* $M^s_{p,q}$.

Proof Boundedness of T_f follows from Proposition 4.10, the proof of Theorem 4.13 and Lemma 4.14. ∎

Remark 4.16 Let $t > 0$ be given. A function $f : \mathbb{R} \to \mathbb{R}$, m-times continuously differentiable, compactly supported and satisfying $f^m \in \mathrm{Lip}\,\alpha$ for some $\alpha \in (0, 1]$, belongs to $B^t_{2,1}(\mathbb{R})$ if $t < m + \alpha$.

4.5 One Example

Ruzhansky, Sugimoto, and Wang [26] suggested to study the operator T_α associated to $f_\alpha(t) := t\,|t|^\alpha$, $t \in \mathbb{R}$, with $\alpha > 0$. This function belongs locally to the Besov space $B^{\alpha+1+1/p}_{p,\infty}(\mathbb{R})$, $1 \leq p \leq \infty$, see [25, Lemma 2.3.1/1] for a related case. Let $\psi \in C_0^\infty(\mathbb{R})$ be a smooth cut-off function such that $\psi(x) = 1$ if $|x| \leq 1$. Then the function

$$\tilde{f}_{\alpha,\lambda}(t) := \psi(t/\lambda) \cdot f_\alpha(t), \qquad t \in \mathbb{R},$$

belongs to $B^{\alpha+1+1/p}_{p,\infty}$ for any p, $1 \leq p \leq \infty$, and any $\lambda > 0$. Applying Corollary 4.15 and

$$u(x)\,|u(x)|^\alpha = \tilde{f}_{\alpha,\lambda}(u(x)), \qquad x \in \mathbb{R}^n, \quad \lambda := \|u\|_{L_\infty},$$

we find the following.

Corollary 4.17 *Let* $1 < p < \infty$, $1 \leq q \leq \infty$ *and* $s > n/q'$. *Let* α *be a positive real number such that*

$$(s + n/q)\Big(1 + \frac{1}{s - n/q'}\Big) < \alpha.$$

Then the composition operator $T_\alpha : u \mapsto u\,|u|^\alpha$ *maps real-valued functions in* $M^s_{p,q}$ *boundedly into* $M^s_{p,q}$.

4.6 The Special Case $p = q = 2$

Finally, we will have a look onto the special case $M^s_{2,2} = H^s$, $s > n/2$. In Bourdaud, Moussai, S. [4] the set of functions f such that $T_f : g \mapsto f \circ g$ maps H^s into itself has been characterized.

Proposition 4.18 *Let* $s > \frac{1}{2}\max(n, 3)$. *For a Borel measurable function* $f : \mathbb{R} \to \mathbb{R}$ *the composition operator* T_f *acts on* H^s *if and only if* $f(0) = 0$ *and* $f \in H^{s,\ell oc}(\mathbb{R})$.

Concerning our example T_α treated above this yields the following: T_α maps H^s into itself if and only if $\alpha > s - 3/2$ (instead of $\alpha > s + \frac{n}{2} + \frac{s+n/2}{s-n/2}$ as required in Corollary 4.17).

Corollary 4.15 and Corollary 4.17 may be understood as first results about sufficient conditions, not more.

4.7 A Final Remark

The method employed here has been used before in connection with composition operators on Gevrey-modulation spaces and modulation spaces of ultradifferentiable functions, see Bourdaud [3], Bourdaud et al. [5], Reich et al. [5], and Reich [24], for Hörmander-type spaces $B_{p,k}$ we refer to Jornet and Oliaro [16]. It would be desirable to develop this method more systematically.

References

1. Bhimani, D.G.: Modulation spaces and nonlinear evolution equations. Ph.D. thesis. Harish-Chandra Research Institute, Allahabad (2015)
2. Bhimani, D.G., Ratnakumar, P.K.: Functions operating on modulation spaces and nonlinear dispersive equations. J. Funct. Anal. **270**(2), 621–648 (2016)
3. Bourdaud, G.: Sur les opérateurs pseudo-différentiels à coefficients peu réguliers. thesis. University of Paris-Sud, Paris (1983)
4. Bourdaud, G., Moussai, M., Sickel, W.: Composition operators acting on Besov spaces on the real line. Annali di Matematica Pura ed Applicata **193**, 1519–1554 (2014)
5. Bourdaud, G., Reissig, M., Sickel, W.: Hyperbolic equations, function spaces with exponential weights and Nemytskij operators. Annali di Matematica Pura ed Applicata **182**(4), 409–455 (2003)
6. Cordero, E., Nicola, F.: Sharpness of some properties of Wiener amalgam and modulation spaces. Bull. Aust. Math. Soc. **80**(01), 105–116 (2009)
7. Feichtinger, H.G.: Modulation Spaces on Locally Compact Abelian Groups. Technical report. University of Vienna (1983)
8. Franke, J.: On the spaces $F^s_{p,q}$ of Triebel-Lizorkin type: pointwise multipliers and spaces on domains. Math. Nachr. **125**, 29–68 (1986)
9. Gol'dman, M.L.: The method of coverings for description of general spaces of Besov type. In: Studies in the Theory of Differentiable Functions of Several Variables and Its Applications, VIII. Trudy Math. Inst. Steklov, vol. 156, pp. 47–81 (1980)
10. Gröchenig, K.: Foundations of Time-Frequency Analysis. Birkhäuser, Boston (2001)
11. Guo, W.C., Fan, D.S., Wu, H.X., Zhao, G.P.: Sharpness of some properties of weighted modulation spaces. Sci. China Math. 1–22 (2015)
12. Helson, H., Kahane, J.-P., Katznelson, Y., Rudin, W.: The functions which operate on Fourier transforms. Acta Math. **102**, 135–157 (1959)
13. Iwabuchi, T.: Navier-Stokes equations and nonlinear heat equations in modulation spaces with negative derivative indices. J. Differ. Equ. **248**(8), 1972–2002 (2010)
14. Johnsen, J.: The stationary Navier-Stokes equations in L_p-related spaces. Kobenhavns University, Ph.D. series. Mat. Inst., Kobenhavn (1993)
15. Johnsen, J.: Pointwise multiplication of Besov and Triebel-Lizorkin spaces. Math. Nachr. **175**, 85–133 (1995)
16. Jornet, D., Oliaro, A.: Functional composition in $B_{p,k}$ spaces and applications. Math. Scand. **99**, 175–203 (2006)

17. Katznelson, Y.: Sur les fonctions opérant sur l'algébre des séries de Fourier absolument convergentes. C.R. Acad.Sci. Paris **247**, 404–406 (1958)
18. Lieb, E.H., Loss, M.: Analysis. Graduate Studies in Mathematics, vol. 14. American Mathematical Society, Providence, RI (1997)
19. Lizorkin, P.I.: Multipliers of Fourier integrals and bounds of convolution in spaces with mixed norms. Appl. Math. Izvestiya **4**, 225–255 (1970)
20. Maz'ya, V.G., Shaposhnikova, T.O.: Theory of Multipliers in Spaces of Differentiable Functions. Springer, Berlin (2011)
21. Nikol'skij, S.M.: Approximation of Functions of Several Variables and Imbedding Theorems. Springer, Berlin (1975)
22. Peetre, J.: New Thoughts on Besov Spaces. Duke University Press, Duke (1976)
23. Reich, M., Reissig, M., Sickel, W.: Non-analytic superposition results on modulation spaces with subexponential weights (2015). arXiv:1510.07521
24. Reich, M.: Superposition in modulation spaces with ultradifferentiable weights (2016). arXiv:1603.08723
25. Runst, T., Sickel, W.: Sobolev Spaces of Fractional Order, Nemytskij Operators, and Nonlinear Partial Differential Equations. de Gruyter, Berlin (1996)
26. Ruzhansky, M., Sugimoto, M., Wang, B.: Modulation spaces and nonlinear evolution equations. In: Evolution Equations of Hyperbolic and Schrödinger Type, pp. 267–283. Springer, Basel (2012)
27. Schmeisser, H.-J., Triebel, H.: Topics in Fourier Analysis and Function Spaces. Geest & Portig K.-g, Leipzig (1987)
28. Strichartz, R.S.: Multipliers on fractional Sobolev spaces. J. Math. Mech. **16**, 1031–1060 (1967)
29. Sugimoto, M., Tomita, N., Wang, B.: Remarks on nonlinear operations on modulation spaces. Integr. Transform. Spec. Funct. **22**, 351–358 (2011)
30. Toft, J.: Continuity properties for modulation spaces, with applications to pseudo-differential calculus-I. J. Funct. Anal. **207**(2), 399–429 (2004)
31. Toft, J.: Convolution and embeddings for weighted modulation spaces. In: Ashino, R., Boggiatto, P., Wong, M.-W. (eds.) Advances in Pseudo-Differential Operators, Birkhäuser, Basel, 165–186 (2004)
32. Toft, J., Johansson, K., Pilipović, S., Teofanov, N.: Sharp convolution and multiplication estimates in weighted spaces. Anal. Appl. **13**(5), 457–480 (2015)
33. Triebel, H.: Theory of Function Spaces. Geest & Portig K.-G., Birkhäuser, Leipzig, Basel (1983)
34. Triebel, H.: Modulation spaces on the Euclidean n-space. ZAA **2**, 443–457 (1983)
35. Wang, B., Han, L., Huang, C.: Global well-posedness and scattering for the derivative nonlinear Schrödinger equation with small rough data. In: Ann. I. H. Poincaré—AN, vol. 26, pp. 2253–2281 (2009)
36. Wang, B., Huang, C.: Frequency-uniform decomposition method for the generalized BO, KdV and NLS equations. J. Differ. Equ. **239**, 213–250 (2007)
37. Wang, B., Hudzik, H.: The global Cauchy problem for the NLS and NLKG with small rough data. J. Differ. Equ. **232**, 36–73 (2007)
38. Wang, B., Lifeng, Z., Boling, G.: Isometric decomposition operators, function spaces $E^{\lambda}_{p,q}$ and applications to nonlinear evolution equations. J. Funct. Anal. **233**(1), 1–39 (2006)

A Reproducing Kernel Theory with Some General Applications

Saburou Saitoh

Abstract In this paper, some essences of the general theory of reproducing kernels from the viewpoint of general applications and general interest will be introduced by our recent results, that are presented in the plenary talk.

Keywords Reproducing kernel · Generalized delta function · Aveiro discretization · Approximation · Tikhonov regularization · General fraction · Division by zero · Integral transform · Initial value problem · Convolution

The theory of reproducing kernels is very fundamental, beautiful, and will have many applications in analysis and numerical analysis.

In this paper, some general essences from the viewpoint of general applications and general interest will be introduced by our recent results, that are presented in the plenary talk. In particular, on this line, reproducing kernels in complex analysis that are typically stated as the Bergman kernels in one and several complex variables will not be referred to here

1. What Is a Reproducing Kernel?
2. Generalized Reproducing Kernels, Generalized Delta Functions, and Generalized Reproducing Kernel Hilbert Spaces
3. Kernel Forms—Connections with Other Fields
4. Inversion Formulas
5. General Integral Transforms
6. The Aveiro Discretization Method
7. Best Approximations—As a Connection
8. The Tikhonov Regularization
9. General Fractional Functions—Division by Zero
10. Convolutions, Integral Transforms, and Integral Equations
11. Eigenfunctions, Initial Value Problems, Integral Transforms, and Reproducing Kernels

---

S. Saitoh (✉)
Institute of Reproducing Kernels, Kiryu, Japan
e-mail: saburou.saitoh@gmail.com

T. Qian and L.G. Rodino (eds.), *Mathematical Analysis, Probability and Applications – Plenary Lectures*, Springer Proceedings in Mathematics & Statistics 177, DOI 10.1007/978-3-319-41945-9_6

1 What Is a Reproducing Kernel?

First of all of the talk, *What is Mathematics?* and *Mystery of Mathematics* were introduced, simply.

Now, let $\mathcal{H}$ be a Hilbert (possibly finite-dimensional) space, and consider E to be an abstract set and $\mathbf{h}$ a Hilbert $\mathcal{H}$-valued function on E. Then, a very general linear transform from $\mathcal{H}$ into the linear space $\mathcal{F}(E)$ consisting of all the complex-valued functions on E will be given by

$$f(p) = (\mathbf{f}, \mathbf{h}(p))_{\mathcal{H}}, \quad \mathbf{f} \in \mathcal{H}, \tag{1.1}$$

in the framework of Hilbert spaces.

In general, a complex-valued function $k : E \times E \to \mathbb{C}$ is called a *positive definite quadratic form function* on the set E, or shortly, *positive definite function*, when it satisfies the property that, for an arbitrary function $X : E \to \mathbb{C}$ and for any finite subset F of E,

$$\sum_{p,q \in F} \overline{X(p)} X(q) k(p,q) \geq 0.$$

In order to investigate the linear mapping (1.1), we form a positive definite quadratic form function $K(p,q)$ on $E \times E$ defined by

$$K(p,q) = (\mathbf{h}(q), \mathbf{h}(p))_{\mathcal{H}} \quad \text{on} \quad E \times E. \tag{1.2}$$

Then, the following fundamental results are valid:

Proposition 1.1 *(I) The range of the linear mapping (1.1) by $\mathcal{H}$ is characterized as the reproducing kernel Hilbert space $H_K(E)$ admitting the reproducing kernel $K(p,q)$ whose characterization is given by the two properties: (i) $K(\cdot, q) \in H_K(E)$ for any $q \in E$ and, (ii) for any $f \in H_K(E)$ and for any $p \in E$, $(f(\cdot), K(\cdot.p))_{H_K(E)} = f(p)$.*

(II) In general, the inequality

$$\|f\|_{H_K(E)} \leq \|\mathbf{f}\|_{\mathcal{H}}$$

holds. Here, for any member f of $H_K(E)$ there exists a uniquely determined $\mathbf{f}^* \in \mathcal{H}$ satisfying

$$f(p) = (\mathbf{f}^*, \mathbf{h}(p))_{\mathcal{H}} \quad \text{on } E$$

and

$$\|f\|_{H_K(E)} = \|\mathbf{f}^*\|_{\mathcal{H}}. \tag{1.3}$$

(III) In general, the inversion formula in (1.1) in the form

$$f \mapsto \mathbf{f}^* \tag{1.4}$$

in (II) holds, by using the reproducing kernel Hilbert space $H_K(E)$.

The typical ill-posed problem (1.1) becomes a well-posed problem, because the image space of (1.1) is characterized as the reproducing kernel Hilbert space $H_K(E)$ with the isometric identity (1.3), which may be considered as a generalization of the *Pythagorean* theorem.

However, this viewpoint is a mathematical one and is not a numerical one and not easy to deal with analytical and numerical problems.

Recently, by a great contribution by *Y. Sawano*, we were able to obtain a general concept of the generalized delta function as a generalized reproducing kernel and, as a general reproducing kernel Hilbert space, we can consider all separable Hilbert spaces consisting of functions. We will refer to the new concepts.

2 Generalized Reproducing Kernels, Generalized Delta Functions, and Generalized Reproducing Kernel Hilbert Spaces

We will consider a family of *any complex-valued functions* $\{U_n(p)\}_{n=0}^{\infty}$ defined on an abstract set E that are linearly independent. Then, we consider the form:

$$K_N(p,q) = \sum_{n=0}^{N} U_n(p)\overline{U_n(q)}. \tag{2.1}$$

Then, $K_N(p,q)$ is a *reproducing kernel* in the following sense:

We will consider the family of all the functions, for arbitrary complex numbers $\{C_n\}_{n=0}^{N}$

$$F(p) = \sum_{n=0}^{N} C_n U_n(p) \tag{2.2}$$

and we introduce the norm

$$\|F\|^2 = \sum_{n=0}^{N} |C_n|^2. \tag{2.3}$$

The function space forms a Hilbert space $H_{K_N}(E)$ determined by the kernel $K_N(p,q)$ with the inner product induced from the norm (2.3), as usual. Then, we note that, for any $y \in E$

$$K_N(\cdot, q) \in H_{K_N}(E) \tag{2.4}$$

and for any $F \in H_{K_N}(E)$ and for any $q \in E$

$$F(q) = (F(\cdot), K_N(\cdot, q))_{H_{K_N}(E)} = \sum_{n=0}^{N} C_n U_n(q). \tag{2.5}$$

The properties (2.4) and (2.5) are called a *reproducing property* of the kernel $K_N(p, q)$ for the Hilbert space $H_{K_N}(E)$.

We introduce a pre-Hilbert space by

$$H_{K_\infty} := \bigcup_{N \geqq 0} H_{K_N}(E).$$

For any $F \in H_{K_\infty}$, there exists a space $H_{K_M}(E)$ containing the function F for some $M \geqq 0$. Then, for any N such that $M < N$,

$$H_{K_M}(E) \subset H_{K_N}(E)$$

and, for the function $F \in H_{K_M}$,

$$\|F\|_{H_{K_M}(E)} = \|F\|_{H_{K_N}(E)}.$$

Therefore, there exists the limit:

$$\|F\|_{H_{K_\infty}} := \lim_{N \to \infty} \|F\|_{H_{K_N}(E)}.$$

Denote by H_∞ the completion of H_{K_∞} with respect to this norm. Then, we obtain:

Theorem 2.1 *Under the above conditions, for any function $F \in H_\infty$ and for F_N defined by*

$$F_N(p) = \langle F, K_N(\cdot, p) \rangle_{H_\infty},$$

$F_N \in H_{K_N}(E)$ for all $N > 0$, and as $N \to \infty$, $F_N \to F$ in the topology of H_∞.

Theorem 2.1 may be looked as a reproducing kernel in the natural topology and by the sense of Theorem 2.1, and the reproducing property may be written as follows:

$$F(p) = \langle F, K_\infty(\cdot, p) \rangle_{H_\infty},$$

with

$$K_\infty(\cdot, p) \equiv \lim_{N \to \infty} K_N(\cdot, p) = \sum_{n=0}^{\infty} U_n(\cdot)\overline{U_n(p)}. \tag{2.6}$$

Here *the limit does, in general, not need to exist*, however, the series are nondecreasing, in the sense: for any $N > M$, $K_N(q,p) - K_M(q,p)$ is a positive definite quadratic form function.

The function (2.6) may be looked as a *generalized Delta function*.

Any reproducing kernel (separable case) may be considered as the form (2.6) by arbitrary linear independent functions $\{U_n(p)\}$ on an abstract set E, here, the sum does not need to converge. Furthermore, the property of linear independence is not essential.

The completion H_∞ may be found, in concrete cases, from the realization of the spaces $H_{K_N}(E)$.

The typical case is that the family $\{U_n(p)\}_{n=0}^{\infty}$ is a complete orthonormal system in a Hilbert space with the norm

$$\|F\|^2 = \int_E |F(p)|^2 dm(p) \tag{2.7}$$

with a dm measurable set E in the usual form $L_2(E, dm)$. Then, the functions (2.2) and the norm (2.3) are realized by this norm and the completion of the space $H_{K_\infty}(E)$ is given by this Hilbert space with the norm (2.7).

For any separable Hilbert space consisting of functions, there exists a complete orthonormal system, so, by our generalized sense, for the Hilbert space there exists an approximating reproducing kernel Hilbert space and therefore, the Hilbert space is the generalized reproducing kernel Hilbert space in the sense of this paper.

This will mean that *we were able to extend the classical reproducing kernels* [1, 2, 29], beautifully and completely.

The fundamental applications to initial value problems using eigenfunctions and reproducing kernels, see [33, 34].

3 Kernel Forms—Connections with Other Fields

For the linear mapping (1.1), to consider the kernel form (1.2) is essentially important, meanwhile any reproducing kernel is given in the form (1.2) and the form will be appeared in natural ways in different theories.

Kolmogorov factorization theorem [22] gives, conversely for any positive definite quadratic form function $K(p,q)$, a factorization representation (1.2) by constructing a Hilbert space $\mathcal{H}$ and a Hilbert $\mathcal{H}$-valued function $\mathbf{h}(p)$ on E. This important result was interestingly derived from the theory of stochastic theory independing of the theory of reproducing kernels. This property is essentially important when we consider a general convolution operator and various operators among abstract Hilbert spaces. See [30] for the details.

Let $(\Omega, \mathcal{B}, P)$ be a probability space and $L_2(\Omega, \mathcal{B}, P)$ the Hilbert space consisting of the second order random variables on Ω with the inner product $E(X\overline{Y})$. Let $X(t)$, t on a set T, be a second order stochastic process defined on the probability space

$(\Omega, \mathcal{B}, P)$. For the mean value function as $m(t) = E(X(t))$, the second moment function

$$R(t,s) = E(X(t)\overline{Y(s)}) \tag{3.1}$$

and the covariance function

$$K(t,s) = E((X(t) - m(t))\overline{(Y(s) - m(s))}) \tag{3.2}$$

are positive definite quadratic form functions on Ω so, both the theories of stochastic processes and reproducing kernels have a fundamental relationship. A typical result is the *Loéve's theorem*: The Hilbert space $H(X)$ generated by the process $X(t)$, t on a set T with the covariance function $R(t,s)$ is *congruent* to the reproducing kernel Hilbert space admitting the kernel $R(t,s)$.

The support vector machine is a powerful computational method for solving learning and function estimating problems such that pattern recognition, density, and regression estimation and operator inversion. See [36] for the details.

From some data input space E we consider a general non linear mapping to a feature space F that is a pre-Hilbert space with the inner product $(\cdot,\cdot)_F$:

$$\Phi : E \longrightarrow F; \qquad x \longrightarrow \Phi(x). \tag{3.3}$$

Then, we form the positive definite quadratic form function

$$K(x,y) = (\Phi(x), \Phi(y))_F. \tag{3.4}$$

The important point of this method is that we can apply this kernel to the problem of construction of the optimal hyperplanes in the space F not by using the explicit values of the transformed data $\Phi(x)$. See [4, 5] for the basic books and their references.

Quite recently a new method is developing known as *kernel method*:

For the transform of the data in the probability space $(\Omega, \mathcal{B}, P)$ for a reproducing kernel Hilbert space H_K admitting a kernel on Ω:

$$\Psi : \Omega \longrightarrow (\cdot,\cdot)_F; \qquad x \longrightarrow K(\cdot, x), \tag{3.5}$$

the theory of reproducing kernels may be applied to the probability problems on the space $(\Omega, \mathcal{B}, P)$. See the basic Refs. [17, 18] and their references.

The Dirac delta function and the Green functions are a family of reproducing kernels, and orthonormal systems and reproducing kernels are the basic tools of *quantum mechanics; for examining of Coherent States*. See the general survey article [37].

4 Inversion Formulas

Consider the inversion in (1.1) formally, however, this idea will be very important for the general inversions and for discretization method.

Following the above general situation, let $\{\mathbf{v}_j\}$ be a complete orthonormal basis for $\mathcal{H}$. Then, for

$$v_j(p) = (\mathbf{v}_j, \mathbf{h}(p))_{\mathcal{H}},$$

$$\mathbf{h}(p) = \sum_j (\mathbf{h}(p), \mathbf{v}_j)_{\mathcal{H}} \mathbf{v}_j = \sum_j \overline{v_j(p)} \mathbf{v}_j.$$

Hence, by setting

$$\overline{\mathbf{h}}(p) = \sum_j v_j(p) \mathbf{v}_j,$$

$$\overline{\mathbf{h}}(\cdot) = \sum_j v_j(\cdot) \mathbf{v}_j.$$

Thus, define

$$(f, \overline{\mathbf{h}}(p))_{H_K} = \sum_j (f, v_j)_{H_K} \mathbf{v}_j.$$

For simplicity, write as follows:

$$H_K = H_K(E).$$

Then, formally, we obtain:

Proposition 4.1 *Assume that for* $f \in H_K$

$$(f, \overline{\mathbf{h}})_{H_K} \in \mathcal{H}$$

and for all $p \in E$,

$$(f, (\mathbf{h}(p), \mathbf{h}(\cdot))_{\mathcal{H}})_{H_K} = ((f, \overline{\mathbf{h}})_{H_K}, \mathbf{h}(p))_{\mathcal{H}}.$$

Then,

$$\|f\|_{H_K} \leqq \|(f, \overline{\mathbf{h}})_{H_K}\|_{\mathcal{H}}.$$

If $\{\mathbf{h}(p);\ p \in E\}$ *is complete in* $\mathcal{H}$, *then equality always holds.*
Furthermore, if

$$(\mathbf{f}_0, (f, \overline{\mathbf{h}})_{H_K})_{\mathcal{H}} = ((\mathbf{f}_0, \mathbf{h})_{\mathcal{H}}, f)_{H_K} \quad for \quad \mathbf{f}_0 \in N(L).$$

Then, for $\mathbf{f}^*$ *in (II) and (III)*

$$\mathbf{f}^* = (f, \overline{\mathbf{h}})_{H_K}.$$

In particular, note that the basic assumption $(f, \overline{\mathbf{h}})_{H_K} \in \mathcal{H}$ in Proposition 4.1, is, in general, not valid and very delicate for many analytical problems and we need some delicate treatment for the inversion.

In order to derive a general inversion formula for (1.1) that is widely applicable in analysis, assume that both the Hilbert spaces $\mathcal{H}$ and H_K are given as $\mathcal{H} = L_2(T, dm)$, $H_K \subset L_2(E, d\mu)$, on the sets T and E, respectively (assume that for $dm, d\mu$ measurable sets T, E, they are the Hilbert spaces consisting of $dm, d\mu - L_2$ integrable complex-valued functions, respectively). Therefore, consider the integral transform

$$f(p) = \int_T F(t)\overline{h(t, p)}dm(t). \tag{4.1}$$

Here, $h(t, p)$ is a function on $T \times E$, $h(\cdot, p) \in L_2(T, dm)$, and $F \in L_2(T, dm)$. The corresponding reproducing kernel for (1.2) is given by

$$K(p, q) = \int_T h(t, q)\overline{h(t, p)}dm(t) \quad \text{on} \quad E \times E.$$

The norm of the reproducing kernel Hilbert space H_K is represented as $L_2(E, d\mu)$. Under these situations:

Proposition 4.2 *Assume that an approximating sequence* $\{E_N\}_{N=1}^{\infty}$ *of* E *satisfies (a)* $E_1 \subset E_2 \subset \cdots \subset \cdots$, *(b)* $\bigcup_{N=1}^{\infty} E_N = E$, *(c)* $\int_{E_N} K(p, p)d\mu(p) < \infty$, $(N = 1, 2, \ldots)$.

Then, for $f \in H_K$ *satisfying* $\int_{E_N} f(p)h(t, p)d\mu(p) \in L_2(T, dm)$ *for any* N, *the sequence*

$$\left\{\int_{E_N} f(p)h(t, p)d\mu(p)\right\}_{N=1}^{\infty} \tag{4.2}$$

converges to F^* *in (1.4) in Proposition 1.1 in the sense of* $L_2(T, dm)$ *norm.*

Practically for many cases, the assumptions in Proposition 4.2 under the condition (4.1) will be satisfied automatically, so Proposition 4.2 may be applied in many cases. Proposition 4.2 will give a new viewpoint and method for the Fredholm integral equation (4.1) of the first kind that is a fundamental integral equation. The method and solutions have the following properties:

(1) The use of the naturally determined reproducing kernel Hilbert space H_K that is determined by the integral kernel.
(2) The solution is given in the sense of $\mathcal{H}$ norm convergence.

(3) The solution (inverse) is given by f^* with minimum norm in Proposition 1.1.
(4) For the ill-posed problem in (4.1), the solution is given as a well-posed solution.

This viewpoint is, however, a mathematical and theoretical one. In practical and physical linear systems, the observation data will be *a finite number of data containing error or noises*, so we will meet to various delicate problems numerically.

5 General Integral Transforms

The basic assumption here for the integral kernels is to belong to some Hilbert spaces. However, as a very typical integral transform, in the case of Fourier integral transform, the integral kernel does not belong to $L_2(\mathbf{R})$ and, however, we can establish the isometric identity and inversion formula in the space $L_2(\mathbf{R})$.

We can develop some general integral transform theory containing the Fourier integral transform case that the integral kernel does not belong to any Hilbert space, based on the recent general concept of generalized reproducing kernels in [33, 34].

When we consider the integral transform

$$LF(p) = \int_T F(\lambda)\overline{h(\lambda, p)}\,dm(\lambda), \quad p \in E \tag{5.1}$$

for $F \in \mathcal{H} = L^2(T, dm)$, indeed, the integral kernel $h(\lambda, p)$ does not need to belong to the space $L^2(T, dm)$ and with the very general assumptions that for any exhausion $\{T_t\}$ of T such that $T_t \subset T_{t'}$ for $t \leq t'$, $\bigcup_{t>0} T_t = T$,

$$h(\lambda, p) \text{ belongs to } L^2(T_t, dm) \text{ for any } p \text{ of } E$$

and

$$\{h(\lambda, p);\, p \in E\} \text{ is complete in } L^2(T_t, dm),$$

we can establish the isometric identity and inversion formula of the integral transform (5.1) by giving the natural interpretation of the integral transform (5.1), as in the Fourier transform by considering the generalized reproducing kernel $K(p, q)$

$$K_t(p, q) = \int_{T_t} h(t, q)\overline{h(t, p)}dm(t) \quad \text{on} \quad E \times E,$$

and

$$K(p, q) = \lim_{t\to\infty} K_t(p, q),$$

which diverses as in the delta function.

6 The Aveiro Discretization Method

Meanwhile, in general, the reproducing kernel Hilbert space H_K has a complicated structure, so we have to consider the approximate realization of the abstract Hilbert space H_K by taking a finite number of points of E. A finite number of data will lead to a discretization principle and practical method, because computers can deal with a finite number of data.

By taking a finite number of points $\{p_j\}_{j=1}^n$, we set

$$K(p_j, p_{j'}) := a_{jj'}. \tag{6.1}$$

Then, if the matrix $A := \| a_{jj'} \|$ is positive definite, then, the corresponding norm in H_A consisting of the vectors $\mathbf{x} = (x_1, x_2, \ldots, x_n)^T$ is determined by

$$\|\mathbf{x}\|_{H_A}^2 = \mathbf{x}^* \widetilde{A}\mathbf{x},$$

where $\widetilde{A} = \overline{A^{-1}} = \|\widetilde{a_{jj'}}\|$.

When we approximate the reproducing kernel Hilbert space H_K by the vector space H_A, then from Proposition 4.1, the following is directly derived:

Proposition 6.1 *In the linear mapping*

$$f(p) = (\mathbf{f}, \mathbf{h}(p))_{\mathcal{H}}, \quad \mathbf{f} \in \mathcal{H} \tag{6.2}$$

for

$$\{p_1, p_2, \ldots, p_n\},$$

the minimum norm inverse $\mathbf{f}_{A_n}^*$ *satisfying*

$$f(p_j) = (\mathbf{f}, \mathbf{h}(p_j))_{\mathcal{H}}, \quad \mathbf{f} \in \mathcal{H} \tag{6.3}$$

is given by

$$\mathbf{f}_{A_n}^* = \sum_{j=1}^{n} \sum_{j'=1}^{n} f(p_j)\widetilde{a_{jj'}}\mathbf{h}(p_{j'}), \tag{6.4}$$

where $\widetilde{a_{jj'}}$ *are assumed to be the elements of the complex conjugate inverse of the positive definite Hermitian matrix* A_n *constituted by the elements*

$$a_{jj'} = (\mathbf{h}(p_{j'}), \mathbf{h}(p_j))_{\mathcal{H}} = K(p_j, p_{j'}).$$

Here, the positive definiteness of A_n *is a basic assumption.*

The following proposition shows the convergence of the approximate inverses in Proposition 6.1.

Proposition 6.2 *Let $\{p_j\}_{j=1}^{\infty}$ be a sequence of distinct points on E, that is the positive definiteness in Proposition 6.1 for any n and a uniqueness set for the reproducing kernel Hilbert space H_K; that is, for any $f \in H_K$, if all $f(p_j) = 0$, then $f \equiv 0$. Then, in the space $\mathcal{H}$*

$$\lim_{n\to\infty} \mathbf{f}^*_{A_n} = \mathbf{f}^*. \tag{6.5}$$

From the result, we can obtain directly the ultimate realization of the reproducing kernel Hilbert spaces and the ultimate sampling theory:

Proposition 6.3 (Ultimate realization of reproducing kernel Hilbert spaces). *In the general situation and for a uniqueness set $\{p_j\}$ of the set E satisfying the linearly independence in Proposition 6.1,*

$$\|f\|^2_{H_K} = \|\mathbf{f}^*\|^2_{\mathcal{H}} = \lim_{n\to\infty} \sum_{j=1}^{n} \sum_{j'=1}^{n} f(p_j)\widetilde{a_{jj'}}\overline{f(p_{j'})}. \tag{6.6}$$

Proposition 6.4 (Ultimate sampling theory). *In the general situation and for a uniqueness set $\{p_j\}$ of the set E satisfying the linearly independence in Proposition 6.1,*

$$f(p) = \lim_{n\to\infty} (\mathbf{f}^*_{A_n}, \mathbf{h}(p))_{\mathcal{H}} = \lim_{n\to\infty} \left(\sum_{j=1}^{n} \sum_{j'=1}^{n} f(p_j)\widetilde{a_{jj'}}\mathbf{h}(p_{j'}), \mathbf{h}(p) \right)_{\mathcal{H}} \tag{6.7}$$

$$= \lim_{n\to\infty} \sum_{j=1}^{n} \sum_{j'=1}^{n} f(p_j)\widetilde{a_{jj'}} K(p, p_{j'}).$$

In Proposition 6.1, for any given finite number $f(p_j)$, $j = 1, 2, \ldots, n$, the result in Proposition 6.1 is valid. Meanwhile, Propositions 6.2 and 6.4 are valid when we consider the sequence $f(p_j)$, $j = 1, 2, \ldots$, for any member f of H_K. The sequence $f(p_j)$, $j = 1, 2, \ldots$, for any member f of H_K is characterized by the convergence of (6.6) in Proposition 6.3. Then, any member f of H_K is represented by (6.7) in terms of the sequence $f(p_j)$, $j = 1, 2, \ldots$, in Proposition 6.4.

In the general setting in Proposition 6.1, assume that we observed the values $f(p_j) = \alpha_j$, $j = 1, 2, \ldots, n$, for a finite number of points $\{p_j\}$, then for the whole value $f(p)$ of the set E, how can we consider it?

One idea is to consider the function $f_1(p)$: among the functions satisfying the conditions $f(p_j) = \alpha_j$, $j = 1, 2, \ldots, n$, we consider the minimum norm member $f_1(p)$ in $H_K(E)$. This function $f_1(p)$ is determined by the formula

$$f_1(p) = \sum_{j=1}^{n} C_j K(p, p_j)$$

where, the constants $\{C_j\}$ are determined by the formula

$$\sum_{j=1}^{n} C_j K(p_{j'}, p_j) = \alpha_{j'}, j' = 1, 2, \ldots, n.$$

(of course, we assume that $\|K(p_{j'}, p_j)\|$ is positive definite).

For this problem, see, Mo and Qian [27], as a new numerical approach by a usual computer system level, we use a special powerful computer system by H. Fujiwara. In particular, they can deal with errorness data.

Meanwhile, by Proposition 1.1 we can consider the function $f_2(p)$ defined by

$$f_2(p) = (\mathbf{f}^*_{A_n}, \mathbf{h}(p))_{\mathcal{H}}$$

in terms of $\mathbf{f}^*_{A_n}$ in Proposition 6.1. This interpolation formula is depending on the linear system.

For analytical problems, we need discretization and using a finite number of data in order to obtain approximate solutions by using computers, the typical methods are finite element method and difference method, however, their practical algorithms will be complicated depending on case by case, depending on the domains and depending on the dimensions, however, the above methods are essentially simple and uniform method in principle, called the *Aveiro discretization method*. However, the hard work part is to solve the linear simultaneous equations, computer powers requested are increasing so, in future, the above simple method may be expected to become a standard method. For the general information and numerical results, see [9, 10].

Many numerical experiments for the sampling theory by Proposition 6.4 were given by [16]. In particular:

We showed a general sampling theorem and the concrete numerical experiments for the simplest and typical examples. We gave the sampling theorem in the Sobolev Hilbert spaces with numerical experiments. For the Sobolev Hilbert spaces, sampling theorems seem to be a new concept.

For the typical Paley-Wiener spaces, the sampling points are automatically determined as the common sense, however, in our general sampling theorem, we can select the sampling points freely so, case by case, following some a priori information of a considering function, we can take the effective sampling points. We showed these properties by the concrete examples, with many Figures by computers.

6.1 A Typical Example of the Aveiro Discretization Method with ODE

Consider a prototype differential operator

$$Ly := \alpha y'' + \beta y' + \gamma y. \tag{6.8}$$

Here, consider a very general situation that the coefficients are *arbitrary* functions essentially and on a general interval I.

For some practical construction of some natural solution of

$$Ly = g \tag{6.9}$$

for a very general function g on a general interval I, we obtain

Proposition 6.5 ([9, 10]) *Let us fix a positive number h and take a finite number of points $\{t_j\}_{j=1}^n$ of I such that*

$$(\alpha(t_j), \beta(t_j), \gamma(t_j)) \neq \mathbf{0}$$

for each j. Then, an optimal solution y_h^A of the Eq. (6.9) is given by

$$y_h^A(t) = \frac{1}{2\pi} \int_{-\pi/h}^{\pi/h} F_h^A(\xi) e^{-it\xi} d\xi$$

in terms of the function $F_h^A \in L_2(-\pi/h, +\pi/h)$ in the sense that F_h^A has the minimum norm in $L_2(-\pi/h, +\pi/h)$ among the functions $F \in L_2(-\pi/h, +\pi/h)$ satisfying, for the characteristic function $\chi_h(t)$ of the interval $(-\pi/h, +\pi/h)$:

$$\frac{1}{2\pi} \int_{\mathbb{R}} F(\xi)[\alpha(t)(-\xi^2) + \beta(t)(-i\xi) + \gamma(t)]\chi_h(\xi) \exp(-it\xi) d\xi = g(t) \tag{6.10}$$

for all $t = t_j$ and for the function space $L_2(-\pi/h, +\pi/h)$.

The best extremal function F_h^A is given by

$$F_h^A(\xi) = \sum_{j,j'=1}^{n} g(t_j)\widetilde{a_{jj'}}\overline{(\alpha(t_{j'})(-\xi^2) + \beta(t_{j'})(-i\xi) + \gamma(t_{j'}))} \exp(it_{j'}\xi). \tag{6.11}$$

Here, the matrix $A = \{a_{jj'}\}_{j,j'=1}^n$ formed by the elements

$$a_{jj'} = K_{hh}(t_j, t_{j'})$$

with

$$K_{hh}(t,t') = \frac{1}{2\pi}\int_{\mathbb{R}} [\alpha(t)(-\xi^2) + \beta(t)(-i\xi) + \gamma(t)]\overline{[\alpha(t')(-\xi^2) + \beta(t')(-i\xi) + \gamma(t')]} \times \chi_h(\xi)\exp(-i(t-t')\xi)d\xi \tag{6.12}$$

is positive definite and the $\widetilde{a_{jj'}}$ *are the elements of the inverse of* $\overline{A}$ *(the complex conjugate of* A*).*

Therefore, the optimal solution y_h^A *of the Eq. (6.9) is given by*

$$y_h^A(t) = \sum_{j,j'=1}^{n} g(t_j)\widetilde{a_{jj'}}\frac{1}{2\pi}[-\overline{\alpha(t_{j'})}\int_{-\frac{\pi}{h}}^{\frac{\pi}{h}} \xi^2 e^{-i(t-t_{j'})\xi}d\xi + i\overline{\beta(t_{j'})}\int_{-\frac{\pi}{h}}^{\frac{\pi}{h}} \xi e^{-i(t-t_{j'})\xi}d\xi + \overline{\gamma(t_{j'})}\int_{-\frac{\pi}{h}}^{\frac{\pi}{h}} e^{-i(t-t_{j'})\xi}d\xi].$$

First, we are considering approximate solutions of the differential equation (6.9) and at this point, we are considering the Paley-Wiener function spaces with parameter h as approximating function spaces; the function spaces are formed by analytic functions of the entire functions of exponential type that are decreasing to zero exponential order. Next, by using the Fourier inversion, the differential equation (6.9) may be transformed into (6.10). However, to solve the integral equation (6.10) is very difficult for the generality of the coefficient functions. So, we assume (6.10) is valid on some finite number of points t_j. This assumption will be very reasonable for the discretization of the integral equation. By this assumption we can obtain an optimal approximate solution in a very simple way.

Here, we assume that Eq. (6.9) is valid on I so, as some practical case we would like to consider the equation in (6.9) on I satisfying some boundary conditions. In the present case, the boundary conditions are given as zero at infinity for $I = \mathbb{R}$.

However, our result gives the approximate general solutions satisfying boundary values. For example, for a finite interval (a, b), we consider $t_1 = a$ and $t_n = b$ and $\alpha(t_1) = \beta(t_1) = \alpha(t_n) = \beta(t_n) = 0$. Then, we can obtain the approximate solution having arbitrary given boundary values $y_h^A(t_1)$ and $y_h^A(t_n)$. In addition, by a simple modification we may give the general approximate solutions satisfying the corresponding boundary values.

For a finite interval case I, following the boundary conditions, we can consider the corresponding reproducing kernels by the Sobolev Hilbert spaces. However, the concrete representations of the reproducing kernels are involved depending on the boundary conditions. However, we can still consider them and we can use them.

Of course, for a smaller h we can obtain a better approximate solution.

For the representation (6.12) of the reproducing kernel $K_{hh}(t, t')$, we can calculate it easily.

The very surprising facts are: for variable coefficients linear differential equations, we can represent their approximate solutions satisfying their boundary conditions without integrals. Approximate function spaces may be considered with the Paley-Wiener spaces and the Sobolev spaces. For many concrete examples and numerical examples, see [9, 10]. We showed Figures of the numerical experiments. See also [28] for some applications to nonlinear partial differential equations.

7 Best Approximations—As a Connection

Let L be any bounded linear operator from a reproducing kernel Hilbert space H_K into a Hilbert space $\mathcal{H}$. Then, the following problem is a classical and fundamental problem known as the best approximate mean square norm problem: For any member $\mathbf{d}$ of $\mathcal{H}$, we would like to find

$$\inf_{f \in H_K} \|Lf - \mathbf{d}\|_{\mathcal{H}}.$$

It is clear that we are considering *operator equations*, generalized solutions and corresponding generalized inverses within the framework of $f \in H_K$ and $\mathbf{d} \in \mathcal{H}$, having in mind

$$Lf = \mathbf{d}. \tag{7.1}$$

However, this problem has a complicated structure, specially in the infinite dimension Hilbert spaces case, leading in fact to the consideration of generalized inverses (in the Moore-Penrose sense). Following the reproducing kernel theory, we can realize its complicated structure. Anyway, the problem turns to be well posed within the reproducing kernels theory framework in the following proposition:

Proposition 7.1 *For any member* $\mathbf{d}$ *of* $\mathcal{H}$, *there exists a function* $\tilde{f}$ *in* H_K *satisfying*

$$\inf_{f \in H_K} \|Lf - \mathbf{d}\|_{\mathcal{H}} = \|L\tilde{f} - \mathbf{d}\|_{\mathcal{H}} \tag{7.2}$$

if and only if, for the reproducing kernel Hilbert space H_k *admitting the kernel defined by* $k(p,q) = (L^*LK(\cdot,q), L^*LK(\cdot,p))_{H_K}$

$$L^*\mathbf{d} \in H_k. \tag{7.3}$$

Furthermore, when there exists a function $\tilde{f}$ *satisfying (7.2), there exists a uniquely determined function that minimizes the norms in* H_K *among the functions satisfying the equality, and its function* $f_{\mathbf{d}}$ *is represented as follows:*

$$f_{\mathbf{d}}(p) = (L^*\mathbf{d}, L^*LK(\cdot,p))_{H_k} \quad \text{on} \quad E. \tag{7.4}$$

Here, the adjoint operator L^* of L, as we see, from

$$(L^*\mathbf{d})(p) = (L^*\mathbf{d}, K(\cdot, p))_{H_K} = (\mathbf{d}, LK(\cdot, p))_{\mathcal{H}}$$

is represented by known $\mathbf{d}$, L, $K(p, q)$, and $\mathcal{H}$. From this Proposition 7.1, we see that the problem is well established by the theory of reproducing kernels that is the existence, uniqueness, and representation of the solutions in the problem are well formulated. In particular, note that the adjoint operator is represented in a good way; this fact will be very important. The extremal function $f_{\mathbf{d}}$ is the *Moore-Penrose generalized inverse* $L^{\dagger}\mathbf{d}$ of the equation $Lf = \mathbf{d}$. The criteria (7.3) is involved so the Moore-Penrose generalized inverse $f_{\mathbf{d}}$ is not good, when the data contain error or noises in some practical cases.

8 The Tikhonov Regularization

We will consider some practical and more concrete representation in the extremal functions involved in the Tikhonov regularization by using the theory of reproducing kernels. Recall that for compact operators the singular values and singular functions representations are popular and in a sense, however the representation may be considered as complicated.

Furthermore, when $\mathbf{d}$ contains error or noises, error estimates are important. For this fundamental problem, we have the following results:

First, we need

Proposition 8.1 *Let $L : H_K \to \mathcal{H}$ be a bounded linear operator, and define the inner product with a positive α*

$$\langle f_1, f_2 \rangle_{H_{K_\alpha}} = \alpha \langle f_1, f_2 \rangle_{H_K} + \langle Lf_1, Lf_2 \rangle_{\mathcal{H}}$$

for $f_1, f_2 \in H_K$. Then $(H_K, \langle \cdot, \cdot \rangle_{H_{K_\alpha}})$ is a reproducing kernel Hilbert space whose reproducing kernel is given by

$$K_\alpha(p, q) = [(\alpha + L^*L)^{-1} K_q](p).$$

Here, $K_\alpha(p, q)$ is the solution $\tilde{K}_\alpha(p, q)$ of the functional equation

$$\tilde{K}_\alpha(p, q) + \frac{1}{\alpha}(L\tilde{K}_q, LK_p)_{\mathcal{H}} = \frac{1}{\alpha} K(p, q), \tag{8.1}$$

that is corresponding to the Fredholm integral equation of the second kind for many concrete cases. Here,

$$\tilde{K}_q = \tilde{K}_\alpha(\cdot, q) \in H_K \ \ \textit{for} \ \ q \in E, \qquad K_p = K(\cdot, p) \ \ \textit{for} \ \ p \in E.$$

Proposition 8.2 *In the Tikhonov functional*

$$f \in H_K \mapsto \left\{\alpha \|f : H_K\|^2 + \|Lf - \mathbf{d} : \mathcal{H}\|^2\right\} \in \mathbb{R}$$

attains the minimum and the minimum is attained only at $f_{\mathbf{d},\alpha} \in H_K$ *such that*

$$(f_{\mathbf{d},\alpha})(p) = \langle \mathbf{d}, LK_\alpha(\cdot, p)\rangle_{\mathcal{H}}.$$

Furthermore, $(f_{\mathbf{d},\alpha})(p)$ *satisfies*

$$|(f_{\mathbf{d},\alpha})(p)| \le \sqrt{\frac{K(p,p)}{2\alpha}}\|\mathbf{d}\|_{\mathcal{H}}. \tag{8.2}$$

This proposition means that in order to obtain good approximate solutions, we must take a sufficiently small α, however, here we have restrictions for them, as we see, when $\mathbf{d}$ moves to $\mathbf{d}'$, by considering $f_{\mathbf{d},\alpha}(p) - f_{\mathbf{d}',\alpha}(p)$ in connection with the relation of the difference $\|\mathbf{d} - \mathbf{d}'\|_{\mathcal{H}}$. This fact is a very natural one, because we cannot obtain good solutions from the data containing errors. Here we wish to know how to take a small α a priori and what is the bound for it. These problems are very important practically and delicate ones, and we have many methods.

The basic idea may be given as follows. We examine for various α tending to zero, the corresponding extremal functions. By examining the sequence of the approximate extremal functions, when it converges to some function numerically and after then when the sequence diverges numerically, it will give the bound for α numerically. See [12–14].

For this important problem and the method of L-curve, see [19, 20], for example.

The Tikhonov regularization is very popular and widely applicable in numerical analysis for its practical power. The application of the theory of reproducing kernels will give more concrete representations of the extremal functions in the Tikhonov regularization.

8.1 A Typical Example with Real and Numerical Inversion of the Laplace Transform

Consider the inversion formula of the Laplace transform

$$(\mathcal{L}F)(p) = f(p) = \int_0^\infty e^{-pt}F(t)dt, \quad p > 0$$

for some natural function spaces.

On the positive real line $\mathbf{R}^+$, consider the norm

$$\left\{ \int_0^\infty |F'(t)|^2 \frac{1}{t} e^t dt \right\}^{1/2}$$

for absolutely continuous functions F satisfying $F(0) = 0$. This space H_K admits the reproducing kernel

$$K(t, t') = \int_0^{\min(t,t')} \xi e^{-\xi} d\xi. \tag{8.3}$$

Then,

$$\int_0^\infty |(\mathcal{L}F)(p)p|^2 dp \le \frac{1}{2} \|F\|^2_{H_K}; \tag{8.4}$$

that is, $(\mathcal{L}F)(p)p$ is a bounded linear operator from H_K into $L_2(\mathbf{R}^+, dp) = L_2(\mathbf{R}^+)$. So the following result holds:

Proposition 8.3 *For any $g \in L_2(\mathbf{R}^+)$ and for any $\alpha > 0$, in the sense*

$$\inf_{F \in H_K} \left\{ \alpha \int_0^\infty |F'(t)|^2 \frac{1}{t} e^t dt + \|(\mathcal{L}F)(p)p - g\|^2_{L_2(\mathbf{R}^+)} \right\} \tag{8.5}$$

$$= \alpha \int_0^\infty |F^{*\prime}_{\alpha,g}(t)|^2 \frac{1}{t} e^t dt + \|(\mathcal{L}F^*_{\alpha,g})(p)p - g\|^2_{L_2(\mathbf{R}^+)}$$

there exists a uniquely determined best approximate function $F^*_{\alpha,g}$ and it is represented by

$$F^*_{\alpha,g}(t) = \int_0^\infty g(\xi) \left(\mathcal{L}K_\alpha(\cdot, t)\right)(\xi)\xi d\xi. \tag{8.6}$$

Here, $K_\alpha(\cdot, t)$ is determined by the functional equation for $K_{\alpha,t'} = K_\alpha(\cdot, t')$, $K_t = K(\cdot, t)$,

$$K_\alpha(t, t') = \frac{1}{\alpha} K(t, t') - \frac{1}{\alpha} ((\mathcal{L}K_{\alpha,t'})(p)p, (\mathcal{L}K_t)(p)p)_{L_2(\mathbf{R}^+)}. \tag{8.7}$$

We calculate the approximate inverse $F^*_{\alpha,g}(t)$ by using (8.6). By taking the Laplace transform of (8.7) with respect to t, by changing the variables t and t'

$$(\mathcal{L}K_\alpha(\cdot, t))(\xi) = \frac{1}{\alpha} (\mathcal{L}K(\cdot, t))(\xi) - \frac{1}{\alpha} ((\mathcal{L}K_{\alpha,t})(p)p, (\mathcal{L}(\mathcal{L}K_\cdot)(p)p))(\xi))_{L_2(\mathbf{R}^+)}. \tag{8.8}$$

Here

$$K(t,t') = \begin{cases} -te^{-t} - e^{-t} + 1 & \text{for } t \le t' \\ -t'e^{-t'} - e^{-t'} + 1 & \text{for } t \ge t'. \end{cases}$$

$$(\mathcal{L}K(\cdot,t'))(p) = e^{-t'p}e^{-t'}\Big[\frac{-t'}{p(p+1)} + \frac{-1}{p(p+1)^2}\Big] + \frac{1}{p(p+1)^2}.$$

$$\int_0^\infty e^{-qt'}(\mathcal{L}K(\cdot,t'))(p)dt' = \frac{1}{pq(p+q+1)^2}.$$

Therefore, by setting as $(\mathcal{L}K_\alpha(\cdot,t))(\xi)\xi = H_\alpha(\xi,t)$, we obtain the *Fredholm integral equation of the second kind*:

$$\alpha H_\alpha(\xi,t) + \int_0^\infty \frac{H_\alpha(p,t)}{(p+\xi+1)^2}dp = -\frac{e^{-t\xi}e^{-t}}{\xi+1}\left(t + \frac{1}{\xi+1}\right) + \frac{1}{(\xi+1)^2}, \quad (8.9)$$

which is corresponding to (7.1). By solving this integral equation, H. Fujiwara derived a very reasonable numerical inversion formula for the integral transform and he expanded very good algorithms for numerical and real inversion formulas of the Laplace transform. For more detailed references and comments for this equation, see [12–14].

In particular, H. Fujiwara solved the integral equation (8.9) with 6000 points discretization with *600 digits precision* based on the concept of *infinite precision* which is in turn based on *multiple-precision arithmetic*. Then, the regularization parameters were $\alpha = 10^{-100}, 10^{-400}$ surprisingly. For the integral equation, he used the *DE formula* by H. Takahashi and M. Mori, using double exponential transforms. H. Fujiwara was successful in deriving numerically the inversion for the Laplace transform of the distribution delta that was proposed by V.V. Kryzhniy as a difficult case. This fact will mean that the above results valid for very general functions approximated by the functions of the reproducing kernel Hilbert space $H_K(\mathbb{R}^+)$.

We can see many Figures for the numerical experiments in the complete version [15] by Professor H. Fujiwara and for the heat conduction problem, by [23].

9 General Fractional Functions

Consider a general fractional function

$$\frac{g}{f} \quad (9.1)$$

for some very general functions g and f on a set E.

In order to consider such fractional functions (9.1), consider the background-related equation

$$f_1(p)f(p) = g(p) \quad \text{on} \quad E \tag{9.2}$$

for some functions f_1 and g on the set E. If the solution f_1 of (9.2) on the set E exists, then the solution f_1 gives the meaning of the fractional function (9.1). So, the problem may be transformed into the very general and popular equation (9.2).

The function f is initially given. So, for analyzing the Eq. (9.2), we introduce a suitable function space containing the function f_1 and then we find the induced function space containing the product $f_1(p)f(p)$. Then, we can consider the solution of the Eq. (9.2).

Here, we note the very interesting fact that the products $f_1(p)f(p)$ determine a natural reproducing kernel Hilbert space that is induced by the reproducing kernel Hilbert space $H_{K_1}(E)$ and by a second reproducing kernel Hilbert space, say $H_K(E)$, containing the function $f(p)$. Note that for an *arbitrary function* f, there exists a reproducing kernel Hillbert space containing the function $f(p)$; indeed the simplest reproducing kernel is given by $f(p)\overline{f(q)}$. Then, the space in question is a reproducing kernel Hilbert space $H_{K_1K}(E)$ that is determined by the product $K_1(p,q)K(p,q)$ and, furthermore, we obtain the inequality

$$\|f_1 f\|_{H_{K_1K}(E)} \leq \|f_1\|_{H_{K_1}(E)} \|f\|_{H_K(E)}. \tag{9.3}$$

This inequality means that for the linear operator $\varphi_f(f_1)$ on $H_{K_1}(E)$ (for a fixed function f), defined by

$$\varphi_f(f_1) = f_1(p)f(p), \tag{9.4}$$

the inequality

$$\|\varphi_f(f_1)\|_{H_{K_1K}(E)} \leq \|f_1\|_{H_{K_1}(E)} \|f\|_{H_K(E)}$$

holds. This means that the mapping φ_f is a bounded linear operator from $H_{K_1}(E)$ into $H_{K_1K}(E)$ (see Sect. 10.1 for the details).

Now we can consider the operator equation (9.2) in this natural framework. We can mathematically analyze this situation in a natural way and can develop a consequent theory, however, the operator problem will be very sensitive on the functions f.

For some reasonable solutions for the operator equation (9.2), we can introduce *approximate fractional functions* and *generalized fractional functions* in correspondence to the usual fractional function by using the above Tikhonov regularization method combined with the theory of reproducing kernels, and as a special case, by the concept of the Moore-Penrose generalized inverses. See [6, 7, 30, 31].

Division by Zero

The general fractional functions may be considered in various general situations. In particular, in the sense of the Moore-Penrose generalized inverse on **R** or **C**, for any real or complex number z,

$$\frac{z}{0} = 0, \tag{9.5}$$

which was derived as the very special case in [32].

For a simple introduction and several physical meanings, see [21].

The division by zero has a long history and great references. See for example Google Site by *division by zero*.

On the division by zero, we believe our mathematics determines that for any complex number z, $z/0 = 0$; here, of course, for the definition $z/0$ of the division by zero, we have to give its definition clearly. At this moment, we have 5 definitions with motivations; that is,

(1) by the generalization of the fractions by the Tikhonov regularization or by the Moore-Penrose generalized inverse,
(2) by the intuitive meaning of the fractions (division) by H. Michiwaki,
(3) by the unique extension of the fractions by S. Takahasi [35],
(4) by the extension of the fundamental function $W = 1/z$ from $\mathbf{C} \setminus \{0\}$ into $\mathbf{C}$ such that $W = 1/z$ is a one to one and onto mapping from $\mathbf{C} \setminus \{0\}$ onto $\mathbf{C} \setminus \{0\}$ and the division by zero $1/0 = 0$ is a one to one and onto mapping extension of the function $W = 1/z$ from $\mathbf{C}$ onto $\mathbf{C}$, and
(5) by considering the values of functions with the mean values of functions.

Further, in order to show the importance of the division by zero, we gave in [25] clear evidences of the reality of the division by zero $z/0 = 0$ with a fundamental algebraic theorem, and physical and geometrical examples; that is, (A) a field structure containing the division by zero, (B) by the gradient of the y axis on the (x, y) plane, (C) by the reflection $1/\overline{z}$ of z with respect to the unit circle with center at the origin on the complex z plane, and (D) by considering rotation of a right circle cone having some very interesting phenomenon from some practical and physical problem.

In particular, by Figure the interpretation of (C) was introduced in the talk.

Furthermore, we were able to find several meanings in the elementary geometry and physical meanings of the division by zero.

For the division by zero in connection with number structures, mathematics logic, and computer sciences, see the paper [3]. However, they state that in the conclusion:

The theory of meadows depends upon the formal idea of a total inverse operator. We do not claim that division by zero is possible in numerical calculations involving the rationals or reals. But we do claim that zero totalized division is logically, algebraically, and computationally useful: for some applications, allowing zero totalized division in formal calculations, based on equations and rewriting, is appropriate because it is conceptually and technically simpler than the conventional concept of partial division.

It seems that the relationship of the division by zero and field structures are abstract.

Anyhow, we have two ideas for the division by zero as follows:

(I) the division by zero is impossible based on the idea that division is an inversion operation of the product (common idea) and

(II) the division by zero is possible as $z/0 = 0$, by the above five basic ideas and four evidences.

Following the idea (II), the idea that division is an inversion operation of the product is incorrect. *Indeed, we think mathematicians made a serious mistake for this point for a very long time.*

Meanwhile, when we take the idea (I), there is no any world to consider the division by zero more; that is, there is no any story for the division by zero, more. That means *the end*. Meanwhile, following the idea (II), we can consider more for the division by zero and we can consider a new mathematics. There exists a new world. We would like to recall *the principle for our existence*. Indeed, for impossible, no more, and the end, for possible, we can expect something, favorable.

10 Convolutions, Integral Transforms, and Integral Equations

We can consider general convolutions, by means of the theory of reproducing kernels.

Consider two systems

$$f_j(p) = (\mathbf{f}_j, \mathbf{h}_j(p))_{\mathcal{H}_j}, \quad \mathbf{f}_j \in \mathcal{H}_j \tag{10.1}$$

as in Sect. 1 by using $\{\mathcal{H}_j, E, \mathbf{h}_j\}_{j=1}^2$. Here, we assume that E is a same set for the two systems in order to have the output functions $f_1(p)$ and $f_2(p)$ on the same set E.

For example, consider the operator

$$f_1(p) f_2(p)$$

in $\mathcal{F}(E)$. Then, consider the following problems: How to represent the product $f_1(p)f_2(p)$ on E in terms of their inputs $\mathbf{f}_1$ and $\mathbf{f}_2$ through one system?

By using the theory of reproducing kernels we can give a natural answer for this problem. Following similar ideas, we can consider various operators among Hilbert spaces. In particular, for the product of two Hilbert spaces, the idea gives generalizations of convolutions and the related natural convolution norm inequalities. These norm inequalities gave various generalizations and applications to forward and inverse problems for linear mappings in the framework of Hilbert spaces, see for example, [8–10]. Furthermore, for some very general nonlinear systems, we can consider similar problems. See [29] for the details. We consider the product case that will give a general concept of convolutions and we refer to applications to integral equations. For this session, see [7].

10.1 Product and Convolution

First, note that in general:

For any two positive definite quadratic form functions $K_1(p,q)$ and $K_2(p,q)$ on E, the usual product $K(p,q) = K_1(p,q)K_2(p,q)$ is again a positive definite quadratic form function on E, and then H_K is the restriction of the tensor product $H_{K_1}(E) \otimes H_{K_2}(E)$ to the diagonal set, and

Let $\{f_j^{(1)}\}_j$ and $\{f_j^{(2)}\}_j$ be some complete orthonormal systems in $H_{K_1}(E)$ and $H_{K_2}(E)$, respectively, then the reproducing kernel Hilbert space H_K is comprised of all functions on E that are represented as, in the sense of absolutely convergence on E

$$f(p) = \sum_{i,j} \alpha_{i,j} f_i^{(1)}(p) f_j^{(2)}(p) \quad \text{on} \quad E, \quad \sum_{i,j} |\alpha_{i,j}|^2 < \infty \tag{10.2}$$

and its norm in H_K is given by $\| f \|_{H_K}^2 = \min \sum_{i,j} |\alpha_{i,j}|^2$ where $\{\alpha_{i,j}\}$ are considered by satisfying (10.2).

By (I), for

$$K_j(p,q) = (\mathbf{h}_j(q), \mathbf{h}_j(p))_{\mathcal{H}_j} \text{ on } E \times E, \tag{10.3}$$

and for $f_1 \in H_{K_1}(E)$ and $f_2 \in H_{K_2}(E)$, we note that for the reproducing kernel Hilbert space $H_{K_1 K_2}(E)$ admitting the reproducing kernel

$$K_1(p,q)K_2(p,q) \text{ on } E,$$

in general, the inequality

$$\| f_1 f_2 \|_{H_{K_1 K_2}(E)} \le \| f_1 \|_{H_{K_1}(E)} \| f_2 \|_{H_{K_2}(E)} \tag{10.4}$$

holds.

For the positive definite quadratic form function $K_1 K_2$ on E, we assume the expression in the form

$$K_1(p,q)K_2(p,q) = (\mathbf{h}_P(q), \mathbf{h}_P(p))_{\mathcal{H}_P} \text{ on } E \times E \tag{10.5}$$

with a Hilbert space $\mathcal{H}_P$-valued function on E and further we assume that

$$\{\mathbf{h}_P(p);\ p \in E\} \text{ is complete in } \mathcal{H}_P. \tag{10.6}$$

Such a representation is, in general, possible by the fundamental result of Kolmogorov. Then, we can consider conversely the linear mapping from $\mathcal{H}_P$ onto $H_{K_1 K_2}(E)$

$$f_P(p) = (\mathbf{f}_P, \mathbf{h}_P(p))_{\mathcal{H}_P}, \ \mathbf{f}_P \in \mathcal{H}_P \tag{10.7}$$

and the isometric identity

$$\|f_P\|_{H_{K_1K_2}(E)} = \|\mathbf{f}_P\|_{\mathcal{H}_P} \tag{10.8}$$

holds.

Hence, for such representations (10.7) with (10.8), we obtain the isometric mapping between the Hilbert spaces $\mathcal{H}_P$ and $H_{K_1K_2}(E)$.

Now, for the product $f_1(p)f_2(p)$ there exists a uniquely determined $\mathbf{f}_P \in \mathcal{H}_P$ satisfying

$$f_1(p)f_2(p) = (\mathbf{f}_P, \mathbf{h}_P(p))_{\mathcal{H}_P} \text{ on } E. \tag{10.9}$$

Then, $\mathbf{f}_P$ will be considered as a product of $\mathbf{f}_1$ and $\mathbf{f}_2$ through these transforms so, we introduce the notation

$$\mathbf{f}_S = \mathbf{f}_1[\times]\mathbf{f}_2. \tag{10.10}$$

This product for the members $\mathbf{f}_1 \in \mathcal{H}_1$ and $\mathbf{f}_2 \in \mathcal{H}_2$ is introduced through the three transforms induced by $\{\mathcal{H}_j,\ E,\ \mathbf{h}_j\}$ $(j = 1, 2)$ and $\{\mathcal{H}_P,\ E,\ \mathbf{h}_P\}$.

The operator $\mathbf{f}_1[\times]\mathbf{f}_2$ is represented in terms of $\mathbf{f}_1$ and $\mathbf{f}_2$ by the inversion formula

$$(\mathbf{f}_1, \mathbf{h}_1(p))_{\mathcal{H}_1}(\mathbf{f}_2, \mathbf{h}_2(p))_{\mathcal{H}_2} \longrightarrow \mathbf{f}_1[\times]\mathbf{f}_2 \tag{10.11}$$

in the sense (II) from $H_{K_1K_2}(E)$ onto $\mathcal{H}_P$. Then, from (II) and (10.6) we have a similar inequality for Schwarz:

The inequality

$$\|\mathbf{f}_1[\times]\mathbf{f}_2\|_{\mathcal{H}_P} \le \|\mathbf{f}_1\|_{\mathcal{H}_1}\|\mathbf{f}_2\|_{\mathcal{H}_2} \tag{10.12}$$

holds.

10.2 Example

A typical application to the convolution inequality is given by:

Proposition 10.1 *Suppose that we are given two positive integrable functions* ρ_1, ρ_2 *on* $\mathbf{R}$. *If* $F_1, F_2 : \mathbf{R} \to [0, \infty]$ *are measurable functions, then the inequality*

$$\int_{\mathbf{R}} \frac{1}{(\rho_1 * \rho_2)(t)} \left| \int_{\mathbf{R}} F_1(\xi)\rho_1(\xi)F_2(t-\xi)\rho_2(t-\xi)\,d\xi \right|^2 dt$$

$$\le \int_{\mathbf{R}} F_1(t)^2\rho_1(t)\,dt \cdot \int_{\mathbf{R}} F_2(t)^2\rho_2(t)\,dt$$

holds for the usual convolution $\rho_1 * \rho_2$.

Proof We define K_j and $L_j : L^2(\mathbf{R}; \rho_j) \to H_{K_j}$ by

$$K_j(x, y) = \int_{\mathbf{R}} \exp(i(x - y) \cdot t)\rho_j(t)\, dt,$$

$$(L_j F)(t) = \frac{1}{2\pi} \int_{\mathbf{R}} F(x)\rho_j(x) \exp(-it \cdot x)\, dx$$

for $j = 1, 2$.

It follows from the Fubini theorem that

$$K_1(x, y)K_2(x, y) = \int_{\mathbf{R}} \exp(i(x - y) \cdot t)(\rho_1 * \rho_2)(t)\, dt.$$

The same can be said for $L_1 F_1 \cdot L_2 F_2$:

$$(L_1 F_1)(x)(L_2 F_2)(x) = \int_{\mathbf{R}} \exp(-ix \cdot t)(F_1\rho_1) * (F_2\rho_2)(t)\, dt.$$

By the property of the product kernel space $H_{K_1K_2}$

$$\|L_1 F_1 \cdot L_2 F_2\|_{H_{K_1K_2}} \leq \|L_1 F_1\|_{H_{K_1}} \cdot \|L_2 F_2\|_{H_{K_2}}.$$

By writing out in full both the sides, the inequality follows. □

This result was expanded for various directions with applications to inverse problems and partial differential equations.

10.3 Applications to Integral Equations

We will assume that the linear transforms in the above

$$L_j : \mathbf{f}_j \in \mathcal{H}_j \longrightarrow f_j \in H_{K_j}(E)$$

and

$$L : \mathbf{f}_P \in \mathcal{H}_P \longrightarrow f_P \in H_{K_1K_2}(E)$$

are isometrical. We now consider the integral equation, for $\mathbf{f}_1 \in \mathcal{H}_1$, $\mathbf{f}_2^{(1)}, \mathbf{f}_2^{(2)} \in \mathcal{H}_2$ and $\mathbf{g} \in \mathcal{H}_P$

$$\mathbf{f}_1[\times]\mathbf{f}_2^{(1)} + \mathbf{f}_1[\times]\mathbf{f}_2^{(2)} = \mathbf{g}.$$

Then, by taking the transform L, we obtain

$$L_1\mathbf{f}_1(L_2\mathbf{f}_2^{(1)} + L_2\mathbf{f}_2^{(2)}) = g(p)$$

so

$$f_1(p)\left(f_2^{(1)}(p)+f_2^{(2)}(p)\right)=g(p) \tag{4.1}$$

on the functions on E. Then, for given $\mathbf{f}_2^{(1)}, \mathbf{f}_2^{(2)} \in \mathcal{H}_2$ and $\mathbf{g} \in \mathcal{H}_P$, when we solve the equation, we wish to consider the following way:

$$\mathbf{f}_1 = L_1^{-1}\left(\frac{g(p)}{f_2^{(1)}(p)+f_2^{(2)}(p)}\right).$$

Here, the essential problem, however, rises how to obtain the solution

$$\frac{g(p)}{f_2^{(1)}(p)+f_2^{(2)}(p)}.$$

The important problems here are that the function $f_2^{(1)}(p)+f_2^{(2)}(p)$ may have many zero points and some properties of the above fractional function for looking the inversion L_1^{-1}. These important problems may be solved effectively by using the Tikhonov regularization in Sect. 8 and the concept in Sect. 10.

11 Eigenfunctions, Initial Value Problems, Integral Transforms, and Reproducing Kernels

Among the very fundamental concepts of eigenfunctions, initial value problems on some general linear PDEs, integral transforms and reproducing kernels on analysis, there exist good relationships. For Sect. 11, see [11].

For some general linear operator L_x for some function space on some domain, consider the initial value problem: For $t > 0$

$$(\partial_t + L_x)u_f(t,x) = 0 \tag{11.1}$$

satisfying the initial value condition with some suitable meaning

$$u_f(0,x) = f(x). \tag{11.2}$$

Furthermore, the relation will give the method how to characterize completely the solutions under the given conditions.

One basic concept is to use some eigenvalues λ on a set I and eigenfunctions W_λ satisfying

$$L_x W_\lambda(x) = \lambda W_\lambda(x). \tag{11.3}$$

Then, the functions

$$\exp\{-\lambda t\}W_\lambda(x) \tag{11.4}$$

are the solutions of the operator equation

$$(\partial_t + L_x)u(t,x) = 0. \tag{11.5}$$

We consider some general solution of (11.5) by a suitable sum of the solutions (11.4). In order to consider a fully good sum, consider the kernel form, with a continuous nonnegative weight function ρ over the interval I

$$K_\lambda(x,y) = \int_I \exp\{-\lambda t\}W_\lambda(x)W_\lambda(y)\rho(\lambda)d\lambda. \tag{11.6}$$

Here, we assume that λ are real-valued and also the eigenfunctions $W_\lambda(x)$ are also real-valued. Then, fully general solutions of the Eq. (11.1) may be represented in the integral form

$$u(t,x) = \int_I \exp\{-\lambda t\}W_\lambda(x)F(\lambda)\rho(\lambda)d\lambda \tag{11.7}$$

for the functions F satisfying

$$\int_I \exp\{-\lambda t\}|F(\lambda)|^2\rho(\lambda)d\lambda < \infty. \tag{11.8}$$

Then, the solution $u(t,x)$ of (11.1) satisfying the initial condition

$$u(0,x) = F(x) \tag{11.9}$$

will be obtained by $t \to +0$ in (11.7) with a natural meaning. However, this point will be very delicate and we will need to consider some deep and beautiful structure. Here, (11.6) is a reproducing kernel and in order to analyze the logic above, we will need the theory of reproducing kernels, essentially and beautiful ways.

We will consider the related reproducing kernel

$$K_0(x,y) = \int_I W_\lambda(x)W_\lambda(y)\rho(\lambda)d\lambda \tag{11.10}$$

and the corresponding reproducing kernel Hilbert space H_{K_0}. Here, the important general property

$$K_t(x,y) \ll K_0(x,y); \tag{11.11}$$

that is, $K_0(x,y) - K_t(x,y)$ is a positive definite quadratic form function and we have

$$H_{K_t} \subset H_{K_0}$$

and for any function $f \in H_{K_t}$

$$\|f\|_{H_{K_0}} = \lim_{t\to 0} \|f\|_{H_{K_t}}$$

in the sense of nondecreasing norm convergence.

Now, the kernel $K_t(x, y)$ will satisfy the operator equation (11.1) for any fixed y as the summation of the solutions of (11.1). Then, consider a further summation in the form, for any given function $f \in H_{K_0}$

$$u_f(t, x) = (f(\cdot), K_t(\cdot, x))_{H_{K_0}}. \tag{11.12}$$

These functions will satisfy the operator equation (11.1) as the summation of the solutions. Then, the initial value condition is given as follows:

$$\begin{aligned} \lim_{t\to 0} u_f(t, x) &= \lim_{t\to 0}(f(\cdot), K_t(\cdot, x))_{H_{K_0}}. \qquad (11.13) \\ &= (f(\cdot), K_0(\cdot, x))_{H_{K_0}} \\ &= f(x). \end{aligned}$$

We construct the solution for the initial value problem satisfying (11.1) and (11.9). The function (11.2) will satisfy the operator equation (11.1) and for the sake of the norm convergence, the limit for $t \to +0$ converges to the function f in a good way. So, the crucial point in our approach is based on the realization of the reproducing kernel Hilbert space H_{K_0}. Of course, these properties depend on the eigenfunctions property.

Furthermore, the complete property of the solutions of (11.1) satisfying the initial value f may be derived from the reproducing kernel Hilbert space admitting the kernel

$$k(x, t; y, \tau) := (K_\tau(\cdot, y), K_t(\cdot, x))_{H_{K_0}}. \tag{11.14}$$

In the method, we see that the existence problem of the initial value problem is based on the eigenfunctions $W_\lambda(x)$ and we are constructing the desired solution satisfying the desired initial condition. For a larger knowledge for the eigenfunctions, we can consider a more general initial value problem. The uniqueness property of the initial value problem is depending on the completeness of the family of functions

$$\{K_t(\cdot, x); x \in \text{ the space}\} \tag{11.15}$$

in the space H_{K_0}.

Furthermore, by considering the linear mapping of (11.7) with various situations, various inverse problems looking for the initial values f from the various output data of $u_f(t, x)$ may be obtained.

11.1 *The Simplest Case Example for the Exponential Function*

For the simplest derivative operator D,

$$De^{\lambda x} = \lambda e^{\lambda x}. \tag{11.16}$$

We can consider the initial value problems with various situations by considering λ and the variable x. The typical cases are the weighted Laplace transforms, the Paley-Wiener spaces and the Sobolev spaces depending on $\lambda > 0$, λ is on a symmetric interval and λ is on the whole real space. The Laplace transform may be considered in many situations by considering the various weights [29], so we will consider the simplest case:

$$K(z, \overline{u}) = \int_0^\infty e^{-\lambda z} e^{-\lambda \overline{u}} d\lambda = \frac{1}{z + \overline{u}}, z = x + iy, \tag{11.17}$$

on the right half complex plane. The reproducing kernel is the Szegö kernel and for the image of the integral transform

$$f(z) = \int_0^\infty e^{-\lambda z} F(\lambda) d\lambda, \tag{11.18}$$

for the $L_2(0, \infty)$ functions $F(\lambda)$, the isometric identity

$$\frac{1}{2\pi} \int_{-\infty}^{+\infty} |f(iy)|^2 dy = \int_0^\infty |F(\lambda)|^2 d\lambda \tag{11.19}$$

is obtained. Here, $f(iy)$ means the Fatou's non-tangential boundary values of analytic functions $f(z)$ of the Szegö space on the right-hand half complex plane.

Now, consider the reproducing kernel $K_t(z, \overline{u})$ and the corresponding reproducing kernel Hilbert space H_{K_t} by

$$K_t(z, \overline{u}) = \int_0^\infty e^{-\lambda t} e^{-\lambda z} e^{-\lambda \overline{u}} d\lambda. \tag{11.20}$$

Note that the reproducing kernel Hilbert space H_{K_t} is the Szegö space on the right-hand complex plane $x > \frac{-t}{2}$.

For any Szegö kernel function space member $f(z)$ on the right half complex plane, the function

$$U_f(t, z) = (f(\cdot), K_t(\cdot, \overline{z}))_{H_K} = \frac{1}{2\pi} \int_{-\infty}^{+\infty} f(iy) \overline{K_t(iy, \overline{z})} dy \tag{11.21}$$

satisfies the partial differential equation

$$(\partial_t - D_z)U(t, z) = 0. \tag{11.22}$$

In order to see the characteristic property of the solutions $U(t, z)$, we will consider the kernel form

$$k(t, z; \tau, \overline{u}) = (K_\tau(\cdot, \overline{u}), K_t(\cdot, \overline{z}))_{H_K} \tag{11.23}$$

$$= \frac{1}{t + \tau + z + \overline{u}}.$$

From this representation we see that, for any fixed $t > 0$, the solutions $U(t, z)$ belong to the Szegö space on the right-hand complex plane

$$Re\, z > -t,$$

and for any fixed z, $Re\, z > 0$, the solutions $U_f(t, z)$ may be continued analytically onto the half complex plane with respect to t

$$Re\, t > -Re\, z.$$

In particular, note that the solutions $U_f(t, z)$ will have meanings on the *negative time*.

For any fixed time t, we can obtain the inversion formula in the complex version in (11.18) by the general formula in Proposition 4.2, because the needed situations are concretely given. Meanwhile, for any fixed space point z, we will be able to see that the situation is similar, and we in a natural way consider the inversion with the *complex time* t. When we wish to establish the real inversion, we can consider the inversion formulas by the Aveiro discretization method [9, 10] or by applying the Tikhonov regularization method [31] as in the numerical real inversion formula of the Laplace transforms. Then, the analytical inversion formula is very deep and complicated.

The above typical example may be expected to have the systematical developments as follows:

(1) Many concrete reproducing kernels may be calculated and the related reproducing kernel Hilbert spaces should be realized with concrete norms.
(2) Eigenfunctions and the related initial value problems in partial differential and integral equations should be examined with their properties of the solutions.
(3) Many new integral transforms and their properties; that is, isometric identities and inversion formulas should be established.
(4) For the associated t kernels and the related small reproducing kernels appeared in the general theory, we can consider the similar problems above.

From the great references by Russian mathematicians containing the special function theory, we can expect new materials and such materials in mathematics are definite values and fundamentals in mathematics.

The general theory in this section was recently extended to the Hilbert space framework by using the generalized reproducing kernels in [33, 34] with Professor *Y. Sawano*.

Meanwhile, the materials except the division by zero in this paper will be published in a book from Springer with the title *Theory of Reproducing Kernels and Applications* with the co-author Y. Sawano in Developments in Mathematics, Vol. 44 (2016).

References

1. Aronszajn, N.: Theory of reproducing kernels. Trans. Am. Math. Soc. **68**, 337–404 (1950)
2. Bergman, S.: 1970 The Kernel Function and Conformal Mapping. American Mathematical Society, Providence, R.I. (1950)
3. Bergstra, J.A., Hirshfeld, Y., Tucker, J.V.: Meadows and the equational specification of division. arXiv:0901.0823v1 [math.RA] 7 Jan 2009 (2009)
4. Berlinet, A., Thomas-Agnan, C.T.: Reproducing Kernel Hilbert Spaces in Probability and Statistics. Kluwer Akademic Publishers, Boston (2004)
5. Berlinet, A.: Reproducing kernels in probability and statistics. In: More Progresses In Analysis, Proceedings of the 5th International ISAAC Congress, pp. 153–162 (2010)
6. Castro, L.P., Saitoh, S.: Fractional functions and their representations. Complex Anal. Oper. Theory **7**(4), 1049–1063 (2013)
7. Castro, L.P., Saitoh, S., Tuan, N.M.: Convolutions, integral transforms and integral equations by means of the theory of reproducing kernels. Opuscula Math. **32**(4), 633–646 (2013)
8. Castro, L.P., Fujiwara, H., Saitoh, S., Sawano, Y., Yamada, A., Yamada, M.: Fundamental error estimates inequalities for the Tikhonov regularization using reproducing kernels. In: International Series of Numerical Mathematics, Inequalities and Applications 2010, vol. 161, pp. 87–101. Springer, Basel (2010)
9. Castro, L.P., Fujiwara, H., Rodrigues, M.M., Saitoh, S., Tuan, V.K.: Aveiro discretization method in mathematics: a new discretization principle. In: Pardalos, P., Rassias, T.M. (eds.) Mathematics Without Boundaries: Surveys in Pure Mathematics, pp. 37–92. Springer (2014)
10. Castro, L.P., Fujiwara, H., Qian, T., Saitoh, S.: How to catch smoothing properties and analyticity of functions by computers? In: Pardalos, P., Rassias, T.M. (eds.) Mathematics Without Boundaries: Surveys in Interdisciplinary Research, pp. 101–116. Springer (2014)
11. Castro, L.P., Rodorigues, M.M., Saitoh, S.: A fundamental theorem on initial value problems by using the theory of reproducing kernels. Complex Anal. Oper. Theory **9**, 87–98 (2015)
12. Fujiwara, H.: Applications of reproducing kernel spaces to real inversions of the Laplace transform. RIMS Koukyuuroku **1618**, 188–209 (2008)
13. Fujiwara, H., Matsuura, T., Saitoh, S., Sawano, Y.: Numerical real inversion of the Laplace transform by using a high-accuracy numerical method. In: Further Progress in Analysis, pp. 574–583. World Science Publisher, Hackensack, NJ (2009)
14. Fujiwara, H.: Numerical real inversion of the Laplace transform by reproducing kernel and multiple-precision arithmetric. In: Proceedings of the 7th International ISAAC Congress Progress in Analysis and Its Applications, pp. 289–295. World Scientific (2010)
15. Fujiwara, H., Higashimori, N.: Numerical inversion of the Laplace transform by using multiple-precision arithmetic. Libertas Mathematica (New Series) **34**(2), 5–21 (2014)

16. Fujiwara, H., Saitoh, S.: The general sampling theory by using reproducing kernels. In: Pardalos, P., Rassias, Th. M. (eds.) Contributions in Mathematics and Engineering in Honor of Constantin Caratheodory Springer (in press)
17. Fukumizu, K., Bach, F.R., Jordan, M.I.: Dimensionality reduction for supervised learning with reproducing kernel Hilbert spaces. J. Mach. Learn. Res. **5**, 73–99 (2004)
18. Fukumizu, K., Bach, F.R., Gretton, A.: Statistical consistency of kernel canonical correlation analysis. J. Mach. Learn. Res. **8**, 361–383 (2007)
19. Hansen, P.C.: Analysis of discrete ill-posed problems by means of the L-curve. SIAM Rev. **34**, 561–580 (1992)
20. Lawson, C.L., Hanson, R.J.: Solving Least Squares Problems. Prentice-Hall, Englewood Cliffs (1974)
21. Kuroda, M., Michiwaki, H., Saitoh, S., Yamane, M.: New meanings of the division by zero and interpretations on $100/0 = 0$ and on $0/0 = 0$. Int. J. Appl. Math. **27**(2), 191–198 (2014)
22. Kolmogorov, A.N.: Stationary sequences in Hilbert's space. Bolletin Moskovskogo Gosudarstvenogo Universiteta, Matematika **2**, 40 (1941) (in Russian)
23. Matsuura, T., Saitoh, S.: Analytical and numerical inversion formulas in the Gaussian convolution by using the Paley-Wiener spaces. Appl. Anal. **85**, 901–915 (2006)
24. Matsuura, T., Saitoh, S.: Matrices and division by zero z/0 = 0. Adv. Linear Algebra Matrix Theory **6**, 51–58 (2016)
25. Michiwaki H, Saitoh S, Yamada M.: Reality of the division by zero $z/0 = 0$. IJAPM Int. J. Appl. Phys. Math. **6**, 1–8 (2016)
26. Michiwaki, H., Okumura, H., Saitoh, S.: Division by zero z/0 = 0 in Euclidean spaces. Int. J. Math. Comput. **28** (2017) (in press)
27. Mo, Y., Qian, T.: Support vector machine adapted Tikhonov regularization method to solve Dirichlet problem. Appl. Math. Comput. **245**, 509–519 (2014)
28. Rocha, E.M.: A reproducing kernel Hilbert discretization method for linear PDEs with nonlinear right-hand side. Libertas Mathematica (New Series) **34**(2), 91–104 (2014)
29. Saitoh, S.: Integral Transforms, Reproducing Kernels and their Applications. Pitman Research Notes in Mathematical Series 369. Addison Wesley Longman, Harlow, CRC Press, Taylor & Francis Group, Boca Raton London, New York (in hard cover) (1997)
30. Saitoh, S.: Various operators in Hilbert space induced by transforms. Int. J. Appl. Math. **1**, 111–126 (1999)
31. Saitoh, S.: Theory of reproducing kernels: applications to approximate solutions of bounded linear operator equations on Hilbert spaces. Am. Math. Soc. Transl. Ser. **2**(230), 107–137 (2010)
32. Saitoh, S.: Generalized inversions of Hadamard and tensor products for matrices. Adv. Linear Algebra Matrix Theory **4**(2), 87–95 (2014)
33. Saitoh, S., Sawano, Y.: Generalized delta functions as generalized reproducing kernels (manuscript)
34. Saitoh, S., Sawano, Y.: General initial value problems using eigenfunctions and reproducing kernels (manuscript)
35. Takahasi, S.E., Tsukada, M.: Kobayashi Y Classification of continuous fractional binary operators on the real and complex fields. Tokyo J. Math. **38**, 369–380 (2015)
36. Vapnik, V.N.: Statistical Learning Theory. Wiley, New York (1998)
37. Vourdas, A.: Analytic representations in quantum mechanics. J. Phys. A: Math. Gen. **39**, 65–141 (2006)

Sparse Approximation by Greedy Algorithms

V. Temlyakov

Abstract It is a survey on recent results in constructive sparse approximation. Three directions are discussed here: (1) Lebesgue-type inequalities for greedy algorithms with respect to a special class of dictionaries, (2) constructive sparse approximation with respect to the trigonometric system, (3) sparse approximation with respect to dictionaries with tensor product structure. In all three cases constructive ways are provided for sparse approximation. The technique used is based on fundamental results from the theory of greedy approximation. In particular, results in the direction (1) are based on deep methods developed recently in compressed sensing. We present some of these results with detailed proofs.

Keywords Sparse · Constructive · Greedy · Lebesgue inequality

1 Introduction

The paper is a survey on recent breakthrough results in constructive sparse approximation. In all cases discussed here the new technique is based on greedy approximation. The main motivation for the study of sparse approximation is that many real-world signals can be well approximated by sparse ones. Sparse approximation automatically implies a need for nonlinear approximation, in particular, for greedy approximation. We give a brief description of a sparse approximation problem and present a discussion of the obtained results and their relation to previous work. In Sect. 2 we concentrate on breakthrough results from [18] and [41]. In these papers we extended a fundamental result of Zhang [48] on the Lebesgue-type inequality for the RIP dictionaries in a Hilbert space (see Theorem 2.2 below) in several directions. We found new more general than the RIP conditions on a dictionary, which still

V. Temlyakov (✉)
University of South Carolina, Columbia, USA
e-mail: temlyak@math.sc.edu

V. Temlyakov
Steklov Institute of Mathematics, Moscow, Russia

T. Qian and L.G. Rodino (eds.), *Mathematical Analysis, Probability and Applications – Plenary Lectures*, Springer Proceedings in Mathematics & Statistics 177, DOI 10.1007/978-3-319-41945-9_7

guarantee the Lebesgue-type inequalities in a Hilbert space setting. We generalized these conditions to a Banach space setting and proved the Lebesgue-type inequalities for dictionaries satisfying those conditions. To illustrate the power of new conditions we applied this new technique to bases instead of redundant dictionaries. In particular, this technique gave very strong results for the trigonometric system.

In a general setting, we are working in a Banach space X with a redundant system of elements $\mathcal{D}$ (dictionary $\mathcal{D}$). There is a solid justification of importance of a Banach space setting in numerical analysis in general and in sparse approximation in particular (see, for instance, [40], Preface, and [29]). An element (function, signal) $f \in X$ is said to be m-sparse with respect to $\mathcal{D}$ if it has a representation $f = \sum_{i=1}^{m} x_i g_i, g_i \in \mathcal{D}, i = 1, \dots, m$. The set of all m-sparse elements is denoted by $\Sigma_m(\mathcal{D})$. For a given element f_0 we introduce the error of best m-term approximation $\sigma_m(f_0, \mathcal{D}) := \inf_{f\in\Sigma_m(\mathcal{D})} \|f_0 - f\|$. We are interested in the following fundamental problem of sparse approximation.

Problem. How to design a practical algorithm that builds sparse approximations comparable to best m-term approximations?

In a general setting, we study an algorithm (approximation method) $\mathcal{A} = \{A_m(\cdot, \mathcal{D})\}_{m=1}^{\infty}$ with respect to a given dictionary $\mathcal{D}$. The sequence of mappings $A_m(\cdot, \mathcal{D})$ defined on X satisfies the condition: for any $f \in X$, $A_m(f, \mathcal{D}) \in \Sigma_m(\mathcal{D})$. In other words, A_m provides an m-term approximant with respect to $\mathcal{D}$. It is clear that for any $f \in X$ and any m we have $\|f - A_m(f, \mathcal{D})\| \ge \sigma_m(f, \mathcal{D})$. We are interested in such pairs $(\mathcal{D}, \mathcal{A})$ for which the algorithm $\mathcal{A}$ provides approximation close to best m-term approximation. We introduce the corresponding definitions.

Definition 1.1 We say that $\mathcal{D}$ is an almost greedy dictionary with respect to $\mathcal{A}$ if there exist two constants C_1 and C_2 such that for any $f \in X$ we have

$$\|f - A_{C_1 m}(f, \mathcal{D})\| \le C_2 \sigma_m(f, \mathcal{D}). \tag{1.1}$$

If $\mathcal{D}$ is an almost greedy dictionary with respect to $\mathcal{A}$ then $\mathcal{A}$ provides almost ideal sparse approximation. It provides $C_1 m$-term approximant as good (up to a constant C_2) as ideal m-term approximant for every $f \in X$. In the case $C_1 = 1$ we call $\mathcal{D}$ a greedy dictionary. We also need a more general definition. Let $\phi(u)$ be a function such that $\phi(u) \ge 1$.

Definition 1.2 We say that $\mathcal{D}$ is a ϕ-greedy dictionary with respect to $\mathcal{A}$ if there exists a constant C_3 such that for any $f \in X$ we have

$$\|f - A_{\phi(m)m}(f, \mathcal{D})\| \le C_3 \sigma_m(f, \mathcal{D}). \tag{1.2}$$

If $\mathcal{D} = \Psi$ is a basis then in the above definitions we replace dictionary by basis. Inequalities of the form (1.1) and (1.2) are called the Lebesgue-type inequalities.

In the above setting, the quality criterion of the algorithm $\mathcal{A}$ is based on the Lebesgue-type inequalities, which hold for every individual $f \in X$. In classical approximation theory very often we use as a quality criterion of the algorithm $\mathcal{A}$ its performance on a given class F. In this case, we compare

$$e_m(F, \mathcal{A}, \mathcal{D}) := \sup_{f \in F} \|f - A_m(f, \mathcal{D})\|$$

with

$$\sigma_m(F, \mathcal{D}) := \sup_{f \in F} \sigma_m(f, \mathcal{D}).$$

We discuss this setting in Sects. 5 and 6.

In the case $\mathcal{A} = \{G_m(\cdot, \Psi)\}_{m=1}^{\infty}$ is the Thresholding Greedy Algorithm (TGA), the theory of greedy and almost greedy bases is well developed (see [40]). We remind that in case of a normalized basis $\Psi = \{\psi_k\}_{k=1}^{\infty}$ of a Banach space X the TGA at the mth iteration gives an approximant

$$G_m(f, \Psi) := \sum_{j=1}^{m} c_{k_j} \psi_{k_j}, \quad f = \sum_{k=1}^{\infty} c_k \psi_k, \quad |c_{k_1}| \geq |c_{k_2}| \geq \cdots.$$

In particular, it is known (see [40], p. 17) that the univariate Haar basis is a greedy basis with respect to TGA for all L_p, $1 < p < \infty$. Also, it is known that the TGA does not work well with respect to the trigonometric system.

We demonstrated in the paper [41] that the Weak Chebyshev Greedy Algorithm (WCGA) which we define momentarily is a solution to the above problem for a special class of dictionaries.

Let X be a real Banach space with norm $\|\cdot\| := \|\cdot\|_X$. We say that a set of elements (functions) $\mathcal{D}$ from X is a dictionary if each $g \in \mathcal{D}$ has norm one ($\|g\| = 1$), and the closure of span $\mathcal{D}$ is X. For a nonzero element $g \in X$ we let F_g denote a norming (peak) functional for g: $\|F_g\|_{X^*} = 1$, $F_g(g) = \|g\|_X$. The existence of such a functional is guaranteed by the Hahn–Banach theorem.

Let $t \in (0, 1]$ be a given weakness parameter. We define the Weak Chebyshev Greedy Algorithm (WCGA) (see [37]) as a generalization for Banach spaces of the Weak Orthogonal Matching Pursuit (WOMP). In a Hilbert space the WCGA coincides with the WOMP. The WOPM is very popular in signal processing, in particular, in compressed sensing. In case $t = 1$, WOMP is called Orthogonal Matching Pursuit (OMP).

Weak Chebyshev Greedy Algorithm (WCGA). Let f_0 be given. Then for each $m \geq 1$ we have the following inductive definition:

(1) $\varphi_m := \varphi_m^{c,t} \in \mathcal{D}$ is any element satisfying

$$|F_{f_{m-1}}(\varphi_m)| \geq t \sup_{g \in \mathcal{D}} |F_{f_{m-1}}(g)|.$$

(2) Define $\Phi_m := \Phi_m^t := \text{span}\{\varphi_j\}_{j=1}^m$, and define $G_m := G_m^{c,t}$ to be the best approximant to f_0 from Φ_m.

(3) Let $f_m := f_m^{c,t} := f_0 - G_m$.

The trigonometric system is a classical system that is known to be difficult to study. In [41] we study among other problems the problem of nonlinear sparse approximation with respect to it. Let $\mathcal{RT}$ denote the real trigonometric system $1, \sin 2\pi x, \cos 2\pi x, \dots$ on $[0, 1]$ and let $\mathcal{RT}_p$ to be its version normalized in $L_p([0, 1])$. Denote $\mathcal{RT}_p^d := \mathcal{RT}_p \times \cdots \times \mathcal{RT}_p$ the d-variate trigonometric system. We need to consider the real trigonometric system because the algorithm WCGA is well studied for the real Banach space. In order to illustrate performance of the WCGA we discuss in this section the above-mentioned problem for the trigonometric system. We proved in [41] the following Lebesgue-type inequality for the WCGA.

Theorem 1.1 *Let $\mathcal{D}$ be the normalized in L_p, $2 \le p < \infty$, real d-variate trigonometric system. Then for any $f_0 \in L_p$ the WCGA with weakness parameter t gives*

$$\|f_{C(t,p,d)m\ln(m+1)}\|_p \le C\sigma_m(f_0, \mathcal{D})_p. \tag{1.3}$$

The Open Problem 7.1 (p. 91) from [38] asks if (1.3) holds without an extra $\ln(m+1)$ factor. Theorem 1.1 is the first result on the Lebesgue-type inequalities for the WCGA with respect to the trigonometric system. It provides a progress in solving the above-mentioned open problem, but the problem is still open.

We note that properties of a given basis with respect to TGA and WCGA could be very different. For instance, the class of quasi-greedy bases (with respect to TGA), that is the class of bases Ψ for which $G_m(f, \Psi)$ converges for each $f \in X$, is a rather narrow subset of all bases. It is close in a certain sense to the set of unconditional bases. The situation is absolutely different for the WCGA. If X is uniformly smooth then WCGA converges for each $f \in X$ with respect to any dictionary in X (see [40], Ch. 6).

Theorem 1.1 shows that the WCGA is very well designed for the trigonometric system. We show in [41] that an analog of (1.3) holds for uniformly bounded orthogonal systems. The proof of Theorem 1.1 uses technique developed in compressed sensing for proving the Lebesgue-type inequalities for redundant dictionaries with special properties. First, results on Lebesgue-type inequalities were proved for incoherent dictionaries (see [40] for a detailed discussion). Then a number of results were proved for dictionaries satisfying the Restricted Isometry Property (RIP) assumption. The incoherence assumption on a dictionary is stronger than the RIP assumption. The corresponding Lebesgue-type inequalities for the Orthogonal Matching Pursuit (OMP) under RIP assumption were not known for a while. As a result new greedy-type algorithms were introduced and exact recovery of sparse signals and the Lebesgue-type inequalities were proved for these algorithms: the Regularized Orthogonal Matching Pursuit (see [22]), Compressive Sampling Matching Pursuit (CoSaMP) (see [21]), and the Subspace Pursuit (SP) (see [3]). The OMP is simpler than CoSaMP and SP, however, at the time of invention of CoSaMP and SP these algorithms provided exact recovery of sparse signals and the Lebesgue-type inequalities for dictionaries satisfying the Restricted Isometry Property (RIP) (see [21] and [3]). The corresponding results for the OMP were not known at that time. Later, a breakthrough result in this direction was obtained by Zhang [48]. In particular, he proved that if $\mathcal{D}$ satisfies RIP then the OMP recovers exactly all m-sparse signals within Cm iterations. In [18] and [41] we developed Zhang's technique to obtain recovery results and the Lebesgue-type inequalities in the Banach space setting.

The above Theorem 1.1 guarantees that the WCGA works very well for each individual function f. It is a constructive method, which provides after $\asymp m \ln m$ iterations an error comparable to $\sigma_m(f, \mathcal{D})$. Here are two important points. First, in order to guarantee a rate of decay of errors $\|f_n\|$ of the WCGA we would like to know how smoothness assumptions on f_0 affect the rate of decay of $\sigma_m(f_0, \mathcal{D})$. Second, if, as we believe, one cannot get rid of $\ln m$ in Theorem 1.1 then it would be nice to find a constructive method, which provides on a certain smoothness class the order of best m-term approximation after m iterations. Thus, as a complement to Theorem 1.1 we would like to obtain results, which relate rate of decay of $\sigma_m(f, \mathcal{T}^d)_p$ to some smoothness type properties of f. In Sect. 5 we concentrate on constructive methods of m-term approximation. We measure smoothness in terms of mixed derivative and mixed difference. We note that the function classes with bounded mixed derivative are not only interesting and challenging object for approximation theory but they are important in numerical computations.

We discuss here the problem of sparse approximation. This problem is closely connected with the problem of recovery of sparse functions (signals). In the sparse recovery problem we assume that an unknown function f is sparse with respect to a given dictionary and we want to recover it. This problem was a starting point for the compressed sensing theory (see [40], Ch. 5). In particular, the celebrated contribution of the work of Candes, Tao, and Donoho was to show that the recovery can be done by the ℓ_1 minimization algorithm. We stress that ℓ_1 minimization algorithm works for the exact recovery of sparse signals. It does not provide sparse approximation. The greedy-type algorithms discussed in this paper provide sparse approximation, satisfying the Lebesgue-type inequalities. It is clear that the Lebesgue-type inequalities (1.1) and (1.2) guarantee exact recovery of sparse signals.

2 Lebesgue-Type Inequalities: General Results

A very important advantage of the WCGA is its convergence and rate of convergence properties. The WCGA is well defined for all m. Moreover, it is known (see [37] and [40]) that the WCGA with weakness parameter $t \in (0, 1]$ converges for all f_0 in all uniformly smooth Banach spaces with respect to any dictionary. That is, when X is a real Banach space and the modulus of smoothness of X is defined as follows:

$$\rho(u) := \frac{1}{2} \sup_{x,y;\|x\|=\|y\|=1} |\|x + uy\| + \|x - uy\| - 2|, \tag{2.1}$$

then the uniformly smooth Banach space is the one with $\rho(u)/u \to 0$ when $u \to 0$.

We discuss here the Lebesgue-type inequalities for the WCGA with weakness parameter $t \in (0, 1]$. This discussion is based on papers [18] and [41]. For notational convenience, we consider here a countable dictionary $\mathcal{D} = \{g_i\}_{i=1}^{\infty}$. The following assumptions **A1** and **A2** were used in [18]. For a given f_0 let sparse element (signal)

$$f := f^{\epsilon} = \sum_{i \in T} x_i g_i, \quad g_i \in \mathcal{D},$$

be such that $\|f_0 - f^{\epsilon}\| \leq \epsilon$ and $|T| = K$. For $A \subset T$ denote

$$f_A := f_A^{\epsilon} := \sum_{i \in A} x_i g_i.$$

A1. We say that $f = \sum_{i \in T} x_i g_i$ satisfies the Nikol'skii-type $\ell_1 X$ inequality with parameter r if

$$\sum_{i \in A} |x_i| \leq C_1 |A|^r \|f_A\|, \quad A \subset T. \tag{2.2}$$

We say that a dictionary $\mathcal{D}$ has the Nikol'skii-type $\ell_1 X$ property with parameters K, r if any K-sparse element satisfies the Nikol'skii-type $\ell_1 X$ inequality with parameter r.

A2. We say that $f = \sum_{i \in T} x_i g_i$ has incoherence property with parameters D and U if for any $A \subset T$ and any Λ such that $A \cap \Lambda = \emptyset$, $|A| + |\Lambda| \leq D$ we have for any $\{c_i\}$

$$\|f_A - \sum_{i \in \Lambda} c_i g_i\| \geq U^{-1} \|f_A\|. \tag{2.3}$$

We say that a dictionary $\mathcal{D}$ is (K, D)-unconditional with a constant U if for any $f = \sum_{i \in T} x_i g_i$ with $|T| \leq K$ inequality (2.3) holds.

The term *unconditional* in **A2** is justified by the following remark. The above definition of (K, D)-unconditional dictionary is equivalent to the following definition. Let $\mathcal{D}$ be such that any subsystem of D distinct elements $e_1, \dots, e_D$ from $\mathcal{D}$ is linearly independent and for any $A \subset [1, D]$ with $|A| \leq K$ and any coefficients $\{c_i\}$ we have

$$\|\sum_{i \in A} c_i e_i\| \leq U \|\sum_{i=1}^{D} c_i e_i\|.$$

It is convenient for us to use the following assumption **A3** introduced in [41] which is a corollary of assumptions **A1** and **A2**.

A3. We say that $f = \sum_{i \in T} x_i g_i$ has ℓ_1 incoherence property with parameters D, V, and r if for any $A \subset T$ and any Λ such that $A \cap \Lambda = \emptyset$, $|A| + |\Lambda| \leq D$ we have for any $\{c_i\}$

$$\sum_{i \in A} |x_i| \leq V |A|^r \|f_A - \sum_{i \in \Lambda} c_i g_i\|. \tag{2.4}$$

A dictionary $\mathcal{D}$ has ℓ_1 incoherence property with parameters K, D, V, and r if for any $A \subset B$, $|A| \le K$, $|B| \le D$ we have for any $\{c_i\}_{i \in B}$

$$\sum_{i \in A} |c_i| \le V |A|^r \| \sum_{i \in B} c_i g_i \|.$$

It is clear that **A1** and **A2** imply **A3** with $V = C_1 U$. Also, **A3** implies **A1** with $C_1 = V$ and **A2** with $U = V K^r$. Obviously, we can restrict ourselves to $r \le 1$.

We give a simple remark that widens the collection of dictionaries satisfying the above properties **A1**, **A2**, and **A3**.

Definition 2.1 Let $\mathcal{D}^1 = \{g_i^1\}$ and $\mathcal{D}^2 = \{g_i^2\}$ be countable dictionaries. We say that $\mathcal{D}^2$ D-dominates $\mathcal{D}^1$ (with a constant B) if for any set Λ, $|\Lambda| \le D$, of indices and any coefficients $\{c_i\}$ we have

$$\| \sum_{i \in \Lambda} c_i g_i^1 \| \le B \| \sum_{i \in \Lambda} c_i g_i^2 \|.$$

In such a case we write $\mathcal{D}^1 \prec \mathcal{D}^2$ or more specifically $\mathcal{D}^1 \le B\mathcal{D}^2$.

In the case $\mathcal{D}^1 \le E_1^{-1} \mathcal{D}^2$ and $\mathcal{D}^2 \le E_2 \mathcal{D}^1$ we say that $\mathcal{D}^1$ and $\mathcal{D}^2$ are D-equivalent (with constants E_1 and E_2) and write $\mathcal{D}^1 \approx \mathcal{D}^2$ or more specifically $E_1 \mathcal{D}^1 \le \mathcal{D}^2 \le E_2 \mathcal{D}^1$.

Proposition 2.1 *Assume $\mathcal{D}^1$ has one of the properties* **A1** *or* **A3**. *If $\mathcal{D}^2$ D-dominates $\mathcal{D}^1$ (with a constant B) then $\mathcal{D}^2$ has the same property as $\mathcal{D}^1$:* **A1** *with $C_1^2 = C_1^1 B$ or* **A3** *with $V^2 = V^1 B$.*

Proof In both cases **A1** and **A3** the proof is the same. We demonstrate the case **A3**. Let $f = \sum_{i \in T} x_i g_i^2$. Then by the **A3** property of $\mathcal{D}^1$ we have

$$\sum_{i \in A} |x_i| \le V^1 |A|^r \| \sum_{i \in A} x_i g_i^1 - \sum_{i \in \Lambda} c_i g_i^1 \| \le V^1 B |A|^r \| \sum_{i \in A} x_i g_i^2 - \sum_{i \in \Lambda} c_i g_i^2 \|.$$

□

Proposition 2.2 *Assume $\mathcal{D}^1$ has the property* **A2**. *If $\mathcal{D}^1$ and $\mathcal{D}^2$ are D-equivalent (with constants E_1 and E_2) then $\mathcal{D}^2$ has property* **A2** *with $U^2 = U^1 E_2 / E_1$.*

Proof Let $f = \sum_{i \in T} x_i g_i^2$. Then by $\mathcal{D}^1 \approx \mathcal{D}^2$ and the **A2** property of $\mathcal{D}^1$ we have

$$\| \sum_{i \in A} x_i g_i^2 - \sum_{i \in \Lambda} c_i g_i^2 \| \ge E_1 \| \sum_{i \in A} x_i g_i^1 - \sum_{i \in \Lambda} c_i g_i^1 \|$$
$$\ge E_1 (U^1)^{-1} \| \sum_{i \in A} x_i g_i^1 \| \ge (E_1/E_2)(U^1)^{-1} \| \sum_{i \in A} x_i g_i^2 \|.$$

□

We now proceed to main results of [18] and [41] on the WCGA with respect to redundant dictionaries. The following Theorem 2.1 from [41] in the case $q = 2$ was proved in [18].

Theorem 2.1 *Let X be a Banach space with $\rho(u) \le \gamma u^q$, $1 < q \le 2$. Suppose K-sparse f^ϵ satisfies* **A1**, **A2** *and $\|f_0 - f^\epsilon\| \le \epsilon$. Assume that $rq' \ge 1$. Then the WCGA with weakness parameter t applied to f_0 provides*

$$\|f_{C(t,\gamma,C_1)U^{q'}\ln(U+1)K^{rq'}}\| \le C\epsilon \quad \textit{for} \quad K + C(t,\gamma,C_1)U^{q'}\ln(U+1)K^{rq'} \le D$$

with an absolute constant C.

It was pointed out in [18] that Theorem 2.1 provides a corollary for Hilbert spaces that gives sufficient conditions somewhat weaker than the known RIP conditions on $\mathcal{D}$ for the Lebesgue-type inequality to hold. We formulate the corresponding definitions and results. Let $\mathcal{D}$ be the Riesz dictionary with depth D and parameter $\delta \in (0, 1)$. This class of dictionaries is a generalization of the class of classical Riesz bases. We give a definition in a general Hilbert space (see [40], p. 306).

Definition 2.2 A dictionary $\mathcal{D}$ is called the Riesz dictionary with depth D and parameter $\delta \in (0, 1)$ if, for any D distinct elements $e_1, \dots, e_D$ of the dictionary and any coefficients $a = (a_1, \dots, a_D)$, we have

$$(1-\delta)\|a\|_2^2 \le \|\sum_{i=1}^{D} a_i e_i\|^2 \le (1+\delta)\|a\|_2^2. \tag{2.5}$$

We denote the class of Riesz dictionaries with depth D and parameter $\delta \in (0, 1)$ by $R(D, \delta)$.

The term Riesz dictionary with depth D and parameter $\delta \in (0, 1)$ is another name for a dictionary satisfying the Restricted Isometry Property (RIP) with parameters D and δ. The following simple lemma holds:

Lemma 2.1 *Let $\mathcal{D} \in R(D, \delta)$ and let $e_j \in \mathcal{D}$, $j = 1, \dots, s$. For $f = \sum_{i=1}^{s} a_i e_i$ and $A \subset \{1, \dots, s\}$ denote*

$$S_A(f) := \sum_{i \in A} a_i e_i.$$

If $s \le D$ then

$$\|S_A(f)\|^2 \le (1+\delta)(1-\delta)^{-1}\|f\|^2.$$

Lemma 2.1 implies that if $\mathcal{D} \in R(D, \delta)$ then it is (D, D)-unconditional with a constant $U = (1+\delta)^{1/2}(1-\delta)^{-1/2}$.

Theorem 2.2 *Let X be a Hilbert space. Suppose K-sparse f^ϵ satisfies* **A2** *and $\|f_0 - f^\epsilon\| \le \epsilon$. Then the WOMP with weakness parameter t applied to f_0 provides*

$$\|f_{C(t,U)K}\| \le C\epsilon \quad for \quad K + C(t,U)K \le D$$

with an absolute constant C.

Theorem 2.2 implies the following corollaries:

Corollary 2.1 *Let X be a Hilbert space. Suppose any K-sparse f satisfies* **A2**. *Then the WOMP with weakness parameter t applied to f_0 provides*

$$\|f_{C(t,U)K}\| \le C\sigma_K(f_0, \mathcal{D}) \quad for \quad K + C(t,U)K \le D$$

with an absolute constant C.

Corollary 2.2 *Let X be a Hilbert space. Suppose $\mathcal{D} \in R(D, \delta)$. Then the WOMP with weakness parameter t applied to f_0 provides*

$$\|f_{C(t,\delta)K}\| \le C\sigma_K(f_0, \mathcal{D}) \quad for \quad K + C(t,\delta)K \le D$$

with an absolute constant C.

We emphasized in [18] that in Theorem 2.1 we impose our conditions on an individual function f^ϵ. It may happen that the dictionary does not have the Nikol'skii $\ell_1 X$ property and (K, D)-unconditionality but the given f_0 can be approximated by f^ϵ which does satisfy assumptions **A1** and **A2**. Even in the case of a Hilbert space the above results from [18] add something new to the study based on the RIP property of a dictionary. First of all, Theorem 2.2 shows that it is sufficient to impose assumption **A2** on f^ϵ in order to obtain exact recovery and the Lebesgue-type inequality results. Second, Corollary 2.1 shows that the condition **A2**, which is weaker than the RIP condition, is sufficient for exact recovery and the Lebesgue-type inequality results. Third, Corollary 2.2 shows that even if we impose our assumptions in terms of RIP we do not need to assume that $\delta < \delta_0$. In fact, the result works for all $\delta < 1$ with parameters depending on δ.

Theorem 2.1 follows from the combination of Theorems 2.3 and 2.4. In case $q = 2$ these theorems were proved in [18] and in general case $q \in (1, 2]$—in [41].

Theorem 2.3 *Let X be a Banach space with $\rho(u) \le \gamma u^q$, $1 < q \le 2$. Suppose for a given f_0 we have $\|f_0 - f^\epsilon\| \le \epsilon$ with K-sparse $f := f^\epsilon$ satisfying* **A3**. *Then for any $k \ge 0$ we have for $K + m \le D$*

$$\|f_m\| \le \|f_k\| \exp\left(-\frac{c_1(m-k)}{K^{rq'}}\right) + 2\epsilon, \quad q' := \frac{q}{q-1},$$

where $c_1 := \frac{t^{q'}}{2(16\gamma)^{\frac{1}{q-1}} V^{q'}}$.

In all theorems that follow we assume $rq' \ge 1$.

Theorem 2.4 *Let X be a Banach space with $\rho(u) \le \gamma u^q$, $1 < q \le 2$. Suppose K-sparse f^ϵ satisfies* **A1**, **A2** *and $\|f_0 - f^\epsilon\| \le \epsilon$. Then the WCGA with weakness parameter t applied to f_0 provides*

$$\|f_{C'U^{q'}\ln(U+1)K^{rq'}}\| \le CU\epsilon \quad \textit{for} \quad K + C'U^{q'}\ln(U+1)K^{rq'} \le D$$

with an absolute constant C *and* $C' = C_2(q)\gamma^{\frac{1}{q-1}}C_1^{q'}t^{-q'}$.

We formulate an immediate corollary of Theorem 2.4 with $\epsilon = 0$.

Corollary 2.3 *Let* X *be a Banach space with* $\rho(u) \le \gamma u^q$. *Suppose* K*-sparse* f *satisfies* **A1** *and* **A2**. *Then the WCGA with weakness parameter* t *applied to* f *recovers it exactly after* $C'U^{q'}\ln(U+1)K^{rq'}$ *iterations under condition* $K + C'U^{q'}\ln(U+1)K^{rq'} \le D$.

We formulate the versions of Theorem 2.4 with assumptions **A1** and **A2** replaced by a single assumption **A3** and replaced by two assumptions **A2** and **A3**. The corresponding modifications in the proofs go as in the proof of Theorem 2.3.

Theorem 2.5 *Let* X *be a Banach space with* $\rho(u) \le \gamma u^q$, $1 < q \le 2$. *Suppose* K*-sparse* f^ϵ *satisfies* **A3** *and* $\|f_0 - f^\epsilon\| \le \epsilon$. *Then the WCGA with weakness parameter* t *applied to* f_0 *provides*

$$\|f_{C(t,\gamma,q)V^{q'}\ln(VK)K^{rq'}}\| \le CVK^r\epsilon \quad \textit{for} \quad K + C(t,\gamma,q)V^{q'}\ln(VK)K^{rq'} \le D$$

with an absolute constant C *and* $C(t,\gamma,q) = C_2(q)\gamma^{\frac{1}{q-1}}t^{-q'}$.

Theorem 2.6 *Let* X *be a Banach space with* $\rho(u) \le \gamma u^q$, $1 < q \le 2$. *Suppose* K*-sparse* f^ϵ *satisfies* **A2**, **A3** *and* $\|f_0 - f^\epsilon\| \le \epsilon$. *Then the WCGA with weakness parameter* t *applied to* f_0 *provides*

$$\|f_{C(t,\gamma,q)V^{q'}\ln(U+1)K^{rq'}}\| \le CU\epsilon \quad \textit{for} \quad K + C(t,\gamma,q)V^{q'}\ln(U+1)K^{rq'} \le D$$

with an absolute constant C *and* $C(t,\gamma,q) = C_2(q)\gamma^{\frac{1}{q-1}}t^{-q'}$.

Theorems 2.5 and 2.3 imply the following analog of Theorem 2.1.

Theorem 2.7 *Let* X *be a Banach space with* $\rho(u) \le \gamma u^q$, $1 < q \le 2$. *Suppose* K*-sparse* f^ϵ *satisfies* **A3** *and* $\|f_0 - f^\epsilon\| \le \epsilon$. *Then the WCGA with weakness parameter* t *applied to* f_0 *provides*

$$\|f_{C(t,\gamma,q)V^{q'}\ln(VK)K^{rq'}}\| \le C\epsilon \quad \textit{for} \quad K + C(t,\gamma,q)V^{q'}\ln(VK)K^{rq'} \le D$$

with an absolute constant C *and* $C(t,\gamma,q) = C_2(q)\gamma^{\frac{1}{q-1}}t^{-q'}$.

The following edition of Theorems 2.1 and 2.7 is also useful in applications. It follows from Theorems 2.6 and 2.3.

Theorem 2.8 *Let* X *be a Banach space with* $\rho(u) \le \gamma u^q$, $1 < q \le 2$. *Suppose* K*-sparse* f^ϵ *satisfies* **A2**, **A3** *and* $\|f_0 - f^\epsilon\| \le \epsilon$. *Then the WCGA with weakness parameter* t *applied to* f_0 *provides*

$$\|f_{C(t,\gamma,q)V^{q'}\ln(U+1)K^{rq'}}\| \le C\epsilon \quad \textit{for} \quad K + C(t,\gamma,q)V^{q'}\ln(U+1)K^{rq'} \le D$$

with an absolute constant C and $C(t,\gamma,q) = C_2(q)\gamma^{\frac{1}{q-1}}t^{-q'}$.

3 Proofs

Proof of Theorem 2.3. We begin with a proof of Theorem 2.3.

Proof Let

$$f := f^\epsilon = \sum_{i\in T} x_i g_i, \quad |T| = K, \quad g_i \in \mathcal{D}.$$

Denote by T^m the set of indices of $g_j \in \mathcal{D}$, $j \in T$, picked by the WCGA after m iterations, $\Gamma^m := T \setminus T^m$. Denote by $A_1(\mathcal{D})$ the closure in X of the convex hull of the symmetrized dictionary $\mathcal{D}^{\pm} := \{\pm g, g \in D\}$. We will bound $\|f_m\|$ from above. Assume $\|f_{m-1}\| \ge \epsilon$. Let $m > k$. We bound from below

$$S_m := \sup_{\phi\in A_1(\mathcal{D})} |F_{f_{m-1}}(\phi)|.$$

Denote $A_m := \Gamma^{m-1}$. Then

$$S_m \ge F_{f_{m-1}}(f_{A_m}/\|f_{A_m}\|_1),$$

where $\|f_A\|_1 := \sum_{i\in A}|x_i|$. Next, by Lemma 6.9, p. 342, from [40] we obtain

$$F_{f_{m-1}}(f_{A_m}) = F_{f_{m-1}}(f^\epsilon) \ge \|f_{m-1}\| - \epsilon.$$

Thus

$$S_m \ge \|f_{A_m}\|_1^{-1}(\|f_{m-1}\| - \epsilon). \tag{3.1}$$

From the definition of the modulus of smoothness we have for any λ

$$\|f_{m-1} - \lambda\varphi_m\| + \|f_{m-1} + \lambda\varphi_m\| \le 2\|f_{m-1}\|\left(1 + \rho\left(\frac{\lambda}{\|f_{m-1}\|}\right)\right) \tag{3.2}$$

and by (1) from the definition of the WCGA and Lemma 6.10 from [40], p. 343, we get

$$|F_{f_{m-1}}(\varphi_m)| \ge t \sup_{g\in\mathcal{D}} |F_{f_{m-1}}(g)| =$$

$$t \sup_{\phi\in A_1(\mathcal{D})} |F_{f_{m-1}}(\phi)| = tS_m.$$

Then either $F_{f_{m-1}}(\varphi_m) \geq t S_m$ or $F_{f_{m-1}}(-\varphi_m) \geq t S_m$. Both cases are treated in the same way. We demonstrate the case $F_{f_{m-1}}(\varphi_m) \geq t S_m$. We have for $\lambda \geq 0$

$$\|f_{m-1} + \lambda \varphi_m\| \geq F_{f_{m-1}}(f_{m-1} + \lambda \varphi_m) \geq \|f_{m-1}\| + \lambda t S_m.$$

From here and from (3.2) we obtain

$$\|f_m\| \leq \|f_{m-1} - \lambda \varphi_m\| \leq \|f_{m-1}\| + \inf_{\lambda \geq 0}(-\lambda t S_m + 2\|f_{m-1}\| \rho(\lambda/\|f_{m-1}\|)).$$

We discuss here the case $\rho(u) \leq \gamma u^q$. Using (3.1) we get

$$\|f_m\| \leq \|f_{m-1}\| \left(1 - \frac{\lambda t}{\|f_{A_m}\|_1} + 2\gamma \frac{\lambda^q}{\|f_{m-1}\|^q}\right) + \frac{\epsilon \lambda t}{\|f_{A_m}\|_1}.$$

Let λ_1 be a solution of

$$\frac{\lambda t}{2\|f_{A_m}\|_1} = 2\gamma \frac{\lambda^q}{\|f_{m-1}\|^q}, \quad \lambda_1 = \left(\frac{t\|f_{m-1}\|^q}{4\gamma \|f_{A_m}\|_1}\right)^{\frac{1}{q-1}}.$$

Our assumption (2.4) gives

$$\|f_{A_m}\|_1 = \|(f^\epsilon - G_{m-1})_{A_m}\|_1 \leq V K^r \|f^\epsilon - G_{m-1}\|$$

$$\leq V K^r (\|f_0 - G_{m-1}\| + \|f_0 - f^\epsilon\|) \leq V K^r (\|f_{m-1}\| + \epsilon). \tag{3.3}$$

Specify

$$\lambda = \left(\frac{t\|f_{A_m}\|_1^{q-1}}{16\gamma (V K^r)^q}\right)^{\frac{1}{q-1}}.$$

Then, using $\|f_{m-1}\| \geq \epsilon$ we get

$$\left(\frac{\lambda}{\lambda_1}\right)^{q-1} = \frac{\|f_{A_m}\|_1^q}{4\|f_{m-1}\|^q (V K^r)^q} \leq 1$$

and obtain

$$\|f_m\| \leq \|f_{m-1}\| \left(1 - \frac{t^{q'}}{2(16\gamma)^{\frac{1}{q-1}} (V K^r)^{q'}}\right) + \frac{\epsilon t^{q'}}{(16\gamma)^{\frac{1}{q-1}} (V K^r)^{q'}}. \tag{3.4}$$

Denote $c_1 := \frac{t^{q'}}{2(16\gamma)^{\frac{1}{q-1}} V^{q'}}$. Then

$$\|f_m\| \leq \|f_k\| \exp\left(-\frac{c_1(m-k)}{K^{rq'}}\right) + 2\epsilon.$$

□

Proof of Theorem 2.4. We proceed to a proof of Theorem 2.4. Modifications of this proof which are in a style of the above proof of Theorem 2.3 give Theorems 2.5 and 2.6.

Proof We begin with a brief description of the structure of the proof. We are given f_0 and $f := f^\epsilon$ such that $\|f_0 - f\| \leq \epsilon$ and f is K-sparse satisfying **A1** and **A2**. We apply the WCGA to f_0 and control how many dictionary elements g_i from the representation of f

$$f := f^\epsilon := \sum_{i \in T} x_i g_i$$

are picked up by the WCGA after m iterations. As above denote by T^m the set of indices $i \in T$ such that g_i has been taken by the WCGA at one of the first m iterations.

Denote $\Gamma^m := T \setminus T^m$. It is clear that if $\Gamma^m = \emptyset$ then $\|f_m\| \leq \epsilon$ because in this case $f \in \Phi_m$.

Our analysis goes as follows. For a residual f_k we assume that Γ^k is nonempty. Then we prove that after $N(k)$ iterations we arrive at a residual $f_{k'}$, $k' = k + N(k)$, such that either

$$\|f_{k'}\| \leq CU\epsilon \tag{3.5}$$

or

$$|\Gamma^{k'}| < |\Gamma^k| - 2^{L-2} \tag{3.6}$$

with some natural number L. An important fact is that for the number $N(k)$ of iterations we have a bound

$$N(k) \leq \beta 2^{aL}, \quad a := rq'. \tag{3.7}$$

Next, we prove that if we begin with $k = 0$ and apply the above argument to the sequence of residuals $f_0, f_{k_1}, \ldots, f_{k_s}$, then after not more than $N := 2^{2a+1}\beta K^a$ iterations, we obtain either $\|f_N\| \leq CU\epsilon$ or $\Gamma^N = \emptyset$, which in turn implies that $\|f_N\| \leq \epsilon$.

We now proceed to the detailed argument. The following corollary of (2.3) will be often used: for $m \leq D - K$ and $A \subset \Gamma^m$ we have

$$\|f_A\| \leq U(\|f_m\| + \epsilon). \tag{3.8}$$

It follows from the fact that $f_A - f + G_m$ has the form $\sum_{i \in \Lambda} c_i g_i$ with Λ satisfying $|A| + |\Lambda| \leq D$, $A \cap \Lambda = \emptyset$, and from our assumption $\|f - f_0\| \leq \epsilon$.

The following lemma plays a key role in the proof. □

Lemma 3.1 *Let f satisfy* **A1** *and* **A2** *and let $A \subset \Gamma^k$ be nonempty. Denote $B := \Gamma^k \setminus A$. Then for any $m \in (k, D-K]$ we have either $\|f_{m-1}\| \le \epsilon$ or*

$$\|f_m\| \le \|f_{m-1}\|(1-u) + 2u(\|f_B\| + \epsilon), \tag{3.9}$$

where

$$u := c_1|A|^{-rq'}, \quad c_1 := \frac{t^{q'}}{2(16\gamma)^{\frac{1}{q-1}}(C_1U)^{q'}},$$

with C_1 and U from **A1** *and* **A2**.

Proof As above in the proof of Theorem 2.3 we bound S_m from below. It is clear that $S_m \ge 0$. Denote $A(m) := A \cap \Gamma^{m-1}$. Then

$$S_m \ge F_{f_{m-1}}(f_{A(m)}/\|f_{A(m)}\|_1).$$

Next,

$$F_{f_{m-1}}(f_{A(m)}) = F_{f_{m-1}}(f_{A(m)} + f_B - f_B).$$

Then $f_{A(m)} + f_B = f^\epsilon - f_\Lambda$ with $F_{f_{m-1}}(f_\Lambda) = 0$. Moreover, it is easy to see that $F_{f_{m-1}}(f^\epsilon) \ge \|f_{m-1}\| - \epsilon$. Therefore,

$$F_{f_{m-1}}(f_{A(m)} + f_B - f_B) \ge \|f_{m-1}\| - \epsilon - \|f_B\|.$$

Thus

$$S_m \ge \|f_{A(m)}\|_1^{-1} \max(0, \|f_{m-1}\| - \epsilon - \|f_B\|).$$

By (2.2) we get

$$\|f_{A(m)}\|_1 \le C_1|A(m)|^r\|f_{A(m)}\| \le C_1|A|^r\|f_{A(m)}\|.$$

Then

$$S_m \ge \frac{\|f_{m-1}\| - \|f_B\| - \epsilon}{C_1|A|^r\|f_{A(m)}\|}. \tag{3.10}$$

From the definition of the modulus of smoothness we have for any λ

$$\|f_{m-1} - \lambda\varphi_m\| + \|f_{m-1} + \lambda\varphi_m\| \le 2\|f_{m-1}\|\left(1 + \rho\left(\frac{\lambda}{\|f_{m-1}\|}\right)\right)$$

and by (1) from the definition of the WCGA and Lemma 6.10 from [40], p. 343, we get

$$|F_{f_{m-1}}(\varphi_m)| \ge t \sup_{g\in\mathcal{D}} |F_{f_{m-1}}(g)| =$$

$$t \sup_{\phi \in A_1(\mathcal{D})} |F_{f_{m-1}}(\phi)|.$$

From here we obtain

$$\|f_m\| \le \|f_{m-1}\| + \inf_{\lambda \ge 0}(-\lambda t S_m + 2\|f_{m-1}\|\rho(\lambda/\|f_{m-1}\|)).$$

We discuss here the case $\rho(u) \le \gamma u^q$. Using (3.10) we get for any $\lambda \ge 0$

$$\|f_m\| \le \|f_{m-1}\| \left(1 - \frac{\lambda t}{C_1|A|^r \|f_{A(m)}\|} + 2\gamma \frac{\lambda^q}{\|f_{m-1}\|^q}\right) + \frac{\lambda t(\|f_B\| + \epsilon)}{C_1|A|^r\|f_{A(m)}\|}.$$

Let λ_1 be a solution of

$$\frac{\lambda t}{2C_1|A|^r\|f_{A(m)}\|} = 2\gamma \frac{\lambda^q}{\|f_{m-1}\|^q}, \quad \lambda_1 = \left(\frac{t\|f_{m-1}\|^q}{4\gamma C_1|A|^r\|f_{A(m)}\|}\right)^{\frac{1}{q-1}}.$$

Inequality (3.8) gives

$$\|f_{A(m)}\| \le U(\|f_{m-1}\| + \epsilon).$$

Specify

$$\lambda = \left(\frac{t\|f_{A(m)}\|^{q-1}}{16\gamma C_1|A|^r U^q}\right)^{\frac{1}{q-1}}.$$

Then $\lambda \le \lambda_1$ and we obtain

$$\|f_m\| \le \|f_{m-1}\| \left(1 - \frac{t^{q'}}{2(16\gamma)^{\frac{1}{q-1}}(C_1U|A|^r)^{q'}}\right) + \frac{t^{q'}(\|f_B\| + \epsilon)}{(16\gamma)^{\frac{1}{q-1}}(C_1|A|^rU)^{q'}}. \quad (3.11)$$

□

For simplicity of notations we consider separately the case $|\Gamma^k| \ge 2$ and the case $|\Gamma^k| = 1$. We begin with the generic case $|\Gamma^k| \ge 2$. We apply Lemma 3.1 with different pairs A_j, B_j, which we now construct. Let n be a natural number such that

$$2^{n-1} < |\Gamma^k| \le 2^n.$$

For $j = 1, 2, \dots, n, n+1$ consider the following pairs of sets A_j, B_j: $A_{n+1} = \Gamma^k$, $B_{n+1} = \emptyset$; for $j \le n$, $A_j := \Gamma^k \setminus B_j$ with $B_j \subset \Gamma^k$ is such that $|B_j| \ge |\Gamma^k| - 2^{j-1}$ and for any set $J \subset \Gamma^k$ with $|J| \ge |\Gamma^k| - 2^{j-1}$ we have

$$\|f_{B_j}\| \le \|f_J\|.$$

We note that the above definition implies that $|A_j| \le 2^{j-1}$ and that if for some $Q \subset \Gamma^k$ we have

$$\|f_Q\| < \|f_{B_j}\| \quad \text{then} \quad |Q| < |\Gamma^k| - 2^{j-1}. \tag{3.12}$$

Set $B_0 := \Gamma^k$. Note that property (3.12) is obvious for $j = 0$.

Let $j_0 \in [1, n]$ be an index such that if $j_0 = 1$ then $B_1 \neq \Gamma^k$ and if $j_0 \ge 2$ then

$$B_1 = B_2 = \cdots = B_{j_0-1} = \Gamma^k, \quad B_{j_0} \neq \Gamma^k.$$

For a given $b > 1$, to be specified later, denote by $L := L(b)$ the index such that $(B_0 := \Gamma^k)$

$$\|f_{B_0}\| < b\|f_{B_{j_0}}\|,$$
$$\|f_{B_{j_0}}\| < b\|f_{B_{j_0+1}}\|,$$
$$\ldots$$
$$\|f_{B_{L-2}}\| < b\|f_{B_{L-1}}\|,$$
$$\|f_{B_{L-1}}\| \ge b\|f_{B_L}\|.$$

Then

$$\|f_{B_j}\| \le b^{L-1-j}\|f_{B_{L-1}}\|, \quad j = j_0, \ldots, L, \tag{3.13}$$

and

$$\|f_{B_0}\| = \cdots = \|f_{B_{j_0-1}}\| \le b^{L-j_0}\|f_{B_{L-1}}\|. \tag{3.14}$$

Clearly, $L \le n + 1$.

Define $m_0 := \cdots m_{j_0-1} := k$ and, inductively,

$$m_j = m_{j-1} + [\beta|A_j|^{rq'}], \quad j = j_0, \ldots, L,$$

where $[x]$ denotes the integer part of x. The parameter β is any, which satisfies the following inequalities with c_1 from Lemma 3.1

$$\beta \ge 1, \qquad e^{-c_1\beta/2} < 1/2, \qquad 16Ue^{-c_1\beta/2} < 1. \tag{3.15}$$

We note that the inequality $\beta \ge 1$ implies that

$$[\beta|A_j|^{rq'}] \ge \beta|A_j|^{rq'}/2.$$

Taking into account that $rq' \ge 1$ and $|A_j| \ge 1$ we obtain

$$m_j \ge m_{j-1} + 1.$$

At iterations from $m_{j-1}+1$ to m_j we apply Lemma 3.1 with $A = A_j$ and obtain from (3.9) that either $\|f_{m-1}\| \le \epsilon$ or

$$\|f_m\| \le \|f_{m-1}\|(1-u) + 2u(\|f_{B_j}\| + \epsilon), \quad u := c_1|A_j|^{-rq'}.$$

Using $1-u \le e^{-u}$ and $\sum_{k=0}^{\infty}(1-u)^k = 1/u$ we derive from here

$$\|f_{m_j}\| \le \|f_{m_{j-1}}\|e^{-c_1\beta/2} + 2(\|f_{B_j}\| + \epsilon). \tag{3.16}$$

We continue it up to $j = L$. Denote $\eta := e^{-c_1\beta/2}$. Then either $\|f_{m_L}\| \le \epsilon$ or

$$\|f_{m_L}\| \le \|f_k\|\eta^{L-j_0+1} + 2\sum_{j=j_0}^{L}(\|f_{B_j}\| + \epsilon)\eta^{L-j}.$$

We bound the $\|f_k\|$. It follows from the definition of f_k that $\|f_k\|$ is the error of best approximation of f_0 by the subspace Φ_k. Representing $f_0 = f + f_0 - f$ we see that $\|f_k\|$ is not greater than the error of best approximation of f by the subspace Φ_k plus $\|f_0 - f\|$. This implies $\|f_k\| \le \|f_{B_0}\| + \epsilon$. Therefore, we continue

$$\le (\|f_{B_0}\| + \epsilon)\eta^{L-j_0+1} + 2\sum_{j=j_0}^{L}(\|f_{B_{L-1}}\|(\eta b)^{L-j}b^{-1} + \epsilon\eta^{L-j})$$

$$\le b^{-1}\|f_{B_{L-1}}\|\left((\eta b)^{L-j_0+1} + 2\sum_{j=j_0}^{L}(\eta b)^{L-j}\right) + \frac{2\epsilon}{1-\eta}.$$

Our choice of β guarantees $\eta < 1/2$. Choose $b = \frac{1}{2\eta}$. Then

$$\|f_{m_L}\| \le \|f_{B_{L-1}}\|8e^{-c_1\beta/2} + 4\epsilon. \tag{3.17}$$

By (3.8) we get

$$\|f_{\Gamma^{m_L}}\| \le U(\|f_{m_L}\| + \epsilon) \le U(\|f_{B_{L-1}}\|8e^{-c_1\beta/2} + 5\epsilon).$$

If $\|f_{B_{L-1}}\| \le 10U\epsilon$ then by (3.17)

$$\|f_{m_L}\| \le CU\epsilon, \qquad C = 44.$$

If $\|f_{B_{L-1}}\| \ge 10U\epsilon$ then by our choice of β we have $16Ue^{-c_1\beta/2} < 1$ and

$$U(\|f_{B_{L-1}}\|8e^{-c_1\beta/2} + 5\epsilon) < \|f_{B_{L-1}}\|.$$

Therefore,

$$\|f_{\Gamma^{m_L}}\| < \|f_{B_{L-1}}\|.$$

This implies

$$|\Gamma^{m_L}| < |\Gamma^k| - 2^{L-2}.$$

In the above proof, our assumption $j_0 \le n$ is equivalent to the assumption that $B_n \ne \Gamma^k$. We now consider the case $B_n = \Gamma^k$ and, therefore, $B_j = \Gamma^k$, $j = 0, 1, \dots, n$. This means that $\|f_{\Gamma^k}\| \le \|f_J\|$ for any J with $|J| \ge |\Gamma^k| - 2^{n-1}$. Therefore, if for some $Q \subset \Gamma^k$ we have

$$\|f_Q\| < \|f_{\Gamma^k}\| \quad \text{then} \quad |Q| < |\Gamma^k| - 2^{n-1}. \tag{3.18}$$

In this case we set $m_0 := k$ and

$$m_1 := k + [\beta|\Gamma^k|^{rq'}].$$

Then by Lemma 3.1 with $A = \Gamma^k$ we obtain as in (3.16)

$$\|f_{m_1}\| \le \|f_{m_0}\|e^{-c_1\beta/2} + 2\epsilon \le \|f_{\Gamma^k}\|e^{-c_1\beta/2} + 3\epsilon. \tag{3.19}$$

By (3.8) we get

$$\|f_{\Gamma^{m_1}}\| \le U(\|f_{m_1}\| + \epsilon) \le U(\|f_{\Gamma^k}\|e^{-c_1\beta/2} + 4\epsilon).$$

If $\|f_{\Gamma^k}\| \le 8U\epsilon$ then by (3.19)

$$\|f_{m_1}\| \le 7U\epsilon.$$

If $\|f_{\Gamma^k}\| \ge 8U\epsilon$ then by our choice of β we have $2Ue^{-c_1\beta/2} < 1$ and

$$\|f_{\Gamma^{m_1}}\| \le U(\|f_{\Gamma^k}\|e^{-c_1\beta/2} + 4\epsilon) < \|f_{\Gamma^k}\|. \tag{3.20}$$

This implies

$$|\Gamma^{m_1}| < |\Gamma^k| - 2^{n-1}.$$

It remains to consider the case $|\Gamma^k| = 1$. By the above argument, where we used Lemma 3.1 with $A = \Gamma^k$ we obtain (3.20). In the case $|\Gamma^k| = 1$ inequality (3.20) implies $\Gamma^{m_1} = \emptyset$, which completes the proof in this case.

We now complete the proof of Theorem 2.4. We begin with f_0 and apply the above argument (with $k = 0$). As a result we either get the required inequality or we reduce the cardinality of support of f from $|T| = K$ to $|\Gamma^{m_{L_1}}| < |T| - 2^{L_1-2}$ (the WCGA picks up at least 2^{L_1-2} dictionary elements g_i from the representation of f), $m_{L_1} \le \beta 2^{aL_1}$, $a := rq'$. We continue the process and build a sequence m_{L_j} such that $m_{L_j} \le \beta 2^{aL_j}$ and after m_{L_j} iterations we reduce the support by at least

2^{L_j-2}. We also note that $m_{L_j} \le \beta 2^{2a} K^a$. We continue this process till the following inequality is satisfied for the first time

$$m_{L_1} + \cdots + m_{L_s} \ge 2^{2a} \beta K^a. \tag{3.21}$$

Then, clearly,

$$m_{L_1} + \cdots + m_{L_s} \le 2^{2a+1} \beta K^a.$$

Using the inequality

$$(a_1 + \cdots + a_s)^\theta \le a_1^\theta + \cdots + a_s^\theta, \quad a_j \ge 0, \quad \theta \in (0, 1]$$

we derive from (3.21)

$$\begin{aligned} 2^{L_1-2} + \cdots + 2^{L_s-2} &\ge \left(2^{a(L_1-2)} + \cdots + 2^{a(L_s-2)}\right)^{\frac{1}{a}} \\ &\ge 2^{-2}\left(2^{aL_1} + \cdots + 2^{aL_s}\right)^{\frac{1}{a}} \\ &\ge 2^{-2}\left((\beta)^{-1}(m_{L_1} + \cdots + m_{L_s})\right)^{\frac{1}{a}} \ge K. \end{aligned}$$

Thus, after not more than $N := 2^{2a+1}\beta K^a$ iterations we either get the required inequality or we recover f exactly (the WCGA picks up all the dictionary elements g_i from the representation of f) and then $\|f_N\| \le \|f_0 - f\| \le \epsilon$.

Proof of Theorem 2.5. We begin with a version of Lemma 3.1 that is used in this proof.

Lemma 3.2 *Let f satisfy* **A3** *and let $A \subset \Gamma^k$ be nonempty. Denote $B := \Gamma^k \setminus A$. Then for any $m \in (k, D-K]$ we have either $\|f_{m-1}\| \le \epsilon$ or*

$$\|f_m\| \le \|f_{m-1}\|(1-u) + 2u(\|f_B\| + \epsilon), \tag{3.22}$$

where

$$u := c_2|A|^{-rq'}, \qquad c_2 := \frac{t^{q'}}{2(16\gamma)^{\frac{1}{q-1}} V^{q'}},$$

with r and V from **A3**.

Proof The proof is a combination of proofs of Theorem 2.3 and Lemma 3.1. As in the proof of Lemma 3.1 we denote $A(m) := A \cap \Gamma^{m-1}$ and get

$$S_m \ge \|f_{A(m)}\|_1^{-1} \max(0, \|f_{m-1}\| - \epsilon - \|f_B\|).$$

From here in the same way as in the proof of Theorem 2.3 we obtain for any $\lambda \ge 0$

$$\|f_m\| \le \|f_{m-1}\| \left(1 - \frac{\lambda t}{\|f_{A(m)}\|_1} + 2\gamma \frac{\lambda^q}{\|f_{m-1}\|^q}\right) + \frac{\lambda t(\|f_B\| + \epsilon)}{\|f_{A(m)}\|_1}.$$

Using definition of $A(m)$ we bound by **A3**

$$\|f_{A(m)}\|_1 = \sum_{i\in A(m)} |x_i| \le V|A(m)|^r \|f_{A(m)} + f - f_{A(m)} - G_{m-1}\|$$

$$\le V|A|^r\|f - G_{m-1}\| \le V|A|^r(\|f_{m-1}\| + \epsilon).$$

This inequality is a variant of inequality (3.3) with K replaced by $|A|$. Arguing as in the proof of Theorem 2.3 with K replaced by $|A|$ we obtain the required inequality, which is the corresponding modification (K is replaced by $|A|$ and ϵ is replaced by $\|f_B\| + \epsilon$) of (3.4).

The rest of the proof repeats the proof of Theorem 2.4 with the use of Lemma 3.2 instead of Lemma 3.1 and with the use of the fact that **A3** implies **A2** with $U = VK^r \le VK$. □

Proof of Theorem 2.6. This proof repeats the proof of Theorem 2.4 with the use of Lemma 3.2 instead Lemma 3.1.

4 Examples

In this section, following [41], we discuss applications of Theorems from Sect. 2 for specific dictionaries $\mathcal{D}$. Mostly, $\mathcal{D}$ will be a basis Ψ for X. Because of that we use m instead of K in the notation of sparse approximation. In some of our examples, we take $X = L_p, 2 \le p < \infty$. Then it is known that $\rho(u) \le \gamma u^2$ with $\gamma = (p-1)/2$. In some other examples, we take $X = L_p, 1 < p \le 2$. Then it is known that $\rho(u) \le \gamma u^p$, with $\gamma = 1/p$.

Proposition 4.1 *Let Ψ be a uniformly bounded orthogonal system normalized in $L_p(\Omega)$, $2 \le p < \infty$, Ω is a bounded domain. Then we have*

$$\|f_{C(t,p,\Omega)m\ln(m+1)}\|_p \le C\sigma_m(f_0,\Psi)_p. \tag{4.1}$$

The proof of Proposition 4.1 is based on Theorem 2.7.

Corollary 4.1 *Let Ψ be the normalized in L_p, $2 \le p < \infty$, real d-variate trigonometric system. Then Proposition 4.1 applies and gives for any $f_0 \in L_p$*

$$\|f_{C(t,p,d)m\ln(m+1)}\|_p \le C\sigma_m(f_0,\Psi)_p. \tag{4.2}$$

We note that (4.2) provides some progress in Open Problem 7.1 (p. 91) from [38].

Proposition 4.2 *Let Ψ be a uniformly bounded orthogonal system normalized in $L_p(\Omega)$, $1 < p \le 2$, Ω is a bounded domain. Then we have*

$$\|f_{C(t,p,\Omega)m^{p'-1}\ln(m+1)}\|_p \le C\sigma_m(f_0,\Psi)_p. \tag{4.3}$$

The proof of Proposition 4.2 is based on Theorem 2.7.

Corollary 4.2 *Let Ψ be the normalized in L_p, $1 < p \le 2$, real d-variate trigonometric system. Then Proposition 4.2 applies and gives for any $f_0 \in L_p$*

$$\|f_{C(t,p,d)m^{p'-1}\ln(m+1)}\|_p \le C\sigma_m(f_0, \Psi)_p. \tag{4.4}$$

Proposition 4.3 *Let Ψ be the normalized in L_p, $2 \le p < \infty$, multivariate Haar basis $\mathcal{H}_p^d = \mathcal{H}_p \times \cdots \times \mathcal{H}_p$. Then*

$$\|f_{C(t,p,d)m^{2/p'}}\|_p \le C\sigma_m(f_0, \mathcal{H}_p^d)_p. \tag{4.5}$$

The proof of Proposition 4.3 is based on Theorem 2.4. Inequality (4.5) provides some progress in Open Problem 7.2 (p. 91) from [38] in the case $2 < p < \infty$.

Proposition 4.4 *Let Ψ be the normalized in L_p, $1 < p \le 2$, univariate Haar basis $\mathcal{H}_p = \{H_{I,p}\}_I$, where $H_{I,p}$ is the Haar function indexed by dyadic intervals of support of $H_{I,p}$ (we index function 1 by $[0, 1]$ and the first Haar function by $[0, 1)$). Then*

$$\|f_{C(t,p)m}\|_p \le C\sigma_m(f_0, \mathcal{H}_p)_p. \tag{4.6}$$

The proof of Proposition 4.4 is based on Theorem 2.8. Inequality (4.6) solves the Open Problem 7.2 (p. 91) from [38] in the case $1 < p \le 2$.

Proposition 4.5 *Let X be a Banach space with $\rho(u) \le \gamma u^2$. Assume that Ψ is a normalized Schauder basis for X. Then*

$$\|f_{C(t,X,\Psi)m^2\ln m}\| \le C\sigma_m(f_0, \Psi). \tag{4.7}$$

The proof of Proposition 4.5 is based on Theorem 2.7. We note that the above bound still works if we replace the assumption that Ψ is a Schauder basis by the assumption that a dictionary $\mathcal{D}$ is $(1, D)$-unconditional with constant U. Then we obtain

$$\|f_{C(t,\gamma,U)K^2\ln K}\| \le C\sigma_K(f_0, \Psi), \quad \text{for} \quad K + C(t, \gamma, U)K^2 \ln K \le D.$$

Proposition 4.6 *Let X be a Banach space with $\rho(u) \le \gamma u^q$, $1 < q \le 2$. Assume that Ψ is a normalized Schauder basis for X. Then*

$$\|f_{C(t,X,\Psi)m^{q'}\ln m}\| \le C\sigma_m(f_0, \Psi). \tag{4.8}$$

The proof of Proposition 4.6 is based on Theorem 2.7. We note that the above bound still works if we replace the assumption that Ψ is a Schauder basis by the assumption that a dictionary $\mathcal{D}$ is $(1, D)$-unconditional with constant U. Then we obtain

$$\|f_{C(t,\gamma,q,U)K^{q'}\ln K}\| \le C\sigma_K(f_0,\mathcal{D}), \quad \text{for} \quad K + C(t,\gamma,q,U)K^{q'}\ln K \le D.$$

We now discuss application of general results of Sect. 2 to quasi-greedy bases. We begin with a brief introduction to the theory of quasi-greedy bases. Let X be an infinite-dimensional separable Banach space with a norm $\|\cdot\| := \|\cdot\|_X$ and let $\Psi := \{\psi_m\}_{m=1}^{\infty}$ be a normalized basis for X. The concept of quasi-greedy basis was introduced in [17].

Definition 4.1 The basis Ψ is called quasi-greedy if there exists some constant C such that

$$\sup_m \|G_m(f,\Psi)\| \le C\|f\|.$$

Subsequently, Wojtaszczyk [46] proved that these are precisely the bases for which the TGA merely converges, i.e.,

$$\lim_{n\to\infty} G_n(f) = f.$$

The following lemma is from [8] (see also [10] and [11] for further discussions).

Lemma 4.1 *Let Ψ be a quasi-greedy basis of X. Then for any finite set of indices Λ we have for all $f \in X$*

$$\|S_\Lambda(f,\Psi)\| \le C\ln(|\Lambda|+1)\|f\|,$$

where for $f = \sum_{k=1}^{\infty} c_k(f)\psi_k$ we denote $S_\Lambda(f,\Psi) := \sum_{k\in\Lambda} c_k(f)\psi_k$.

We now formulate a result about quasi-greedy bases in L_p spaces. The following theorem is from [45]. We note that in the case $p = 2$ Theorem 4.1 was proved in [46]. Some notations first. For a given element $f \in X$ we consider the expansion

$$f = \sum_{k=1}^{\infty} c_k(f)\psi_k$$

and the decreasing rearrangement of its coefficients

$$|c_{k_1}(f)| \ge |c_{k_2}(f)| \ge \dots .$$

Denote

$$a_n(f) := |c_{k_n}(f)|.$$

Theorem 4.1 *Let $\Psi = \{\psi_m\}_{m=1}^{\infty}$ be a quasi-greedy basis of the L_p space, $1 < p < \infty$. Then for each $f \in X$ we have*

$$C_1(p) \sup_n n^{1/p} a_n(f) \le \|f\|_p \le C_2(p) \sum_{n=1}^{\infty} n^{-1/2} a_n(f), \quad 2 \le p < \infty;$$

$$C_3(p) \sup_n n^{1/2} a_n(f) \le \|f\|_p \le C_4(p) \sum_{n=1}^{\infty} n^{1/p-1} a_n(f), \quad 1 < p \le 2.$$

Proposition 4.7 *Let Ψ be a normalized quasi-greedy basis for L_p, $2 \le p < \infty$. Then*

$$\|f_{C(t,p)m^{2(1-1/p)}\ln(m+1)}\| \le C\sigma_m(f_0, \Psi). \tag{4.9}$$

The proof of Proposition 4.7 is based on Theorem 2.7.

Proposition 4.8 *Let Ψ be a normalized quasi-greedy basis for L_p, $1 < p \le 2$. Then*

$$\|f_{C(t,p)m^{p'/2}\ln(m+1)}\| \le C\sigma_m(f_0, \Psi). \tag{4.10}$$

The proof of Proposition 4.8 is based on Theorem 2.7.

Proposition 4.9 *Let Ψ be a normalized uniformly bounded orthogonal quasi-greedy basis for L_p, $2 \le p < \infty$ (for existence of such bases see [23]). Then*

$$\|f_{C(t,p,\Psi)m\ln\ln(m+3)}\|_p \le C\sigma_m(f_0, \Psi)_p. \tag{4.11}$$

The proof of Proposition 4.9 is based on Theorem 2.8.

Proposition 4.10 *Let Ψ be a normalized uniformly bounded orthogonal quasi-greedy basis for L_p, $1 < p \le 2$ (for existence of such bases see [23]). Then*

$$\|f_{C(t,p,\Psi)m^{p'/2}\ln\ln(m+3)}\|_p \le C\sigma_m(f_0, \Psi)_p. \tag{4.12}$$

The proof of Proposition 4.10 is based on Theorem 2.8.

Proposition 4.4 is the first result about almost greedy bases with respect to WCGA in Banach spaces. It shows that the univariate Haar basis is an almost greedy basis with respect to the WCGA in the L_p spaces for $1 < p \le 2$. Proposition 4.1 shows that uniformly bounded orthogonal bases are ϕ-greedy bases with respect to WCGA with $\phi(u) = C(t, p, \Omega)\ln(u+1)$ in the L_p spaces for $2 \le p < \infty$. We do not know if these bases are almost greedy with respect to WCGA. They are good candidates for that.

It is known (see [40], p. 17) that the univariate Haar basis is a greedy basis with respect to TGA for all L_p, $1 < p < \infty$. Proposition 4.3 only shows that it is a ϕ-greedy basis with respect to WCGA with $\phi(u) = C(t, p)u^{1-2/p}$ in the L_p spaces for $2 \le p < \infty$. It is much weaker than the corresponding results for the $\mathcal{H}_p$, $1 < p \le 2$, and for the trigonometric system, $2 \le p < \infty$ (see Corollary 4.1). We do not know if this result on the Haar basis can be substantially improved. At the level of our today's technique we can observe that the Haar basis is ideal (greedy basis) for the TGA in L_p, $1 < p < \infty$, almost ideal (almost greedy basis) for the WCGA in L_p, $1 < p \le 2$, and that the trigonometric system is very good for the WCGA in L_p, $2 \le p < \infty$.

Corollary 4.2 shows that our results for the trigonometric system in $L_p, 1 < p < 2$, are not as strong as for $2 \le p < \infty$. We do not know if it is a lack of appropriate technique or it reflects the nature of the WCGA with respect to the trigonometric system.

We note that Propositions 2.1 and 2.2 can be used to formulate the above Propositions for a more general bases. In these cases, we use Propositions 2.1 and 2.2 with $D = \infty$. In Propositions 4.1, 4.2, 4.7, and 4.8, where we used Theorem 2.7, we can replace the basis Ψ by a basis Φ, which dominates the basis Ψ. In Propositions 4.3, 4.4, 4.9, and 4.10, where we used either Theorem 2.4 or 2.8, we can replace the basis Ψ by a basis Φ, which is equivalent to the basis Ψ.

It is interesting to compare Theorem 2.3 with the following known result. The following theorem provides rate of convergence (see [40], p. 347). We denote by $A_1(\mathcal{D})$ the closure in X of the convex hull of the symmetrized dictionary $\mathcal{D}^{\pm} := \{\pm g : g \in \mathcal{D}\}$.

Theorem 4.2 *Let X be a uniformly smooth Banach space with modulus of smoothness $\rho(u) \le \gamma u^q$, $1 < q \le 2$. Take a number $\epsilon \ge 0$ and two elements f_0, f^ϵ from X such that*

$$\|f_0 - f^\epsilon\| \le \epsilon, \quad f^\epsilon/A(\epsilon) \in A_1(\mathcal{D}),$$

with some number $A(\epsilon) > 0$. Then, for the WCGA we have

$$\|f_m^{c,t}\| \le \max\left(2\epsilon, C(q,\gamma)(A(\epsilon)+\epsilon)t^{-1}(1+m)^{1/q-1}\right).$$

Both Theorem 4.2 and Theorem 2.3 provide stability of the WCGA with respect to noise. In order to apply them for noisy data we interpret f_0 as a noisy version of a signal and f^ϵ as a noiseless version of a signal. Then, assumption $f^\epsilon/A(\epsilon) \in A_1(\mathcal{D})$ describes our smoothness assumption on the noiseless signal and assumption $f^\epsilon \in \Sigma_K(\mathcal{D})$ describes our structural assumption on the noiseless signal. In fact, Theorem 4.2 simultaneously takes care of two issues: noisy data and approximation in an interpolation space. Theorem 4.2 can be applied for approximation of f_0 under assumption that f_0 belongs to one of the interpolation spaces between X and the space generated by the $A_1(\mathcal{D})$-norm (atomic norm).

5 Sparse Trigonometric Approximation

Sparse trigonometric approximation of periodic functions began by the paper of Stechkin [28], who used it in the criterion for absolute convergence of trigonometric series. Ismagilov [14] found nontrivial estimates for m-term approximation of functions with singularities of the type $|x|$ and gave interesting and important applications to the widths of Sobolev classes. He used a deterministic method based on number theoretical constructions. His method was developed by Maiorov [19], who used a method based on Gaussian sums. Further strong results were obtained in [7] with the help of a nonconstructive result from finite-dimensional Banach spaces due to Gluskin [12]. Other powerful nonconstructive method, which is based on a probabilistic argument, was used by Makovoz [20] and by Belinskii [2]. Different methods were created in [16, 31, 36, 42] for proving lower bounds for function classes. It was discovered in [9] and [39] that greedy algorithms can be used for constructive m-term approximation with respect to the trigonometric system. We demonstrate in [43] how greedy algorithms can be used to prove optimal or best-known upper bounds for m-term approximation of classes of functions with mixed smoothness. It is a simple and powerful method of proving upper bounds. The reader can find a detailed study of m-term approximation of classes of functions with mixed smoothness, including small smoothness, in the paper [24] by Romanyuk and in [43, 44]. We note that in the case $2 < p < \infty$ the upper bounds in [24] are not constructive.

We discuss some approximation problems for classes of functions with mixed smoothness. We define these classes momentarily. We will begin with the case of univariate periodic functions. Let for $r > 0$

$$F_r(x) := 1 + 2\sum_{k=1}^{\infty} k^{-r}\cos(kx - r\pi/2) \tag{5.1}$$

and

$$W_p^r := \{f : f = \varphi * F_r, \quad \|\varphi\|_p \le 1\}. \tag{5.2}$$

It is well known that for $r > 1/p$ the class W_p^r is embedded into the space of continuous functions $C(\mathbb{T})$. In a particular case of W_1^1 we also have embedding into $C(\mathbb{T})$.

In the multivariate case for $\mathbf{x} = (x_1, \dots, x_d)$ denote

$$F_r(\mathbf{x}) := \prod_{j=1}^{d} F_r(x_j)$$

and

$$\mathbf{W}_p^r := \{f : f = \varphi * F_r, \quad \|\varphi\|_p \le 1\}.$$

For $f \in \mathbf{W}_p^r$ we will denote $f^{(r)} := \varphi$ where φ is such that $f = \varphi * F_r$.

The main results of Sect. 2 of [43] are the following two theorems. We use the notation $\beta := \beta(q,p) := 1/q - 1/p$ and $\eta := \eta(q) := 1/q - 1/2$. In the case of trigonometric system $\mathcal{T}^d$ we drop it from the notation:

$$\sigma_m(\mathbf{W})_p := \sigma_m(\mathbf{W}, \mathcal{T}^d)_p.$$

Theorem 5.1 *We have*

$$\sigma_m(\mathbf{W}^r_q)_p \asymp \begin{cases} m^{-r+\beta}(\log m)^{(d-1)(r-2\beta)}, & 1 < q \le p \le 2, \quad r > 2\beta, \\ m^{-r+\eta}(\log m)^{(d-1)(r-2\eta)}, & 1 < q \le 2 \le p < \infty, \quad r > 1/q, \\ m^{-r}(\log m)^{r(d-1)}, & 2 \le q \le p < \infty, \quad r > 1/2. \end{cases}$$

Theorem 5.2 *We have*

$$\sigma_m(\mathbf{W}^r_q)_\infty \ll \begin{cases} m^{-r+\eta}(\log m)^{(d-1)(r-2\eta)+1/2}, & 1 < q \le 2, \quad r > 1/q, \\ m^{-r}(\log m)^{r(d-1)+1/2}, & 2 \le q < \infty, \quad r > 1/2. \end{cases}$$

The case $1 < q \le p \le 2$ in Theorem 5.1, which corresponds to the first line, was proved in [31] (see also [30], Ch. 4). The proofs from [31] and [30] are constructive. In [43] we concentrate on the case $p \ge 2$. We use recently developed techniques on greedy approximation in Banach spaces to prove Theorems 5.1 and 5.2. It is important that greedy approximation allows us not only to prove the above theorems but also to provide a constructive way for building the corresponding m-term approximants. We give a precise formulation from [43].

Theorem 5.3 *For $p \in (1,\infty)$ and $\mu > 0$ there exist constructive methods $A_m(f,p,\mu)$, which provide for $f \in \mathbf{W}^r_q$ an m-term approximation such that*

$$\|f - A_m(f,p,\mu)\|_p$$

$$\ll \begin{cases} m^{-r+\beta}(\log m)^{(d-1)(r-2\beta)}, & 1 < q \le p \le 2, \quad r > 2\beta + \mu, \\ m^{-r+\eta}(\log m)^{(d-1)(r-2\eta)}, & 1 < q \le 2 \le p < \infty, \quad r > 1/q + \mu, \\ m^{-r}(\log m)^{r(d-1)}, & 2 \le q \le p < \infty, \quad r > 1/2 + \mu. \end{cases}$$

Similar modification of Theorem 5.2 holds for $p = \infty$. We do not have matching lower bounds for the upper bounds in Theorem 5.2 in the case of approximation in the uniform norm L_∞.

As a direct corollary of Theorems 1.1 and 5.1 we obtain the following result.

Theorem 5.4 *Let $p \in [2,\infty)$. Apply the WCGA with weakness parameter $t \in (0,1]$ to $f \in L_p$ with respect to the real trigonometric system $\mathcal{RT}^d_p$. If $f \in \mathbf{W}^r_q$, then we have*

$$\|f_m\|_p \ll \begin{cases} m^{-r+\eta}(\log m)^{(d-1)(r-2\eta)+r-\eta}, & 1 < q \le 2, \quad r > 1/q, \\ m^{-r}(\log m)^{rd}, & 2 \le q < \infty, \quad r > 1/2. \end{cases}$$

The reader can find results on best m-term approximation as well as results on constructive m-term approximation for Besov-type classes in the paper [43]. Some new results in the case of small smoothness are contained in [44].

6 Tensor Product Approximations

In the paper [1] we study multilinear approximation (nonlinear tensor product approximation) of functions. For a function $f(x_1, \dots, x_d)$ denote

$$\Theta_M(f)_X := \inf_{\{u^i_j\}, j=1,\dots,M, i=1,\dots,d} \|f(x_1, \dots, x_d) - \sum_{j=1}^{M} \prod_{i=1}^{d} u^i_j(x_i)\|_X$$

and for a function class F define

$$\Theta_M(F)_X := \sup_{f\in F} \Theta_M(f)_X.$$

In this section we use the notation M instead of m for the number of terms in an approximant because this notation is a standard one in the area. In the case $X = L_p$ we write p instead of L_p in the notation. In other words we are interested in studying M-term approximations of functions with respect to the dictionary

$$\Pi^d := \{g(x_1, \dots, x_d) : g(x_1, \dots, x_d) = \prod_{i=1}^{d} u^i(x_i)\}$$

where $u^i(x_i)$ are arbitrary univariate functions. We discuss the case of 2π-periodic functions of d variables and approximate them in the L_p spaces. Denote by Π^d_p the normalized in L_p dictionary Π^d of 2π-periodic functions. We say that a dictionary $\mathcal{D}$ has a tensor product structure if all its elements have a form of products $u^1(x_1)\cdots u^d(x_d)$ of univariate functions $u^i(x_i)$, $i = 1, \dots, d$. Then any dictionary with tensor product structure is a subset of Π^d. The classical example of a dictionary with tensor product structure is the d-variate trigonometric system $\{e^{i(\mathbf{k},\mathbf{x})}\}$. Other examples include the hyperbolic wavelets and the hyperbolic wavelet-type system $\mathcal{U}^d$ defined in [35].

Modern problems in approximation, driven by applications in biology, medicine, and engineering, are being formulated in very high dimensions, which brings to the fore new phenomena. For instance, partial differential equations in a phase space of large spacial dimensions (e.g., Schrödinger and Fokker–Plank equations) are very important in applications. It is known (see, for instance, [4]) that such equations involving large number of spacial variables pose a serious computational challenge because of the so-called *curse of dimensionality*, which is caused by the use of classical notions of smoothness as the regularity characteristics of the solution. The authors of [4] show that replacing the classical smoothness assumptions by structural assumptions in terms of sparsity with respect to the dictionary Π^d, they overcome the above computational challenge. They prove that the solutions of certain high-dimensional equations inherit sparsity, based on tensor product decompositions, from given data. Thus, our algorithms, which provide good sparse approximation with respect to Π^d for individual functions might be useful in applications to PDEs of the above type. The nonlinear tensor product approximation is very important in numerical applications. We refer the reader to the monograph [13] which presents the state of the art on the topic. Also, the reader can find a very recent discussion of related results in [27].

In the case $d = 2$ the multilinear approximation problem is the classical problem of bilinear approximation. In the case of approximation in the L_2 space the bilinear approximation problem is closely related to the problem of singular value decomposition (also called Schmidt expansion) of the corresponding integral operator with the kernel $f(x_1, x_2)$. There are known results on the rate of decay of errors of best bilinear approximation in L_p under different smoothness assumptions on f. We only mention some known results for classes of functions with mixed smoothness. We study the classes $\mathbf{W}^r_q$ of functions with bounded mixed derivative defined above in Sect. 5.

The problem of estimating $\Theta_M(f)_2$ in case $d = 2$ (best M-term bilinear approximation in L_2) is a classical one and was considered for the first time by E. Schmidt [26] in 1907. For many function classes F an asymptotic behavior of $\Theta_M(F)_p$ is known. For instance, the relation

$$\Theta_M(\mathbf{W}^r_q)_p \asymp M^{-2r+(1/q-\max(1/2,1/p))_+} \tag{6.1}$$

for $r > 1$ and $1 \leq q \leq p \leq \infty$ follows from more general results in [32]. In the case $d > 2$ almost nothing is known. There is (see [33]) an upper estimate in the case $q = p = 2$

$$\Theta_M(\mathbf{W}^r_2)_2 \ll M^{-rd/(d-1)}. \tag{6.2}$$

Results of [1] are around the bound (6.2). First of all we discuss the lower bound matching the upper bound (6.2). In the case $d = 2$ the lower bound

$$\Theta_M(W^r_p)_p \gg M^{-2r}, \qquad 1 \leq p \leq \infty, \tag{6.3}$$

follows from more general results in [32] (see (6.1) above). A stronger result

$$\Theta_M(W^r_\infty)_1 \gg M^{-2r} \tag{6.4}$$

follows from Theorem 1.1 in [34].

We could not prove the lower bound matching the upper bound (6.2) for $d > 2$. Instead, we proved in [1] a weaker lower bound. For a function $f(x_1, \dots, x_d)$ denote

$$\Theta_M^b(f)_X := \inf_{\{u_j^i\}, \|u_j^i\|_X \le b\|f\|_X^{1/d}} \|f(x_1, \dots, x_d) - \sum_{j=1}^{M} \prod_{i=1}^{d} u_j^i(x_i)\|_X$$

and for a function class F define

$$\Theta_M^b(F)_X := \sup_{f \in F} \Theta_M^b(f)_X.$$

In [1] we proved the following lower bound (see Corollary 2.2)

$$\Theta_M^b(\mathbf{W}_\infty^r)_1 \gg (M \ln M)^{-\frac{rd}{d-1}}.$$

This lower bound indicates that probably the exponent $\frac{rd}{d-1}$ is the right one in the power decay of the $\Theta_M(\mathbf{W}_p^r)_p$.

Second, we discuss some upper bounds which extend the bound (6.2). The relation (6.1) shows that for $2 \le p \le \infty$ in the case $d = 2$ one has

$$\Theta_M(\mathbf{W}_2^r)_p \ll M^{-2r}. \tag{6.5}$$

In [1] we extend (6.5) for $d > 2$.

Theorem 6.1 *Let $2 \le p < \infty$ and $r > (d-1)/d$. Then*

$$\Theta_M(\mathbf{W}_2^r)_p \ll \left(\frac{M}{(\log M)^{d-1}}\right)^{-\frac{rd}{d-1}}.$$

The proof of Theorem 6.1 in [1] is not constructive. It goes by induction and uses a nonconstructive bound in the case $d = 2$, which is obtained in [33]. The corresponding proof from [33] uses the bounds for the Kolmogorov width $d_n(W_2^r, L_\infty)$, proved by Kashin [15]. Kashin's proof is a probabilistic one, which provides existence of a good linear subspace for approximation, but there is no known explicit constructions of such subspaces. This problem is related to a problem from compressed sensing on construction of good matrices with Restricted Isometry Property. It is an outstanding difficult open problem. In [1] we discuss constructive ways of building good multilinear approximations. The simplest way would be to use known results about M-term approximation with respect to special systems with tensor product structure. However, this approach (see [35]) provides error bounds, which are not as good as best m-term approximation with respect to Π^d (we have exponent r instead of $\frac{rd}{d-1}$ for Π^d). It would be very interesting to provide a constructive multilinear approximation method with the same order of the error as the best m-term approximation.

As we pointed out in a discussion of Theorem 6.1 the upper bound in Theorem 6.1 is proved with a help of probabilistic results. There is no known deterministic constructive methods (theoretical algorithms), which provide the corresponding upper bounds. In [1] we apply greedy-type algorithms to obtain upper estimates of $\Theta_M(\mathbf{W}^r_2)_p$. The important feature of our proof is that it is deterministic and moreover it is constructive. Formally, the optimization problem

$$\Theta_M(f)_X := \inf_{\{u^i_j\}, j=1,\dots,M, i=1,\dots,d} \|f(x_1,\dots,x_d) - \sum_{j=1}^{M}\prod_{i=1}^{d} u^i_j(x_i)\|_X$$

is deterministic: one needs to minimize over u^i_j. However, minimization by itself does not provide any upper estimate. It is known (see [5]) that simultaneous optimization over many parameters is a very difficult problem. Thus, in nonlinear M-term approximation we look for methods (algorithms), which provide approximation close to best M-term approximation and at each step solve an optimization problem over only one parameter ($\prod_{i=1}^d u^i_j(x_i)$ in our case). In [1] we provide such an algorithm for estimating $\Theta_M(f)_p$. We call this algorithm *constructive* because it provides an explicit construction with feasible one parameter optimization steps. We stress that in the setting of approximation in an infinite-dimensional Banach space, which is considered in [1], the use of term *algorithm* requires some explanation. In that paper we discuss only theoretical aspects of the efficiency (accuracy) of M-term approximation and possible ways to realize this efficiency. The *greedy algorithms* used in [1] give a procedure to construct an approximant, which turns out to be a good approximant. The procedure of constructing a greedy approximant is not a numerical algorithm ready for computational implementation. Therefore, it would be more precise to call this procedure a *theoretical greedy algorithm* or *stepwise optimizing process*. Keeping this remark in mind we, however, use the term *greedy algorithm* in this paper because it has been used in previous papers and has become a standard name for procedures used in [1] and for more general procedures of this type (see for instance [6, 40]). Also, the theoretical algorithms, which we use in [1], become algorithms in a strict sense if instead of an infinite-dimensional setting we consider a finite-dimensional setting, replacing, for instance, the L_p space by its restriction on the set of trigonometric polynomials. We note that the greedy-type algorithms are known to be very efficient in numerical applications (see, for instance, [47] and [25]).

In [1] we use two very different greedy-type algorithms to provide a constructive multilinear approximant. The first greedy-type algorithm is based on a very simple dictionary consisting of shifts of the de la Vallée Poussin kernels. The algorithm uses function (dyadic blocks of a function) evaluations and picks the largest of them. The second greedy-type algorithm is more complex. It is based on the dictionary Π^d and uses the Weak Chebyshev Greedy Algorithm with respect to Π^d to update the approximant. Surprisingly, these two algorithms give the same error bound. For instance, Theorems 4.3 and 4.4 from [1] give for big enough r the following constructive upper bound for $2 \le p < \infty$

$$\Theta_M(\mathbf{W}^r_2)_p \ll \left(\frac{M}{(\ln M)^{d-1}}\right)^{-\frac{rd}{d-1}+\frac{\beta}{d-1}}, \quad \beta := \frac{1}{2} - \frac{1}{p}.$$

This constructive upper bound has an extra term $\frac{\beta}{d-1}$ in the exponent compared to the best M-term approximation. It would be interesting to find a constructive proof of Theorem 6.1.

Acknowledgments University of South Carolina and Steklov Institute of Mathematics. Research was supported by NSF grant DMS-1160841.

References

1. Bazarkhanov, D., Temlyakov, V.: Nonlinear tensor product approximation of functions. J. Complexity **31**, 867–884 (2015). arXiv:1409.1403v1 [stat.ML] 4 Sep 2014 (to appear in J. Complex. 2015)
2. Belinskii, E.S.: Decomposition theorems and approximation by a "floating" system of exponentials. Trans. Am. Math. Soc. **350**, 43–53 (1998)
3. Dai, W., Milenkovic, O.: Subspace pursuit for compressive sensing signal reconstruction. IEEE Trans. Inf. Theory **55**, 2230–2249 (2009)
4. Dahmen, W., DeVore, R., Grasedyck, L., Süli, E.: Tensor-sparsity of solutions to high-dimensional elliptic Partial Differential Equations, Manuscript, 23 July 2014
5. Davis, G., Mallat, S., Avellaneda, M.: Adaptive greedy approximations. Constr. Approx. **13**, 57–98 (1997)
6. DeVore, R.A.: Nonlinear approximation. Acta Numerica **7**, 51–150 (1998)
7. DeVore, R.A., Temlyakov, V.N.: Nonlinear approximation by trigonometric sums. J. Fourier Anal. Appl. **2**, 29–48 (1995)
8. Dilworth, S.J., Kalton, N.J.: and Denka Kutzarova. On the existence of almost greedy bases in Banach spaces, Studia Math. **158**, 67–101 (2003)
9. Dilworth, S.J., Kutzarova, D., Temlyakov, V.N.: Convergence of some Greedy Algorithms in Banach spaces. J. Fourier Anal. Appl. **8**, 489–505 (2002)
10. Dilworth, S.J., Soto-Bajo, M., Temlyakov, V.N.: Quasi-greedy bases and Lebesgue-type inequalities. Stud. Math. **211**, 41–69 (2012)
11. Garrigós, G., Hernández, E., Oikhberg, T.: Lebesgue type inequalities for quasi-greedy bases. Constr. Approx. **38**, 447–479 (2013)
12. Gluskin, E.D.: Extremal properties of orthogonal parallelpipeds and their application to the geometry of Banach spaces. Math USSR Sbornik **64**, 85–96 (1989)
13. Hackbusch, W.: Tensor Spaces and Numerical Tensor Calculus. Springer, Heidelberg (2012)
14. Ismagilov, R.S.: Widths of sets in normed linear spaces and the approximation of functions by trigonometric polynomials. Uspekhi Mat. Nauk **29**, 161–178 (1974); English transl. in Russian Math. Surv. **29** (1974)
15. Kashin, B.S.: Widths of certain finite-dimensional sets and classes of smooth functions. Izv. AN SSSR **41**, 334–351 (1977); English transl. in Math. Izv. **11** (1977)
16. Kashin, B.S., Temlyakov, V.N.: On best m-term approximations and the entropy of sets in the space L^1. Math. Notes **56**, 1137–1157 (1994)
17. Konyagin, S.V., Temlyakov, V.N.: A remark on greedy approximation in Banach spaces. East. J. Approx. **5**, 365–379 (1999)
18. Livshitz, E.D., Temlyakov, V.N.: Sparse approximation and recovery by greedy algorithms, IEEE Trans. Inf. Theory **60**, 3989–4000 (2014). arXiv:1303.3595v1 [math.NA] 14 Mar 2013
19. Maiorov, V.E.: Trigonometric diameters of the Sobolev classes W^r_p in the space L_q. Math. Notes **40**, 590–597 (1986)

20. Makovoz, Y.: On trigonometric n-widths and their generalizations. J. Approx. Theory **41**, 361–366 (1984)
21. Needell, D., Tropp, J.A.: CoSaMP: iterative signal recovery from incomplete and inaccurate samples. Appl. Comput. Harmon. Anal. **26**, 301–321 (2009)
22. Needell, D., Vershynin, R.: Uniform uncertainty principle and signal recovery via orthogonal matching pursuit. Found. Comput. Math. **9**, 317–334 (2009)
23. Nielsen, M.: An example of an almost greedy uniformly bounded orthonormal basis for $L_p(0, 1)$. J. Approx. Theory **149**, 188–192 (2007)
24. Romanyuk, A.S.: Best M-term trigonometric approximations of Besov classes of periodic functions of several variables, Izvestia RAN, Ser. Mat. 67 (2003), 61–100; English transl. in. Izvestiya: Mathematics **67**(2), 265 (2003)
25. Shalev-Shwartz, S., Srebro, N., Zhang, T.: Trading accuracy for sparsity in optimization problems with sparsity constrains. SIAM J. Optim. **20**(6), 2807–2832 (2010)
26. Schmidt, E.: Zur Theorie der linearen und nichtlinearen Integralgleichungen. I. Math. Ann. **63**, 433–476 (1907)
27. Schneider, R., Uschmajew, A.: Approximation rates for the hierarchical tensor format in periodic Sobolev spaces. J. Complex. **30**, 56–71 (2014)
28. Stechkin, S.B.: On absolute convergence of orthogonal series. Dokl. AN SSSR **102**, 37–40 (1955) (in Russian)
29. Savu, D., Temlyakov, V.N.: Lebesgue type inequalities for greedy approximation in Banach Spaces. IEEE Trans. Inf. Theory **58**, 1098–1106 (2013)
30. Temlyakov, V.N.: Approximation of functions with bounded mixed derivative. Trudy MIAN **178**, 1–112 (1986). English transl. in Proc. Steklov Inst. Math. **1** (1989)
31. Temlyakov, V.N.: Approximation of periodic functions of several variables by bilinear forms. Izvestiya AN SSSR **50**, 137–155 (1986); English transl. in Math. USSR Izvestija **28**, 133–150 (1987)
32. Temlyakov, V.N.: Estimates of the best bilinear approximations of functions of two variables and some of their applications. Mat. Sb. **134**, 93–107 (1987); English transl. in Math. USSR-Sb **62**, 95–109 (1989)
33. Temlyakov, V.N.: Estimates of best bilinear approximations of periodic functions. Trudy Mat. Inst. Steklov **181**, 250–267 (1988); English transl. Proc. Steklov Inst. Math. **4**, 275–293 (1989)
34. Temlyakov, V.N.: Estimates of best bilinear approximations of functions and approximation numbers of integral operators. Matem. Zametki **51**, 125–134 (1992); English transl. in Math. Notes **51**, 510–517 (1992)
35. Temlyakov, V.N.: Greedy algorithms with regard to multivariate systems with special structure. Constr. Approx. **16**, 399–425 (2000)
36. Temlyakov, V.N.: Nonlinear Kolmogorov's widths. Matem. Zametki **63**, 891–902 (1998)
37. Temlyakov, V.N.: Greedy algorithms in Banach spaces. Adv. Comput. Math. **14**, 277–292 (2001)
38. Temlyakov, V.N.: Nonlinear method of approximation. Found. Comput. Math. **3**, 33–107 (2003)
39. Temlyakov, V.N.: Greedy-type approximation in Banach Spaces and applications. Constr. Approx. **21**, 257–292 (2005)
40. Temlyakov, V.N.: Greedy Approximation. Cambridge University Press (2011)
41. Temlyakov, V.N.: Sparse approximation and recovery by greedy algorithms in Banach Spaces. Forum Math. Sigma **2**, e12, e26 (2014). arXiv:1303.6811v1 [stat.ML] 27 Mar 2013, 1–27
42. Temlyakov, V.N.: An inequality for the entropy numbers and its application. J. Approx. Theory **173**, 110–121 (2013)
43. Temlyakov, V.N.: Constructive sparse trigonometric approximation and other problems for functions with mixed smoothness. Math. Sbornik **206**, 131–160 (2015). arXiv:1412.8647v1 [math.NA] 24 Dec 2014, 1–37 (to appear in Math. Sbornik, 2015)
44. Temlyakov, V.N.: Constructive sparse trigonometric approximation for functions with small mixed smoothness. arXiv:1503.0282v1 [math.NA] 1 Mar 2015, 1–30
45. Temlyakov, V.N., Yang, M., Ye, P.: Greedy approximation with regard to non-greedy bases. Adv. Comput. Math. **34**, 319–337 (2011)

46. Wojtaszczyk, P.: Greedy algorithm for general biorthogonal systems. J. Approx. Theory **107**, 293–314 (2000)
47. Zhang, T.: Sequential greedy approximation for certain convex optimization problems. IEEE Trans. Inf. Theory **49**(3), 682–691 (2003)
48. Zhang, T.: Sparse recovery with orthogonal matching pursuit under RIP. IEEE Trans. Inf. Theory **57**, 6215–6221 (2011)

The Bi-free Extension of Free Probability

Dan-Virgil Voiculescu

Abstract Free probability is a noncommutative probability theory adapted to variables with the highest degree of noncommutativity. The theory has connections with random matrices, combinatorics, and operator algebras. Recently, we realized that the theory has an extension to systems with left and right variables, based on a notion of bi-freeness. We provide a look at the development of this new direction. The paper is an expanded version of the plenary lecture at the 10th ISAAC Congress in Macau.

Keywords Two-faced pair · Bi-free probability · Bi-free convolution

2000 Mathematics Subject Classification. Primary: 46L54 · Secondary: 46L53

1 Introduction

Free probability is now in its early thirties. After such a long time I became aware that the theory has a natural extension to a theory with two kinds of variables: left and right. This is not the same as passing from a theory of modules to a theory of bimodules, since our left and right variables will not commute in general, a noncommutation which will appear already when we shall take a look at what the Gaussian variables of the theory are.

We call the theory with left and right variables bi-free probability and the independence relation that underlies it is called bi-freeness. This new type of independence does not contradict the theorems of Muraki [15] and Speicher [20] about the possible types of independence in noncommutative probability with all the nice properties ("classical", free, Boolean and if we give up symmetry also monotonic and anti-monotonic), the reason being that we play a new game here, by replacing the usual sets of random variables by sets with two types of variables.

D.-V. Voiculescu (✉)
Department of Mathematics, University of California at Berkeley,
Berkeley, CA 94720-3840, USA
e-mail: dvv@math.berkeley.edu

T. Qian and L.G. Rodino (eds.), *Mathematical Analysis, Probability and Applications – Plenary Lectures*, Springer Proceedings in Mathematics & Statistics 177, DOI 10.1007/978-3-319-41945-9_8

The observation about the possibility of left and right variables could have been made at the very beginning of free probability. At present it becomes necessary to look back and think about the problems which would have appeared earlier had we been aware of the possibility. Several advances on this road have been made. Developments are happening faster since the lines along which free probability developed are serving often as a guide.

2 Free Probability Background

Free probability is a noncommutative probability framework adapted for variables with the highest degree of noncommutativity. By "highest" noncommutativity we mean the kind of noncommutativity one encounters, for instance, in free groups, free semigroups or in the creation and destruction operators on a full Fock space. At this heuristic level, Bosonic and Fermionic creation and destruction operators are "less noncommutative" because the commutation or anticommutation relations they satisfy represent restrictions on the noncommutativity. These are the reasons for the adjective "free" in the name of free probability. It should also be noticed that heuristically, when noncommutativity is at the highest, a certain homogeneity appears which simplifies matters.

The distinguishing feature of free probability among noncommutative probability theories is the independence relation, called free independence or freeness, on which it is based.

Thus the notions of random variables and of expectation values are the usual ones in noncommutative probability that is quantum mechanical observables, and their expectation values or some purely algebraic version of these. So our random variables will be operators T on some complex Hilbert space $\mathcal{H}$ and there will be a unit vector $\xi \in \mathcal{H}$ so that the expectation of T is $\langle T\xi, \xi\rangle$ (we use the mathematician's scalar product which is linear in the first and conjugate linear in the second variable). The purely algebraic version is a "noncommutative probability space" which is a unital algebra $\mathcal{A}$ over $\mathbb{C}$ with a linear expectation functional $\varphi : \mathcal{A} \to \mathbb{C}$, $\varphi(1) = 1$ and the elements $a \in \mathcal{A}$ are the noncommutative random variables (in the Hilbert space setting $\mathcal{A}$ is $\mathcal{L}(\mathcal{H})$ the linear operators on $\mathcal{H}$ and $\varphi(a) = \langle a\xi, \xi\rangle$).

The distribution μ_α of a family $\alpha = (a_i)_{i\in I}$ of noncommutative random variables in a noncommutative probability space $(\mathcal{A}, \varphi)$ is the information provided by the collection of noncommutative moments $\varphi(a_{i_1} \dots a_{i_n})$ when $n \in \mathbb{N}$ and $i_1, \dots, i_n \in I$. This information can be structured in better ways. For instance in the case of just one variable a, which is a bounded hermitian operator on a Hilbert space $\mathcal{H}$ then $\varphi(a^n) = \langle a^n\xi \mid \xi\rangle$ are precisely the moments of a probability measure, which we shall also denote by μ_a and which is given by $\mu_a(\omega) = \langle E(a; \omega)\xi, \xi\rangle$ where $E(a; \omega)$ is the spectral project of a for the Borel set $\omega \subset \mathbb{R}$.

The definition of freeness, which is the notion of independence, for a family of subalgebras $(\mathcal{A}_i)_{i\in I}$ which contain the unit $1 \in \mathcal{A}_i$ $(i \in I)$ of $(\mathcal{A}, \varphi)$ is that:

$$\varphi(a_1 \dots a_n) = 0$$

whenever $\varphi(a_j) = 0$, $a_j \in \mathcal{A}_{i_j}$ $1 \le j \le n$ and $i_j \ne i_{j+1}$ if $1 \le j < n$. A family of sets $\alpha_i \subset (\mathcal{A}, \varphi)_{i \in I}$ is free if the algebras $\mathcal{A}_i$ generated by $\{1\} \cup \alpha_i$ are freely independent. When compared with the usual notion of independence (often called "classical" or "tensor product" independence) in quantum mechanics, a key difference is the noncommutation of independent variables in the free case. In the usual definition, at least in distribution, the independent variables commute. So if $a, b \in (\mathcal{A}, \varphi)$ are classically independent and $\varphi(a) = \varphi(b) = 0$ then $\varphi(abab) = \varphi(a^2)\varphi(b^2)$ which may be $\ne 0$ while $\varphi(abab) = 0$ when a, b are freely independent.

Once random variables, distributions, and independence of random variables in a noncommutative probability theory have been defined, one can imitate classical probability theory and look at the limit processes which give rise to basic types of variables: Gaussian, Poisson, etc. For instance, a free central limit process will consider a sequence $a_k, k \in \mathbb{N}$ of freely independent variables in some $(\mathcal{A}, \varphi)$ which are identically distributed, i.e., $\varphi(a_k^p) = m_p$ for all $k \in \mathbb{N}$, and assuming the variables are centered $\varphi(a_k) = m_1 = 0$ one looks at the limits as $n \to +\infty$ of the distributions of $n^{-1/2}(a_1 + \dots + a_n)$. One finds that the limit distribution if $m_2 > 0$ is a semicircle law, that is if we normalize $m_2 = 1$ the moments are those of a probability measure on $\mathbb{R}$ with support $[-2, 2]$ and density $\frac{1}{2\pi}\sqrt{4 - t^2}$ with respect to Lebesgue measure.

Similarly, to find a free Poisson distribution one takes for each n freely independent variables $a_k^{(n)}$, $1 \le k \le n$ which have identical distributions corresponding to Bernoulli measures $\left(1 - \frac{\alpha}{n}\right)\delta_0 + \frac{\alpha}{n}\delta_1$ and then looks at the limit distribution of $a_1^{(n)} + \dots + a_n^{(n)}$ as $n \to \infty$. The distribution one finds (in the case of $\alpha > 0$) looks like a tilted shifted semicircle with the possibility of an additional atom at 0, that is the probability measure $(1 - a)\delta_0 + \nu$ if $0 \le a \le 1$ and only ν if $a \ge 1$ where ν has support $[(1 - \sqrt{a})^2, (1 + \sqrt{a})^2]$ and density $(2\pi t)^{-1}\sqrt{4a - (t - (1 + a))^2}$. This law is quite different from the classical Poisson law which is

$$\sum_{n \ge 0} e^{-a} \frac{a^n}{n!} \delta_n$$

a measure concentrated on the natural numbers.

Actually, the free Gaussian and free Poisson laws are the same as well-known limit eigenvalue distribution laws in random matrix theory: the Wigner semicircle law (the limit distribution for eigenvalues of a suitably normalized hermitian Gaussian random matrix with i.i.d. entries) and the Marchenko–Pastur law. Clearly, the fact that the free Gauss law and the free Poisson law appear in random matrix theory provided a strong indication of a connection between free probability and random matrices.

The explanation for the connection between random matrices and free probability I found in [25] is that free independence appears asymptotically among large random matrices under suitable conditions. The algebra of $N \times N$ random matrices with entries p-integrable for all $1 \le p < \infty$ over a probability space (Ω, Σ, μ) can be endowed with an expectation functional φ_N, where $\varphi_N(Y) = N^{-1}E(\mathrm{Tr}_N Y)$.

Thus random matrices become noncommutative random variables and by doing this their entries are forgotten and only noncommutative moments are remembered. The simple occurrence of asymptotic freeness is that an m-tuple $(Y_1^{(N)}, \ldots, Y_m^{(N)})$ of Hermitian Gaussian random matrices with i.i.d. suitably normalized entries is asymptotically free as $N \to \infty$. This generalizes to sets of independent random matrices with distributions invariant under unitary conjugation and even, using a combination of concentration and operator algebra techniques to a kind of generic asymptotic freeness result. The fact that free probability can be modeled by random matrices in the limit $N \to \infty$, was the source of the applications of free probability to the operator algebras of free groups (see [26, 28]), a subject we will not discuss here.

The computations of noncommutative distributions in free probability has evolved in two directions. On one hand my initial analytic approach using complex analysis and a bit of operator algebras evolved toward an analytic approach and connections with noncommutative analysis. On the other hand the streamlining of the computation of moments led to Roland Speicher's combinatorial approach to free probability. In essence free probability from the point of view of moments and cumulants can be viewed as replacing the lattice of partitions of $\{1, \ldots, n\}$ which underlies classical probability by the lattice of noncrossing partitions. A partition of $\{1, \ldots, n\}$ is noncrossing if there are no $1 \leq a < b < c < d \leq n$ so that $\{a, c\}$ and $\{b, d\}$ lie in different blocks of the partition.

All this seems to be connected to the discovery of t'Hooft [12] about the large N-limit of gauge theory: that in the large N-limit of the gauge group, ($U(N)$ with $N \to \infty$) the contribution to the expectation values concentrates on planar diagrams. Given a partition of $\{1, \ldots, n\}$ if we draw limits connecting the elements in the same block, the resulting diagram is planar precisely when the partition is noncrossing. On the other hand, the large N limit of random matrix models, was early on recognized by physicists as a kind of simplified large N-limit of gauge theories situation. Perhaps the connection to the large N limit of gauge theories is also the answer to the question: if free probability is a successful noncommutative probability theory, why is the noncommutative probability which underlies quantum mechanics the one based on "classical" independence? The answer seems to be that free probability is related to another region of quantum theory, to the large N limit of gauge theories.

Imitation of basic classical probability for the corresponding notions of free probability has developed in many directions. The free parallel to classical probability goes quite far and after more than 30 years one should wonder about the extent of this parallelism. A partial list of items which appear in the free/classical parallel includes: limit laws, stochastic processes with independent increments, convolution operations corresponding to addition or multiplication of independent random variables, combinatorics of cumulants, continuous entropy, extreme values, exchangeability.

Some comments are here in order. This is only a rough parallel, to be taken with a grain of salt sometimes. For instance, there is a free entropy theory [30], resembling more classical entropy than von Neumann's quantum entropy of states. On the other hand the free entropy is a "continuous" entropy, an analogue of Shannon's differential entropy, in contrast with the fact that in classical probability the fundamental entropy notion is the discrete one.

The parallelism for infinitely divisible and stable laws (see [2]) is quite close, but there are a few surprises. The $\mathcal{X}^2$-law, the distribution of the square of a Gaussian variable in the free setting coincides unexpectedly with a free Poisson law (i.e., a Marchenko–Pastur law). Another interesting detail is perhaps that free and classical Cauchy laws are the same.

A quite unexpected rather recent addition to the parallel was the discovery by Koestler and Speicher [13] that there is a free analogue to de Finneti's exchangeability theorem. At first sight such a result is quite unlikely, since invariance under permutations is too weak a condition for joint distributions of noncommutative variables, the number of noncommutative moments of monomials grows exponentially with the degree compared with the polynomial growth for commuting variables. The discovery was that instead of classical permutations the appropriate symmetry is provided by Wang's C^*-algebraic universal quantum permutation groups.

Other rather unexpected items with free analogues are, for instance, extreme values [1] and optimal transportation [3, 10].

An important feature of free probability is that conditional probabilities have a quite natural free counterpart: a "base change" from the complex field $\mathbb{C}$ to some unital algebra $\mathcal{B}$ over $\mathbb{C}$. This works especially nice in the setting of von Neumann algebras with faithful trace states, where there are unique state-preserving conditional expectations onto von Neumann subalgebras. We should note that because of the noncommutativity, "conditional free" is a much more complex matter than in the classical commutative setting. For instance, conditional independence when dealing with group examples amounts to free products of groups with amalgamation over a subgroup. Also, in the setting of von Neumann algebras of type II_1, one may have to face the complexity of a subfactor inclusion $\mathcal{B} \subset \mathcal{A}$, to describe the position of $\mathcal{B}$ in $\mathcal{A}$.

3 Bi-free Independence

The framework for dealing with left and right variables is that of *a pair of faces in a noncommutative probability space* $(\mathcal{A}, \varphi)$, which is a pair $1 \in \mathcal{B} \subset \mathcal{A}$, $1 \in \mathcal{C} \subset \mathcal{A}$ of subalgebras in $\mathcal{A}$, the first one $\mathcal{B}$ being the left face and the second $\mathcal{C}$ the right face. Often such a structure arises from $((z_i)_{i\in I}, (z_j)_{j\in J})$ a two-faced set of noncommutative random variables in $(\mathcal{A}, \varphi)$, where $(z_i)_{i\in I}$ are the left and $(z_j)_{j\in J}$ the right variables, the algebras $\mathcal{B}$ and $\mathcal{C}$ being then the algebras generated by $\{1\} \cup \{(z_i)_{i\in I}\}$ and $\{1\} \cup \{(z_j)_{j1J}\}$, respectively. The distribution of such a system is that of the left and right variables taken together $(z_i)_{i\in I} \cup (z_j)_{j\in J}$ in $(\mathcal{A}, \varphi)$, that is expectation values of monomials in left and right variables.

To explain why left and right variables are natural in free probability we shall revisit how classical independence is defined from the tensor product of Hilbert spaces and then try to imitate this in the free setting.

Let $(T_i)_{i\in I}$ be operators on a Hilbert space $\mathcal{H}$ with the state vector $\xi \in \mathcal{H}$, $\|\xi\| = 1$ and $(S_j)_{j\in J}$ operators on $\mathcal{K}$ with state vector η. Two sets of noncommutative random variables are classically independent if they have the same joint distribution as two sets of the form $(T_i \otimes I_{\mathcal{K}})_{i \in I}$, $(I_{\mathcal{H}} \otimes S_j)_{j \in J}$ on $\mathcal{H} \otimes \mathcal{K}$ with the state vector $\xi \otimes \eta$ (to make this quite general the operators are not bounded and the tensor product is not completed).

The free analogue of the tensor product of the Hilbert spaces with state vector, for a family $(\mathcal{H}_i, \xi_i)_{i \in I}$ of such spaces is $(\mathcal{H}, \xi) = \underset{i \in I}{*} (\mathcal{H}_i, \xi_i)$ where

$$\mathcal{H} = \mathbb{C}\xi \oplus \bigoplus_{n \geq 1} \bigoplus_{i_1 \neq i_2 \neq i_3 \neq \cdots \neq i_n} \overset{\circ}{\mathcal{H}}_{i_1} \otimes \cdots \otimes \overset{\circ}{\mathcal{H}}_{i_n}$$

and $\overset{\circ}{\mathcal{H}}_i = \mathcal{H}_i \ominus \mathbb{C}\xi_i$. The moment left and right make their appearance is when we want to lift operators acting on the spaces $\mathcal{H}_i$ to operators acting on $\mathcal{H}$: this can be done in two ways, as left and as right operators, respectively. Indeed there are left and right factorizations, identifications via isomorphisms

$$V_i : \mathcal{H}_i \otimes \left(\mathbb{C}\xi \oplus \bigoplus_{n \geq 1} \bigoplus_{i \neq i_1 \neq \cdots \neq i_n} \overset{\circ}{\mathcal{H}}_{i_1} \otimes \cdots \otimes \overset{\circ}{\mathcal{H}}_{i_n} \right) \to \mathcal{H}$$

$$W_i : \left(\mathbb{C}\xi \oplus \bigoplus_{n \geq 1} \bigoplus_{i_1 \neq \cdots \neq i_n \neq i} \overset{\circ}{\mathcal{H}}_{i} \otimes \cdots \otimes \overset{\circ}{\mathcal{H}}_{i_n} \right) \otimes \mathcal{H}_i \to \mathcal{H},$$

where we view $\mathcal{H}_i$ as $\mathbb{C}\xi_i \oplus \overset{\circ}{\mathcal{H}}_i$ and $\xi_i \otimes$ or $\otimes \xi_i$ acting as a blank. If T is an operator on $\mathcal{H}_i$ we define $\lambda_i(T) = V_i(T \otimes I)V_i^{-1}$ and $\rho_i(T) = W_i(I \otimes T)W_i^{-1}$ the left and right operators, respectively.

When defining an independence based on the free products of Hilbert spaces we have a choice between left and right. If we choose all operators to be left operators or all operators to be right operators we get the usual free independence. Bi-freeness arises when we combine left and right.

Two two-faced systems in $(\mathcal{A}, \varphi)$, $((b'_i)_{i \in I'}, (c'_j)_{j \in J'})$ *and* $((b''_i)_{i \in I''}, (c''_j)_{j \in J''})$ *are bi-free* if there are $(\mathcal{H}_1, \xi_1)$ and $(\mathcal{H}_2, \xi_2)$ Hilbert spaces with state vectors and operators $((T'_i)_{i \in I'}, (S'_j)_{j \in J'})$ on $\mathcal{H}_1$ and $((T''_i)_{i \in I'}, (S''_j)_{j \in J''})$ on $\mathcal{H}_2$ so that the distribution of $((b'_i)_{i \in I'}, (c'_j)_{j \in J'}, (b''_i)_{i \in I''}, (c''_j)_{j \in J''})$ is the same as that of

$$((\lambda_1(T'_i))_{i \in I'}, (\rho_1(S'_j))_{j \in J'}, (\lambda_2(T''_i)_{i \in I''}, \rho_2(S''_j)_{j \in J''}))$$

on $\mathcal{H}$ w.r.t. the state vector ξ where $(\mathcal{H}, \xi) = (\mathcal{H}_1, \xi_1) * (\mathcal{H}_2, \xi_2)$. (To achieve full generality in this the operators will not be bounded and all completions left out.)

Bi-freeness defined in this way has the right properties to serve as a noncommutative independence relation for a new type of systems of noncommutative random variables with two faces (two kinds of variables, left and right variables).

Among the consequences of this definition, if $((b'_i)_{i\in I'}, (c'_j)_{j\in J'})$ and $((b''_i)_{i\in I''}, (c''_j)_{j\in J''})$ are bi-free then $(b'_i)_{i\in I'}$ and $(b''_i)_{i\in I''}$ are free and similarly $(c'_j)_{j\in J'}$ and $(c''_j)_{j\in J''}$ are free. On the other hand $(b'_i)_{i\in I'}$ and $(c''_j)_{j\in J''}$ are classically independent and also $(c'_j)_{j\in J'}$ and $(b''_i)_{i\in I''}$ are classically independent. This bi-freeness involves some freeness (for the same kind of variables and classical independence for different kinds). However bi-freeness does not reduce just to this combination of free and classical independences.

We should also point out that on a free product of Hilbert spaces $(\mathcal{H}_i, \xi_i)$ if T is an operator on $\mathcal{H}_i$ and S an operator on $\mathcal{H}_j$ then $\lambda_i(T)$ and $\rho_j(S)$ commute when $i \neq j$ but if $i = j$ then $[\lambda_i(T), \rho_i(S)] = [T, S] \oplus O$ where $\mathcal{H}_i$ is identified with the subspace $\mathbb{C}\xi \oplus \overset{\circ}{\mathcal{H}}_i$ of the free product space.

Two-faced systems of variables where the left and right variables commute will be called bipartite.

We have chosen to call the left and right variables "faces" of a system in reference to Janus. In roman mythology, the two faces of Janus were used to look into the past and into the future and combining these two kinds of observations to deal with the transition. It would be interesting if models involving some past/future interface could be developed in the noncommutative probability setting which we discuss here.

To conclude this section, we will mention a few basic examples of bi-freeness.

If $(\mathcal{X}_i, \xi_i)$ are Hilbert spaces with state vectors and $\mathcal{L}(\mathcal{X}_i)$ denotes the algebra of linear operators on $\mathcal{X}_i$ and $\varphi_i(\cdot) = \langle \cdot\xi_i, \xi_i\rangle$ is the expectation functional on $\mathcal{L}(\mathcal{H}_i)$ let $(\mathcal{X}, \xi)$ be the free product of the $(\mathcal{X}_i, \xi_i)$ and $\varphi(\cdot) = \langle \cdot\xi, \xi\rangle$ the expectation functional on $\mathcal{L}(\mathcal{X})$. It is almost tautological then that $(\lambda_i(\mathcal{L}(\mathcal{X}_i)), \rho_i(\mathcal{L}(\mathcal{X}_i)))_{i\in I}$ are bi-free in $(\mathcal{L}(\mathcal{X}), \varphi)$.

If $\mathcal{H}$ is a complex Hilbert space and $\mathcal{T}(\mathcal{H}) = \bigoplus_{n\geq 1} \mathcal{H}^{\otimes n} \oplus \mathbb{C}1$ is the full Fock space, let $l(h)$ be the left and $r(h)$ be the right creation operator: $l(h)\xi = h \otimes \xi$ and $r(h)\xi = \xi \otimes h$. If $\omega_i \subset \mathcal{H}$ are subsets, $i \in I$ which are pairwise orthogonal $i \neq j \Rightarrow \omega_i \perp \omega_j$ then in $(\mathcal{L}(\mathcal{T}(\mathcal{H})), \langle \cdot 1, 1\rangle)$ the family of two-faced sets of noncommutative random variables $((l(\omega_i) \cup l^*(\omega_i)), (r(\omega_i) \cup r^*(\omega_i)))_{i\in I}$ is bi-free.

Similarly, let $(\mathcal{G}_i)_{i\in I}$ be groups and $\mathcal{G} = \underset{i\in I}{*} \mathcal{G}$ their free product. If $g \in \mathcal{G}$ we shall denote by $L(g)$ the left shift by g on $l^2(\mathcal{G})$ and by $R(g)$ the right shift on $l^2(\mathcal{G})$ by g and let τ on $\mathcal{L}(l^2(\mathcal{G}))$ be the functional $\varphi(\cdot) = \langle \cdot\delta_e, \delta_e\rangle$ where δ_g where $g \in \mathcal{G}$ is the canonical basis of $l^2(\mathcal{G})$. Then the family $(L(g))_{g\in\mathcal{G}_i}, (R(g))_{g\in\mathcal{G}_i})_{i\in I}$ of two-faced sets in $(\mathcal{L}(l^2(\mathcal{G})), \varphi)$ is bi-free.

4 Generalities on Operations on Bi-free Systems of Variables

Like for other types of independences operations on bi-free systems of variables give rise to corresponding convolution operations on the distributions. For instance, if $z' = ((z'_i)_{i\in I}, (z'_j)_{j\in J})$, $z'' = ((z''_i)_{i\in I}, (z''_j)_{j\in J})$ are bi-free two-faced systems of noncommutative random variables in $(\mathcal{A}, \varphi)$ and if $z' + z'' = ((z'_i + z''_i)_{i\in I}, (z'_j + z''_j)_{j\in J})$ then the distribution $\mu_{z'+z''}$ depends only on the distributions $\mu_{z'}, \mu_{z''}$. This yields an operation on the distributions of systems of variables with these index sets, so that

$$\mu_{z'} \boxplus\boxplus \mu_{z''} = \mu_{z'+z''}.$$

The operation $\boxplus\boxplus$ will be called additive bi-free convolution, in analogy with the additive free convolution $\boxplus$ in free probability. Clearly many kinds of such convolution operations can be defined, like in the case of free probability and it is also possible to combine different operations on left and right variables. For instance, passing from z', z'' to $((z'_i + z''_i)_{i\in I}, (z'_j z''_j)_{j\in J})$ defines an additive–multiplicative bi-free convolution which we shall denote by $\boxplus\boxtimes$.

In classical probability the linearizing map for the additive convolution is provided by the logarithm of the Fourier transform, in particular for probability measures with compact support the sequence of derivatives of all orders of the logarithm of the Fourier transform at zero, is a sequence of polynomials in the moments of the measure which add when the probability measures are convolved. Roughly up to normalization these polynomials are the classical cumulants of the probability measure.

The bi-free cumulants can be described as follows. Let $z = ((z_i)_{i\in I}, (z_j)_{j\in J})$ be a two-faced system of noncommutative random variables with index see $\mathcal{I}, \mathcal{J}$. Given a map $\alpha : \{1, \dots, n\} \to \mathcal{I} \coprod \mathcal{J}$ the bi-free cumulant R_α is a polynomial in variables $X_{\alpha(k_1),\dots,\alpha(k_r)}$ where $\{k_1 < \cdots < k_r\} \subset \{1, \dots, n\}$ so that the quantity $R_\alpha(\mu_z)$ obtained under the substitution

$$X_{\alpha(k_1),\dots,\alpha(k_r)} \to \varphi(z_{\alpha(k_1)} \dots z_{\alpha(k_r)})$$

has the property that $R_\alpha(\mu_{z'} \boxplus\boxplus \mu_{z''}) = R_\alpha(\mu_{z'}) + R_\alpha(\mu_{z''})$ and moreover R_α is homogeneous of degree n when $X_{\alpha(k_1),\dots,\alpha(k_r)}$ is assigned degree r and the coefficient of $X_{\alpha(1),\dots,\alpha(n)}$ is 1. *The simplest result about cumulants is their existence and uniqueness* which we proved in [31]. This is a consequence of quite general considerations about the bi-free convolution being an operation which can be described by polynomials at the level of moments and which then yields an inverse limit of simply connected abelian complex Lie groups. Such Lie groups are isomorphic to their Lie algebras via the exponential maps. Roughly, the bi-free cumulants are the result of using the inverse of these isomorphisms.

Such general considerations were also used in free probability in [22], at the very beginning of the theory, to prove existence and uniqueness of cumulants, before the effective results about cumulants which were the result of later developments. This very primitive result about existence and uniqueness was sufficient to prove an algebraic central limit theorem [22] and find that the semicircle law plays the role of the Gauss law in free probability. In the bi-free setting a development along similar lines has taken place and will be the subject of the next section.

5 Bi-free Central Limit and Bi-free Gaussian Distributions [31]

From the existence and uniqueness of bi-free cumulants one immediately finds that the bi-free cumulants of degrees 1 and 2, when $z = ((z_i)_{i\in I}, (z_j)_{j\in J})$ are $R_\alpha(\mu_z) = \varphi(z_{\alpha(1)})$, $\alpha : \{1\} \to I \coprod J$ and, respectively, $R_\alpha(\mu_z) = \varphi(z_{\alpha(1)}z_{\alpha(2)}) - \varphi(z_{\alpha(1)})\varphi(z_{\alpha(2)})$, $\alpha : \{1, 2\} \to I \coprod J$. With this at hand one then easily proves an *algebraic bi-free central limit theorem* [31] if $z^{(k)} = ((z_i^{(k)})_{i\in I}, (z_j^{(k)})_{k\in J})$, $k \in \mathbb{N}$, *are bi-free and their distributions coincide* $\mu_z(k) = \mu$, $k \in \mathbb{N}$ *then assuming* $\varphi(z_p) = 0$ *for all* $p \in I \coprod J$, *the central limit process*

$$S^{(N)} = \left(\left(N^{-1/2} \sum_{1\le k\le N} z_i^{(k)}\right)_{i\in I}, \left(N^{-1/2} \sum_{1\le k\le N} z_j^{(k)}\right)_{j\in J}\right)$$

has a limit distribution γ.

Indeed, the cumulants of $S^{(N)}$ are easily seen to converge and this is equivalent with the convergence of the moments, that is the convergence of the distributions. Moreover, one easily sees that the limit distributions one finds, which are the centered bi-free Gaussian distributions, are precisely those for which $R_\alpha(\mu_z) = 0$, where $\alpha : \{1, \ldots, n\} \to I \coprod J$ and $n \neq 2$. To find all these centered bi-free Gaussian distributions is equivalent to finding for each covariance matrix $(C_{pq})_{p,q\in I\coprod J}$ a distribution μ_z so that $\varphi(z_p) = 0$, $\varphi(z_p z_q) = C_{pq}$ and μ_z is equal to the limit distribution of a central limit process.

Using these simple remarks we found [31] that in case I and J are finite sets, the bi-free Gaussian distributions are the distributions of $((l(h_i') + l^*(h_i''))_{i\in I}, (r(h_j') + r^*(h_j''))_{j\in J}$ in $(\mathcal{L}(\mathcal{T}(\mathcal{H})), \langle \cdot 1 \mid 1\rangle)$ where $h_i', h_i'', h_j', h_j'' \in \mathcal{H}$. Here l, l^*, r, r^* are left and, respectively, right creation and destruction operators on the full Fock space $\mathcal{T}(\mathcal{H})$. The covariance matrix which determines the distribution depends only on the scalar products of the vectors h_i', h_i'', h_j', h_j'' or $i \in I$, $j \in J$.

It should be noted that the left and right operators which realize the bi-free Gaussian distribution do not commute in general, indeed:

$$[l(h_i') + l^*(h_i''), r(h_j') + r^*(h_j'')] = (\langle h_j', h_i''\rangle - \langle h_i', h_j''\rangle)\mathcal{P}$$

where $\mathcal{P}$ is the rank one projector operator $\mathcal{P} = \langle \cdot 1 \mid 1\rangle 1$.

6 The Combinatorics of Bi-freeness

Knowing that bi-free cumulants, which linearize the additive bi-free convolution exist, immediately leads to the question of extending to the bi-free setting what is known in free probability about computing free convolutions in free probability. This meant looking for bi-free extensions on one hand of the analytic machinery and on the other hand of the combinatorial machinery of free probability. This section will briefly deal with the extension of Speicher's noncrossing partitions approach [21] and I will turn later to my initial analytic approach.

A first step in the combinatorial approach to bi-freeness was the paper of Mastnak and Nica [14], who found a connection to the combinatorics of double-ended queues and identified some of the basic objects. This beginning was then brought to fruition in a work of Charlesworth et al. [4] and also carried further to go beyond the field of scalars $\mathbb{C}$ to a general algebra $\mathcal{B}$ in [5].

Instead of noncrossing partitions one considers so-called bi-noncrossing partitions [4], that is for each $n \in \mathbb{N}$ and each map $\chi : \{1, \dots, n\} \to \{L, R\}$ the set $BNC(\chi)$ is the set of partitions π of $\{1, \dots, n\}$ so that $s_\chi^{-1}\pi$ is noncrossing, where s_χ is a permutation of $\{1, \dots, n\}$ defined as follows. If $\chi^{-1}(L) = \{i_1 < \cdots < i_p\}$ and $\chi^{-1}(R) = \{j_1 < \cdots < j_q\}$ then

$$s_\chi(k) = \begin{cases} i_k & \text{if } 1 \leq k \leq p \\ j_{n+1-k} & \text{if } p < k \leq n. \end{cases}$$

The role of the map χ in formulae connecting noncommutative moments and cumulants is to indicate in an ordered product which are the factors which are left and, respectively, right variables. With this change of lattices of partitions, the connection between moments and cumulants is of the same kind as in the free setting, that is based on an incidence algebra and the corresponding Möbius function. Note, that since $BNC(\chi)$ and the lattice of noncrossing partitions $NC(n)$ of $\{1, \dots, n\}$ are isomorphic via the permutation s_χ, the Möbius functions are also related via s_χ. However, since the role of s_χ is to change the order of factors in a product and all possible χ are to be considered, the resulting cumulant formulae are a quite nontrivial generalization.

Like in the free setting, also in the bi-free generalization the independence relation corresponds to the vanishing of mixed cumulants.

To illustrate these results in one of the simplest cases: the formulae expressing moments of Gaussian variables in terms of covariances using pair-partitions is the relation between moments and cumulants (in the Gaussian case the only nonzero ones are covariances). If $((z_i)_{i\in I}, (z_j)_{j\in J})$ is a bi-free Gaussian two-faced system in $(\mathcal{A}, \varphi)$ and $BNC_2(\chi)$ denotes the bi-noncrossing pair-partitions for a given $\chi : \{1, \dots, n\} \to \{L, R\}$ then:

$$\varphi(z_{\alpha(1)}, \dots, z_{\alpha(n)}) = \sum_{(\{a_k, b_k\})_{1\leq k\leq m} \in BNC_2(\chi)} \prod_{1\leq k\leq m} \varphi(z_{\alpha(a_k)} z_{\alpha(b_k)})$$

where $\chi(\alpha^{-1}(I)) \subset \{L\}$, $\chi(\alpha^{-1}(J)) \subset \{R\}$.

7 One-Variable Free Convolutions of Free Probability

Before we discuss the simplest bi-free convolutions, we need to briefly recall their free probability precursors.

If $a, b \in (\mathcal{A}, \varphi)$ are freely independent noncommutative random variables, then $\mu_{a+b} = \mu_a \boxplus \mu_b$ which is the definition of the additive free convolution $\boxplus$. The transform which linearizes additive free convolution and which can be used for its computation is the R-transform [23], $R_a(z)$ defined by the formulae

$$G_z(z) = \sum_{n \geq 0} z^{-n-1}\varphi(a^n) = \varphi((z1 - a)^{-1})$$

$$G_a(K_a(z)) = z, \;\; R_a(z) = K_a(z) - z^{-1}$$

and which satisfies

$$R_{a+b} = R_a + R_b.$$

Here R_a is either a formal power series if we work purely algebraically or the germ of a holomorphic function at 0 in the case of a Banach algebra $\mathcal{A}$. Note that in the case of a hermitian operator a, μ is a compactly supported probability measure on $\mathbb{R}$ and G_a is its Cauchy–Stieltjes transform. The inversion used to define K_a is for z near 0. Note also that since the sum of hermitian operators is a hermitian operator, $\boxplus$ is an operation on probability measures on $\mathbb{R}$. In this case one uses the above result from [23] to get R_{a+b} and then by the same formulae used backward one finds G_{a+b} and gets μ_{a+b} by the solution of a moment problem which boils down to finding the distributional boundary values of $-\text{Im } G_{a+b}(x + i\epsilon)$ as $\epsilon \downarrow 0$.

In [24] we found also another transform which computes the multiplicative convolution. If $a, b \in (\mathcal{A}, \varphi)$ are free then $\mu_{ab} = \mu_a \boxtimes \mu_b$. Assuming $\varphi(a) \neq 0$ one considers the moment-generating series

$$\psi_a(z) = \sum_{n \geq 1} z^n \varphi(a^n) = \varphi((1 - za)^{-1}) - 1$$

and defines $\chi_a(\psi_a(z)) = z$, $S_a(z) = \frac{z+1}{z}\chi_a(z)$ which then satisfies

$$S_{\mu_a \boxtimes \mu_b}(z) = S_{\mu_a}(z) S_{\mu_b}(z).$$

The map $\mu \to S_\mu$ is a free analogue of the Mellin transform. Surprisingly, if μ_a, μ_b are compactly supported probability measures on $(0, \infty)$, then so is $\mu_a \boxtimes \mu_b$.

The proofs of these results in [23, 24] were analytic, using operator theory and complex analysis. Later alternative combinatorial proofs were found. For more references about free convolution see [29].

8 Partial Bi-free Transforms and the Computation of the Simplest Bi-free Convolutions

The simplest bi-free convolutions arise from operations on two bi-free two-faced pairs (a, b) and (c, d) in some noncommutative probability space $(\mathcal{A}, \varphi)$. The three operations which we can consider combining addition and multiplication give rise to bi-additive, additive–multiplicative, and bi-multiplicative convolutions

$$\begin{aligned} \mu_{a+c,b+d} &= \mu_{a,b} \boxplus\boxplus \mu_{c,d} \\ \mu_{a+c,bd} &= \mu_{a,b} \boxplus\boxtimes \mu_{c,d} \\ \mu_{ac,bd} &= \mu_{a,b} \boxtimes\boxtimes \mu_{c,d}. \end{aligned}$$

Looking for the simplest situations, we may restrict our attention to "two-bands moments, starting left" that is to moments of the form $\varphi(a^p b^q)$ for a two-faced pair (a, b). Note that in case the pairs (a, b), (c, d) satisfy the additional simplifying assumption $[a, b] = 0$, $[c, d] = 0$ and since we may also find realizations of the bi-freeness so that $[a, d] = [b, c] = 0$ we will get in all three cases pairs where left and right commute (i.e., bipartite pairs):

$$[a + c, b + d] = [a + c, bd] = [ac, bd] = 0.$$

Thus the convolution operation at the level of two-bands moments actually completely describes the bi-free convolution operations in the case of bipartite pairs.

We found in [32, 33] three transforms which together with the one-variable transforms which we discussed in Sect. 7 provide the solution to computing these simplest bi-free convolutions.

If (a, b) is a two-faced pair in $(\mathcal{A}, \varphi)$, in the Banach algebra setting, the moment-generating functions we use can be written:

$$\begin{aligned} G_{a,b}(z, w) &= \varphi((z1 - a)^{-1}(w1 - b)^{-1}) \\ H_{a,b}(z, w) &= \varphi((1 - za)^{-1}(1 - wb)^{-1}) \\ F_{a,b}(z, w) &= \varphi((z1 - a)^{-1}(1 - wb)^{-1}) \end{aligned}$$

which of course have also formal power series versions. In case a, b are commuting hermitian operators and φ is given by a probability measure on $\mathbb{R}^2$, $G_{a,b}(z, w)$ is a double Cauchy–Stieltjes transform.

The reduced partial transforms are defined by the formulae

$$\begin{aligned} \tilde{R}_{a,b}(z, w) &= 1 - \frac{zw}{G_{a,b}(K_a(z), K_b(w))} \\ \tilde{S}_{a,b}(z, w) &= \frac{z+1}{z}\frac{w+1}{w}\left(1 - \frac{1+z+w}{H_{a,b}(\chi_a(z), \chi_b(w))}\right) \\ \tilde{T}_{a,b}(z, w) &= \frac{w+1}{w}\left(1 - \frac{z}{F_{a,b}(K_a(z), \chi_b(w))}\right) \end{aligned}$$

where K_a, χ_a are according to the notation used in 7. We called these transforms, "reduced" because in case $\varphi(a^p b^q) = \varphi(a^p)\varphi(b^q)$ for all $p \geq 0$, $q \geq 0$ we have $\tilde{R}_{a,b} = 0$, $\tilde{S}_{a,b} = 1$, $\tilde{T}_{a,b} = 1$.

The key properties of these transforms are that if (a, b) and (c, d) are bi-free in $(\mathcal{A}, \varphi)$ we have

$$\begin{aligned}\tilde{R}_{a+c,b+d} &= \tilde{R}_{a,b} + \tilde{R}_{c,d}\\ \tilde{S}_{ac,bd} &= \tilde{S}_{a,b}\tilde{S}_{c,d}\\ \tilde{T}_{a+c,bd} &= \tilde{T}_{a,b}\tilde{T}_{c,d}.\end{aligned}$$

To compute the bi-free convolutions at the level of two-bands moments the reduced transforms are used in conjunction with the one-variable free transforms applied to the marginals. For instance, to compute $\boxplus\boxtimes$ one uses $(R_a, S_b, \tilde{T}_{a,b})$ and $(R_c, S_d, \tilde{T}_{c,d})$ to get $R_{a+c} = R_a + R_c$, $S_{bd} = S_b S_d$ and $\tilde{T}_{a+c,bd} = \tilde{T}_{a,b}\tilde{T}_{c,d}$. Note that the computation of $\tilde{T}_{a+c,bd}$ requires the knowledge of K_{a+c} and χ_{bd} which are obtained from R_{a+c} and S_{bd}. Then from $(R_{a+c}, S_{bd}, \tilde{T}_{a+c,bd})$ one finds G_{a+c}, ψ_{bd} and then one can recover from $\tilde{T}_{a+c,bd}$ the moment-generating function $F_{a+c,bd}$.

Our work [32, 33] about the $\tilde{R}$, $\tilde{S}$, and $\tilde{T}$ transforms is analytic. It takes as starting point our one-variable results in free probability about the R and S-transforms, but instead of our original proofs, the alternative proofs of Uffe Haagerup [11] turned out to be better suited for approaching the bi-free generalization. Soon after these results were obtained analytically, Paul Skoufranis [16, 17] was able to find also alternative combinatorial proofs.

9 Bi-free Extreme Values

In classical probability theory, the max of two independent random variables has as distribution function the product of the distribution functions of the random variables. The realization that there is a free probability analogue of this basic observation was the starting point of our joint work with Gerard Ben Arous [1] on free extreme values. We showed in [34] how to extend this basic fact also to the bi-free setting, which opens the way to study basic bi-free extreme value questions. We will explain the free "dictionary" and go on to explain the bi-free "dictionary" for extreme values.

The noncommutative probability framework $(\mathcal{A}, \varphi)$ will be that of a von Neumann algebra $\mathcal{A}$ and φ a normal state. If $P, Q \in \mathcal{A}$ are hermitian projections, then $P \wedge Q$, $P \vee Q \in \mathcal{A}$ denote the projections unto $P\mathcal{H} \cap Q\mathcal{H}$ and $\overline{P\mathcal{H} + Q\mathcal{H}}$ where $\mathcal{H}$ is the Hilbert space on which $\mathcal{A}$ acts. If $X = X^*$, $Y = Y^*$ are hermitian operators then $X \vee Y$ is defined with respect to Ando's spectral order, that is

$$E(X \vee Y; (-\infty, a]) = E(X; (-\infty, a]) \wedge E(Y; (-\infty, a]), \ a \in \mathbb{R}$$

where $E(X; \omega)$ is the spectral projection of X for the Borel set $\omega \subset \mathbb{R}$. There is a similar definition of $X \wedge Y$.

If $(X_i)_{i\in I}$, $(Y_i)_{i\in I}$ are families of hermitian elements in $(\mathcal{A}, \varphi)$, one defines a free max-convolution for distributions of such families, so that

$$\mu_{(X_i)_{i\in I}} \boxed{\vee} \mu_{(Y_i)_{i\in I}} = \mu_{(X_i \vee Y_i)_{i\in I}}.$$

In the one-variable case the distribution corresponds to a probability measure μ on $\mathbb{R}$ with compact support. To compute the free max-convolution, it is convenient to pass to the distribution function

$$F_\mu(a) = \mu((-\infty, a])$$

and to consider the corresponding operation, denoted also by $\boxed{\vee}$, on distribution functions. Then one has

$$(F \boxed{\vee} G)(t) = (F(t) + G(t) - 1)_+.$$

This is what replaces in free probability in this case the multiplication of the distribution functions.

For the bi-free extension one considers a bi-free pair of faces of hermitian elements in $(\mathcal{A}, \varphi)$

$$z' = ((z'_i)_{i\in I}, (z'_j)_{j\in J})$$
$$z'' = ((z''_i)_{i\in I}, (z''_j)_{j\in J})$$

and

$$z' \vee z'' = ((z'_i \vee z''_i)_{i\in I}, (z''_j \vee z''_j)_{j\in J}).$$

This gives then rise to a bi-free max-convolution operation on the distributions

$$\mu_{z'} \boxed{\vee}\boxed{\vee} \mu_{z''} = \mu_{z' \vee z''}.$$

The simplest case of bi-free max-convolutions to consider is that of the distributions of two-faced pairs of hermitian operators which commute (i.e., the bipartite case). The distribution of such a pair is described by a probability measure μ with compact support on $\mathbb{R}^2$. The bivariate distribution function is then

$$F_\mu(s, t) = \mu((-\infty, s] \times (-\infty, t]).$$

The marginals of a bivariate distribution function $F(s, t)$ will be denoted by $F_1(s)$ and $F_2(t)$. The operation on the bivariate distribution functions is also denoted by $\boxed{\vee}\boxed{\vee}$. The result of [34] is that if F, G are bivariate distribution functions and $H = F \boxed{\vee}\boxed{\vee} G$ the $H_j = (F_j + G_j - 1)_+$, $j = 1, 2$ and

$$\frac{H_1(s)H_2(t)}{H(s,t)} - 1 = \left(\frac{F_1(s)F_2(t)}{F(s,t)} - 1\right) + \left(\frac{G_1(s)G_2(t)}{G(s,t)} - 1\right)$$

if $F(s,t) > 0$, $G(s,t) > 0$, $H_1(s) > 0$, $H_2(t) > 0$ and $H(s,t) = 0$ otherwise.

This opens the way to finding the bi-freely max-stable and max-infinitely-divisible laws, which is reduced to a classical analysis problem. Note that the determination of the univariate free max-stable laws in [1] showed that these are generalized Pareto laws, related to "peaks over thresholds" in classical extreme values theory. One may wonder whether the bi-free max-stable and max-infinitely divisible laws will also turn out to be similarly related to classical bivariate extreme values questions.

10 Concluding Remarks

The replacement of the complex field $\mathbb{C}$ by a general algebra B in bi-freeness was briefly sketched in [31] and then developed in detail together with the corresponding combinatorics of cumulants in [5]. In free probability B-valued R-transforms were initially found analytically in [27], the bi-free B-valued transforms have now been developed using combinatorics in [19].

General infinite divisibility results for the simplest cases have already been obtained [8, 9]. See also [6] about the operator theory side of bi-free Gaussian pairs.

In free probability random matrix realizations in the large N limit play a key role. The question about random matrix realizations for bi-free probability has not yet been clarified. On one hand realizations for certain bipartite situations, that is when left and right commute, are easy to construct but do not seem to add much to what we already know from free probability. Going beyond the bipartite case the best results at present are in [18].

The question whether there are de Finetti type theorems for bi-freeness was considered in [7]. It is not clear at present whether the Köstler–Speicher theorem [13] has a complete bi-free analogue.

Acknowledgments Research supported in part by NSF Grant DMS-1301727.

References

1. Ben Arous, G., Voiculescu, D.V.: Free extreme values. Ann. Probab. **34**(5), 2037–2059 (2006)
2. Bercovici, H., Pata, V.: Stable laws and domains of attraction in free probability, with an appendix by P. Biane, Ann. Math. (2) **149**(3), 1023–1060 (1999)
3. Biane, P., Voiculescu, D.V.: A free probability analogue of the Wasserstein metric on the trace-state space. Geom. Funct. Anal. **11**(6), 1125–1138 (2001)
4. Charlesworth, I., Nelson, B., Skoufranis, P.: On two-faced families of non-commutative random variables. Can. J. Math. **67**(6), 1290–1325 (2015)
5. Charlesworth, I., Nelson, B., Skoufranis, P.: Combinatorics of bi-freeness with amalgamation. Commun. Math. Phys. **338**(2), 801–847 (2015)
6. Dykema, K.J., Na, W.: Principal functions for bi-free central limit distributions. arXiv:1510.03328

7. Freslon, A., Weber, M.: On bi-free De Finetti theorems. arXiv:1501.05124
8. Gao, M.: Two-faced families of non-commutative random variables having bi-free infinitely divisible distributions. arXiv:1507.08270
9. Gu, Y., Huang, H.-W., Mingo, J.A.: An analogue of the Levy–Hincin formula for bi-free infinitely divisible distributions. arXiv:1501.05369
10. Guionnet, A., Shlyakhtenko, D.: Free monotone transport. Invent. Math. **197**(3), 613–661 (2014)
11. Haagerup, U.: On Voiculescu's R- and S-transforms for free non-commuting random variables. In: Free Probability. Waterloo, ON (1995). Fields Institute Communications, vol. 12, pp. 127–148. American Mathematical Society, Providence, RI (1997)
12. 't Hooft, G.: A planar diagram theory for strong interactions. Nucl. Phys. **B72**, 461–473 (1974)
13. Köstler, C., Speicher, R.: A non-commutative de Finetti theorem: invariance under quantum permutations is equivalent to freeness with amalgamation. Commun. Math. Phys. **291**(2), 473–490 (2009)
14. Mastnak, M., Nica, A.: Double-ended queues and joint moments of left–right canonical operators on full Fock space. arXiv:1312.0269
15. Muraki, N.: The five independences as natural products. Infin. Dimen. Anal. Quantum Probab. Relat. Top. **06**, 337–371 (2003)
16. Skoufranis, P.: Independences and partial R-transforms in bi-free probability. arXiv:1410.4265
17. Skoufranis, P.: A combinatorial approach to Voiculescu's bi-free partial transforms. arXiv:1504.06005
18. Skoufranis, P.: Some bi-matrix models for bi-free limit distributions. arXiv:1506.03896
19. Skoufranis, P.: On operator-valued bi-free distributions. arXiv:1510.03896
20. Speicher, R.: On universal products. In: Free Probability Theory. Waterloo ON (1995). Fields Institute Communication, vol. 12, pp. 257–266. American Mathematical Society, Providence, RI (1997)
21. Speicher, R.: Combinatorial theory of the free product with amalgamation and operator-valued free probability theory. Mem. Am. Math. Soc. **132**(627) (1998)
22. Voiculescu, D.V.: Symmetries of some reduced free product C^*-algebras. In: Operator algebras and their connections with topology and ergodic theory, Busteni (1983). Lecture Notes in Mathematics, vol. 132, pp. 556–588. Springer (1985)
23. Voiculescu, D.V.: Addition of certain non-commuting random variables. J. Funct. Anal. **66**(3), 323–346 (1986)
24. Voiculescu, D.V.: Multiplication of certain non-commuting random variables. J. Oper. Theory **18**(2), 223–235 (1987)
25. Voiculescu, D.V.: Limit laws for random matrices and free products. Invent. Math. **104**(1), 201–220 (1991)
26. Voiculescu, D.V., Dykema, K.J., Nica, A.: Free random variables, A non-commutative probability approach to free products with applications to random matrices, operator algebras and harmonic analysis on free groups, CRM Monograph Series 1. American Mathematical Society, Providence, RI (1992)
27. Voiculescu, D.V.: Operations on certain non-commutative operator-valued random variables, Recent advances in operator algebras. Orléans (1995). Asterisque, No. 232, 243–273 (1992)
28. Voiculescu, D.V.: Free probability theory: random matrices and von Neumann algebras. In: Proceedings of the International Congress of Mathematicians, vol. 1, **2**, pp. 227–241. Zürich (1994), Birkhäuser, Basel (1995)
29. Voiculescu, D.V.: Lectures on free probability. In: Lectures on probability theory and statistics. Saint–Flour (1998). Lecture Notes in Mathematics, vol. 1738, pp. 279–349. Springer, Berlin (2000)
30. Voiculescu, D.V.: Free entropy. Bull. London Math. Soc. **34**(3), 257–278 (2002)
31. Voiculescu, D.V.: Free probability for pairs of faces I. Commun. Math. Phys. **332**(3), 955–980 (2014)

32. Voiculescu, D.V.: Free probability for pairs of faces II: 2-variables bi-free partial R-transform and systems with rank ≤ 1 commutation. Ann. Inst. Henri Poincare Probab. Stat. **52**(1), 1–15 (2016)
33. Voiculescu, D.V.: Free probability for pairs of faces III: 2-variables bi-free partial S- and T-transforms. arXiv:1504.03765
34. Voiculescu, D.V.: Free probability for pairs of faces IV: bi-free extremes in the plane. arXiv:1505.05020

Stability of the Prandtl Boundary Layers

Y.-G. Wang

Abstract This note is to survey our recent study on the stability and instability of the Prandtl boundary layers in incompressible viscous flows near a physical boundary. Both of the two-dimensional and three-dimensional problems are considered. First, we present an energy method for studying the well-posedness of the two-dimensional Prandtl boundary layer equations in Sobolev spaces under the Oleinik monotonicity condition on the tangential velocity field. Then, we give an instability result for the Prandtl equations in three space variables, which shows that the monotonicity condition of tangential velocity fields is not sufficient for the well-posedness of the three-dimensional Prandtl equations. Later, we present a well-posedness result of the three-dimensional Prandtl equations for a special structured flow. These results show that a shear flow is linearly and nonlinearly stable for the three-dimensional Prandtl equations, if and only if, the tangential velocity field direction is invariant with respect to the normal variable.

Keywords Prandtl boundary layers · Energy method · Monotonic flow · Three-dimensional stable and unstable flow

Mathematics Subject Classication (2010). 35M13 · 35Q35 · 76D10 · 76D03 · 76N20

Y.-G. Wang (✉)
School of Mathematical Sciences, MOE-LSC and SHL-MAC,
Shanghai Jiao Tong University, Shanghai 200240, People's Republic of China
e-mail: ygwang@sjtu.edu.cn

T. Qian and L.G. Rodino (eds.), *Mathematical Analysis, Probability and Applications – Plenary Lectures*, Springer Proceedings in Mathematics & Statistics 177, DOI 10.1007/978-3-319-41945-9_9

1 Introduction

Incompressible flows occupied in a domain Ω of $\mathbb{R}^d$ $(d = 2, 3)$ can be described by the following well-known Navier-Stokes equations:

$$\begin{cases} \partial_t \mathbf{u}^\epsilon + (\mathbf{u}^\epsilon \cdot \nabla)\mathbf{u}^\epsilon + \nabla p^\epsilon - \epsilon\Delta\mathbf{u}^\epsilon = 0, & t > 0,\ x \in \Omega \\ \nabla \cdot \mathbf{u}^\epsilon = 0 \end{cases} \tag{1.1}$$

where $\mathbf{u}^\epsilon = (u_1^\epsilon, \dots, u_d^\epsilon)^T$ is the velocity field, p^ϵ is the pressure, ∇ and Δ are the gradient and Laplacian in the space variables $x = (x_1, \dots, x_d)^T$, and ϵ is the viscosity. The boundary condition on $\partial\Omega$ plays an important role in determining the behavior of flow near the physical boundary. One classical type is the so-called the nonslip condition,

$$\mathbf{u}^\epsilon|_{\partial\Omega} = 0. \tag{1.2}$$

To understand the behavior of a flow near the physical boundary is a classical and challenging problem, not only in developing mechanical theory but also in application. It was Prandtl [23] in 1904 who first noted that away from the physical boundary, the flow is mainly driven by the convection while the viscous force can be neglected approximately due to friction being tiny, and only in a small neighborhood of the boundary, the forces arising from the convection and viscosity of the flow are important simultaneously. Mathematically, away from $\partial\Omega$, the flow given by (1.1) and (1.2) can be approximated by the following incompressible Euler equations:

$$\begin{cases} \partial_t \mathbf{u} + (\mathbf{u} \cdot \nabla)\mathbf{u} + \nabla p = 0, & t > 0,\ x \in \Omega \\ \nabla \cdot \mathbf{u} = 0 \end{cases} \tag{1.3}$$

with the boundary condition

$$\mathbf{u} \cdot \vec{n}|_{\partial\Omega} = 0 \tag{1.4}$$

a simple consequence of (1.2), where $\vec{n}$ is the unit outward normal vector on $\partial\Omega$.

The inconsistence between the boundary conditions (1.2) and (1.4) generates a thin layer near the physical boundary $\partial\Omega$, in which the tangential velocity fields change rapidly. This thin transition layer was first studied by Prandtl in his seminal work [23], it was called later as the boundary layer. As Prandtl claimed, both of the forces arose from the convection and the viscosity are important in the layer, so by multi-scale analysis one can deduce that the boundary layer thickness is $\sqrt{\epsilon}$. Near a regular point of the physical boundary $\partial\Omega$, one can transform $\partial\Omega$ into the flat one, e.g., see [13], and can suppose Ω to be the half space $\Omega = \{x = (x', x_d) | x' \in \mathbb{R}^{d-1}, x_d > 0\}$ with the tangential variables $x' = (x_1, \dots, x_{d-1})^T$.

Take the following ansatz of the asymptotic expansions:

$$\begin{cases} u_j(t, x) = u_j^p(t, x', \frac{x_d}{\sqrt{\epsilon}}) + o(1), & j = 1, \dots, d-1 \\ u_d(t, x) = \sqrt{\epsilon} u_d^p(t, x', \frac{x_d}{\sqrt{\epsilon}}) + o(\sqrt{\epsilon}) \end{cases} \tag{1.5}$$

for the solutions of the problem (1.1) near the boundary $\{x_d = 0\}$, where $u_j^p(t, x', \eta)$ $(1 \le j \le d)$ approach the Euler flow on the boundary $\{x_d = 0\}$ rapidly as $\eta = \frac{x_d}{\sqrt{\epsilon}} \to +\infty$. Plugging the ansatz (1.5) into the problem (1.1), it follows that the pressure does not change in the layer, and the boundary layer profiles $\mathbf{u}^p = (u_1^p, \dots, u_d^p)^T$ satisfy the following d−dimensional Prandtl equations in $\{(t, x', \eta) | t > 0, x' \in \mathbb{R}^{d-1}, \eta > 0\}$:

$$\begin{cases} \partial_t u_j^p + (\mathbf{u}_\tau^p \cdot \nabla_{x'}) u_j^p + u_d^p \partial_\eta u_j^p + \partial_{x_j} P = \partial_\eta^2 u_j^p, \quad j = 1, \dots, d-1 \\ \nabla_{x'} \cdot \mathbf{u}_\tau^p + \partial_{x_d} u_d^p = 0 \\ \mathbf{u}^p|_{\eta=0} = 0, \\ \lim\limits_{\eta \to +\infty} u_j^p(t, x', \eta) = u_j^E(t, x', 0), \quad j = 1, \dots, d-1 \end{cases} \tag{1.6}$$

where $\mathbf{u}_\tau^p = (u_1^p, \dots, u_{d-1}^p)^T$ are tangential velocity fields, $\nabla_{x'} = (\partial_{x_1}, \dots, \partial_{x_{d-1}})^T$, $P(t, x') = p^E(t, x', 0)$, and $(\mathbf{u}^E, p^E)$ is the Euler flow determined by the problem (1.3) and (1.4).

The first theoretic study of the Prandtl equation problem (1.6) was developed by Oleinik and her collaborators in [21, 22], in which they had obtained the well-posedness of the two-dimensional Prandtl boundary layer equations, $d = 2$, in the class of tangential velocity u_1^p being strictly monotonic with respect to the normal variable $\eta > 0$, by introducing the Crocco transformation. Under the Oleinik monotonicity assumption and an additional favorite condition of pressure, $\partial_{x_1} P < 0$, Xin and Zhang [29] obtained a global weak solution to the two-dimensional Prandtl equations.

Without the monotonicity assumption of the tangential velocity, many works show that the two-dimensional Prandtl equations are linearly and nonlinearly unstable in the Sobolev spaces in general. E and Engquist in [27] constructed a finite time blowup solution to the Prandtl equation. In [6], Grenier showed that the unstable Euler shear flow $(u_s(y), 0)$ with $u_s(y)$ having an inflection point implies instability for the two-dimensional Prandtl equations by a spectral argument and the WKB method. In the spirit of Grenier's approach, Gérard-Varet and Dormy [4] showed that if the shear flow profile $(u^s(t, y), 0)$ of the two-dimensional Prandtl equation has a nondegenerate critical point, then it leads to a strong linear ill-posedness of the Prandtl equation in the Sobolev framework. Guo and Nguyen in [8] proved that the nonlinear two-dimensional Prandtl equation is ill-posed near nonstationary and non-monotonic shear flows, and showed that the asymptotic boundary layer expansion is not valid for non-monotonic shear layer flows in Sobolev spaces. For the related mathematical results and discussions, also see the review papers [2, 28]. Another approach for studying the Prandtl equations is in the frame of analytic functions. In [24, 25], Sammartino and Caflisch obtained the local existence of analytic solutions to the two-dimensional and three-dimensional Prandtl equation, and a rigorous theory on the stability of boundary layers with analytic data in the framework of the abstract

Cauchy-Kowaleskaya theory. This result was extended to the function space which is only analytic in the tangential variable in [3, 14].

A rigorous mathematical theory of Prandtl's boundary layer theory, the solutions of the Navier-Stokes equations can be approximately decomposed into the solutions of the Euler equations away from the boundary and the solutions of the Prandtl boundary layer equations in the small viscosity limit is almost completely unknown, except a few special cases, such as circularly symmetric flows [15], in the space of analytic solutions [25], under the assumption of the support of vortices in the Euler flow being away from boundary [16], and a steady flow over a moving plate [7] in the two-dimensional problem. It is well known that the energy method works well for both of the Navier-Stokes equations and the Euler equations, as proposed in [2], it is very interesting to develop an energy method for studying the well-posedness of the Prandtl equations in Sobolev spaces.

The first part of this note shall review our recent study of a direct energy method for the two-dimensional Prandtl equations under the Oleinik monotonicity assumption, which was collaborated with Alexandre, Xu and Yang in [1].

The strict monotonicity of the tangential velocity implies the positivity of the vorticity of the flow in the two-dimensional boundary layer, as shown in the existing works, this positivity of vorticity is crucial to have the stability of the boundary layer in the existing literature. It is known that the vorticity of the flow is much more complicated in the three-dimensional flow in general, as shown in [18], extra instability could be generated by the secondary flow in the three-dimensional boundary layers. Till now, there is basically no any well-posedness theory for the three-dimensional Prandtl equations, except the analytic case [24]. The well-posedness of the Prandtl equations in three space variables is one of the important open questions proposed by Oleinik and Samokhin in their classical monograph [21].

The second part of this note is to review our recent rigorous study on the stability and instability of the three-dimensional Prandtl equations, collaborated with Liu and Yang in [10–12]. From these works, we found that the three-dimensional shear flow is stable in the Prandtl equations if and only if the direction of the tangential velocity field is invariant with respect to the normal vector to the boundary, this special structure avoids the appearance of the complicated secondary flow in the three-dimensional Prandtl boundary layers.

The remainder of this note is arranged as follows. In Sect. 2, we present the well-posedness result on the two-dimensional Prandtl equations by the energy method, and in Sect. 3, we study the stability and instability of the three-dimensional Prandtl equations.

2 Well-Posedness of the Two-Dimensional Prandtl Equations

The proposal of this section is to study the following initial boundary value problem for the two-dimensional Prandtl equations in $\{(t,x,z)|t>0, x\in\mathbb{R}, z>0\}$:

$$\begin{cases} u_t + uu_x + vu_z - u_{zz} + P_x = 0, \quad t>0, \quad x\in\mathbb{R}, \quad z>0, \\ u_x + v_z = 0, \\ u|_{z=0} = v|_{z=0} = 0, \quad \lim\limits_{z\to+\infty} u(t,x,z) = U(t,x), \\ u|_{t=0} = u_0(x,z) \end{cases} \tag{2.1}$$

under the assumption that

$$u_0(x,z) \text{ is strictly monotonic in } z, \tag{2.2}$$

where $P(t,x) = p^E(t,x,0)$ and $U(t,x) = u^E(t,x,0)$ satisfy the Bernoulli law

$$U_t + UU_x + P_x = 0.$$

For simplicity of presentation, we shall only consider the case of an uniform outflow, i.e., $U(t,x) = 1$.

Linearized Prandtl Equations

Let $(\tilde{u},\tilde{v})$ be a smooth background state of (2.1) satisfying

$$\partial_z\tilde{u}(t,x,z) > 0, \qquad \partial_x\tilde{u} + \partial_z\tilde{v} = 0. \tag{2.3}$$

It is easy to know the linearized problem of (2.1) at $(\tilde{u},\tilde{v})$ can be written as

$$\begin{cases} \partial_t u + \tilde{u}\partial_x u + \tilde{v}\partial_z u + u\partial_x\tilde{u} + v\partial_z\tilde{u} - \partial_z^2 u = f, \\ \partial_x u + \partial_z v = 0, \\ u|_{z=0} = v|_{z=0} = 0, \quad \lim\limits_{z\to+\infty} u(t,x,z) = 0, \\ u|_{t\le 0} = 0\,. \end{cases} \tag{2.4}$$

It is easy to represent $v(t,x,z)$ in term of u by

$$v(t,x,y) = -\int_0^z \partial_x u(t,x,\tilde{z})d\tilde{z} \tag{2.5}$$

from the divergence-free constraint and the boundary condition $v(t,x,0) = 0$ given in (2.4).

Plugging the representation (2.5) into the Prandtl equation, it follows that $u(t, x, z)$ satisfies the following equation:

$$\partial_t u + \tilde{u}\partial_x u + \tilde{v}\partial_z u + u\partial_x \tilde{u} - \partial_z \tilde{u} \int_0^z \partial_x u(t, x, \tilde{z})d\tilde{z} - \partial_z^2 u = f. \tag{2.6}$$

The crucial difficult term for estimating the energy of u from the Eq. (2.6) is the integral term

$$\int_0^z \partial_x u(t, x, \tilde{z})d\tilde{z}.$$

Noting that the coefficient $\partial_z \tilde{u}$ of this integral term in (2.6) has a good sign due to monotonicity assumption (2.3), by introducing the transformation

$$w(t, x, z) = \left(\frac{u}{\partial_z \tilde{u}}\right)_z (t, x, z), \quad \text{i.e.} \quad u(t, x, z) = (\partial_z \tilde{u}) \int_0^z w(t, x, \tilde{z})d\tilde{z}. \tag{2.7}$$

it follows that the problem (2.4) is equivalent to the following one for the unknown $w(t, x, z)$:

$$\begin{cases} \partial_t w + \partial_x(\tilde{u}w) + \partial_z(\tilde{v}w) - 2\partial_z(\eta w) + \partial_z(\zeta \int_0^z w(t, x, \tilde{z})d\tilde{z}) - \partial_z^2 w = \partial_z \tilde{f}, \\ (\partial_z w + 2\eta w)|_{z=0} = -\tilde{f}|_{z=0}, \\ w|_{t\leq 0} = 0, \end{cases} \tag{2.8}$$

where

$$\eta = \frac{\partial_z^2 \tilde{u}}{\partial_z \tilde{u}}, \quad \zeta = \frac{(\partial_t + \tilde{u}\partial_x + \tilde{v}\partial_z - \partial_z^2)\partial_z \tilde{u}}{\partial_z \tilde{u}}, \quad \tilde{f} = \frac{f}{\partial_z \tilde{u}}.$$

One can obtain a priori estimates in the weighted Sobolev spaces for the solution to the problem (2.8), but with a loss of regularity with respect to the background states $(\tilde{u}, \tilde{v})$. Returning back the transformation (2.7), we deduce

Theorem 2.1 ([1], Theorem 3.1) *Under certain regularity assumption on the background state $(\tilde{u}, \tilde{v})$ and the decay rate of $\tilde{u}(t, x, z)$ as $z \to +\infty$, and compatibility conditions for the problem* (2.4), *we have the energy estimates of the solution (u, v) to the problem* (2.4) *in the weighted Sobolev spaces, with a fixed order loss of regularity with respect to $(\tilde{u}, \tilde{v})$.*

Remark 2.1 (1) By using the energy estimates for the linearized problem (2.4), and adapting the Nash–Moser iteration scheme [9, 19, 20] for the nonlinear problem (2.1), we have obtained the existence of a classical solution local time to the problem (2.1) under certain order compatibility conditions and the monotonicity assumption (2.2). One can find the detail in [1].

(2) In collaboration with Xie and Yang in [26], by using a similar energy method we have obtained the well-posedness of a classical solution in the class of monotonic tangential velocity to the compressible Prandtl boundary layer equations in two space variables, which is derived from the compressible isentropic Navier-Stokes

equations with nonslip boundary condition in the small viscosity limit. Recently, collaborating with Gong and Guo, we have also studied the incompressible limit for the compressible Prandtl equations in [5].

(3) Another energy method has been introduced by Masmoudi and Wong in [17] for the two-dimensional Prandtl equations in the class of monotonic tangential velocity.

3 The Three-Dimensional Prandtl Equations

In this section, we study the well-posedness and ill-posedness of the following initial boundary value problem for the three-dimensional Prandtl equations in $\{(t,x,y,z)|t>0,(x,y)\in D, z>0\}$ for a fixed domain D of $\mathbb{R}^2$:

$$\begin{cases} \partial_t u + (u\partial_x + v\partial_y + w\partial_z)u - \partial_z^2 u = -\partial_x P, \\ \partial_t v + (u\partial_x + v\partial_y + w\partial_z)v - \partial_z^2 v = -\partial_y P, \\ \partial_x u + \partial_y v + \partial_z w = 0, \\ u|_{z=0} = w|_{z=0} = 0, \quad \lim\limits_{z\to+\infty}(u,v) = (U(t,x,y), V(t,x,y)), \\ (u,v)|_{\Gamma^-} = (u_1(t,x,y,z), v_1(t,x,y,z)), \\ (u,v)|_{t=0} = (u_0(x,y,z), v_0(x,y,z)), \end{cases} \tag{3.1}$$

where $\Gamma^- = \mathbb{R}_t^+ \times \gamma_- \times \mathbb{R}_z^+$ with $\gamma_- = \{(x,y)\in\partial D | n_x u_1 + n_y v_1 < 0\}$, $\vec{n} = (n_x, n_y)^T$ being the outward unit normal vector of ∂D, $P(t,x,y) = p^E(t,x,y,0)$ and $U(t,x,y) = u^E(t,x,y,0)$, $V(t,x,y) = v^E(t,x,y,0)$ are the Euler flow on the boundary.

As we pointed out in Introduction, since the tangential flow given by (3.1) are two-dimensional, the vorticity of the flow is very complicated, so to study the stability of the three-dimensional Prandtl boundary layer is very challenging. We shall first study the stability mechanism for a shear flow, then consider the stability of a general flow.

3.1 *Instability of a Shear Flow*

Let $(u^s(t,z), v^s(t,z), 0)$ be a given shear flow solution to the three-dimensional Prandtl equations, then from (1.6) we know that u^s and v^s should satisfy

$$\begin{cases} \partial_t u^s - \partial_z^2 u^s = 0, \qquad \partial_t v^s - \partial_z^2 v^s = 0, \\ (u^s, v^s)|_{z=0} = 0, \qquad \lim\limits_{z\to+\infty}(u^s, v^s) = (U_0, V_0), \\ (u^s, v^s)|_{t=0} = (U_s, V_s)(z) \end{cases} \tag{3.2}$$

with U_0, V_0 being constants.

Consider the following problem for the three-dimensional linearized Prandtl equations at $(u^s, v^s, 0)$ in $\Omega \triangleq \{(t, x, y, z) : t > 0, (x, y) \in \mathbb{T}^2, z \in \mathbb{R}^+\}$:

$$\begin{cases} \partial_t u + (u^s \partial_x + v^s \partial_y)u + w\partial_z u^s - \partial_z^2 u = 0 \\ \partial_t v + (u^s \partial_x + v^s \partial_y)v + w\partial_z v^s - \partial_z^2 v = 0 \\ \partial_x u + \partial_y v + \partial_z w = 0 \\ (u, v, w)|_{z=0} = 0, \quad \lim_{z\to+\infty} (u, v) = (0, 0), \\ (u, v)|_{t=0} = (u_0(x, y, z), v_0(x, y, z)) . \end{cases} \tag{3.3}$$

It is not difficult to obtain the following results:

Proposition 3.1 ([12], Sect. 2) *(1) When the initial data (u_0, v_0) are analytic in (x, y), then the problem* (3.3) *has a local solution (u, v, w) analytic in (x, y).*

(2) When the shear flow $u^s(t, z)$ satisfies $\partial_z u^s(t, z) > 0$, and the initial data (u_0, v_0) are analytic in y and H^m in x, then the problem (3.3) *has a local classical solution (u, v, w) which is analytic in y as well.*

A natural and interesting question is whether the well-posedness of the linearized Prandtl equations (3.3) is still true in the Sobolev spaces if one imposes monotonicity condition on both tangential velocity components u^s, v^s but without analyticity assumption anymore.

From the following discussion, we shall see that the answer to the above question is no in general. To state our instability result, we first introduce notations: Denote by $T(t, s)\big((u_0, v_0)\big) = (u, v)(t, \cdot)$ the solution operator of the linearized Prandtl equations (3.3) with $(u, v)|_{t=s} = (u_0, v_0)$, and

$$H^m_\alpha := H^m\big(\mathbb{T}^2_{x,y}; L^2_\alpha(\mathbb{R}^+_z)\big),$$

with $L^2_\alpha(\mathbb{R}^+_z) = \{u | e^{\alpha z} u \in L^2\}$.

Theorem 3.1 ([12], Theorem 1) *(1) If the initial data of the shear flow $(u^s, v^s, 0)$ satisfy that there is $z_0 > 0$ such that,*

$$V_s'(z_0)U_s''(z_0) \neq U_s'(z_0)V_s''(z_0) \tag{3.4}$$

Then there exists $\sigma > 0$ such that for all $\delta > 0$,

$$\sup_{0\le s\le t\le\delta} \left\| e^{-\sigma(t-s)\sqrt{|\partial_{\mathcal{T}}|}} T(t, s) \right\|_{L(H^m_\alpha, H^{m-\mu}_\alpha)} = +\infty \tag{3.5}$$

for all $m > 0$ and $\mu \in [0, \frac{1}{4})$, with the operator $\partial_{\mathcal{T}} = \partial_x$ or ∂_y.

(2) There exists an initial shear layer (U_s, V_s) *and* $\sigma > 0$*, such that for all* $\delta > 0$ *and* $m_1, m_2 > 0$,

$$\sup_{0 \le s \le t \le \delta} \left\| e^{-\sigma(t-s)\sqrt{|\partial_T|}} T(t,s) \right\|_{L(H_\alpha^{m_1}, H_\alpha^{m_2})} = +\infty. \tag{3.6}$$

Discussion of the Proof Idea:
To study the instability mechanism in the problem (3.3), as the assertions (3.5) and (3.6) hold for any time interval $[0, \delta]$, so as in [4], one can first consider the following linear problem by frozing the coefficient functions of the equations (3.3) at $t = 0$:

$$\begin{cases} \partial_t u + (U_s \partial_x + V_s \partial_y) u + w U_s' - \partial_z^2 u = 0, & \text{in } \Omega, \\ \partial_t v + (U_s \partial_x + V_s \partial_y) v + w V_s' - \partial_z^2 v = 0, & \text{in } \Omega, \\ \partial_x u + \partial_y v + \partial_z w = 0, & \text{in } \Omega, \\ (u, v, w)|_{z=0} = 0, \quad \lim\limits_{z \to +\infty} (u, v) = 0. \end{cases} \tag{3.7}$$

A crucial fact for the problem (3.7) is that, first from the assumption (3.4) we know that $U_s'(z_0)$ and $V_s'(z_0)$ can not vanish simultaneously, for example, we assume that $U_s'(z_0) \neq 0$, then from the assumption (3.4), we get that the tangential velocity of the background state

$$V_s(z) - \frac{V_s'(z_0)}{U_s'(z_0)} U_s(z) \ \ \text{has a non} - \text{degenerate critical point at} \ \ z = z_0. \tag{3.8}$$

Next, we construct solutions of the problem (3.7) in the form of planar waves in (x, y),

$$(u, v, w)(t, x, y, z) = e^{ik(y - ax + \lambda(k)t)} (\hat{u}^k, \hat{v}^k, \hat{w}^k)(z),$$

with $a = \frac{V_s'(z_0)}{U_s'(z_0)}$.

Set $\epsilon = \frac{1}{k}$. By using the divergence-free condition given in (3.7), we know that the solution of the above form can be rewritten as

$$\begin{cases} (u, v)(t, x, y, z) = e^{i\epsilon^{-1}(y - ax + \lambda_\epsilon t)} (u_\epsilon', v_\epsilon')(z), \\ w(t, x, y, z) = -i\epsilon^{-1} e^{i\epsilon^{-1}(y - ax + \lambda_\epsilon t)} (v_\epsilon - a u_\epsilon)(z). \end{cases} \tag{3.9}$$

Plugging the solution representation (3.9) into the equations given in (3.7), it follows that $w_\epsilon(z) \triangleq (v_\epsilon - a u_\epsilon)(z)$ satisfies the following problem in $\{z > 0\}$:

$$\begin{cases} (\lambda_\epsilon + W_s) w_\epsilon' - W_s' w_\epsilon + i\epsilon w_\epsilon^{(3)} = 0, \\ w_\epsilon(0) = w_\epsilon'(0) = 0 \end{cases} \tag{3.10}$$

with $W_s(z) \triangleq (V_s - a U_s)(z)$. Note that the above problem of $w_\epsilon(z)$ is the same one studied by Gérard-Varet and Dormy in [4] for the linearized two-dimensional Prandtl equations, in which they had obtained the following results

Lemma 3.1 ([4]) *For the problem* (3.10), *there is a complex number* τ *with* $\mathrm{Im}\tau < 0$, *such that*

$$\begin{cases} \lambda_\epsilon \sim -W_s(z_0) + \epsilon^{\frac{1}{2}}\tau, \\ w_\epsilon(z) \sim H(z-z_0)\big[W_s(z) - W_s(z_0) + \epsilon^{\frac{1}{2}}\tau\big] + \epsilon^{\frac{1}{2}} W(\frac{z-z_0}{\epsilon^{\frac{1}{4}}}), \end{cases} \tag{3.11}$$

where $W(Z)$ *is the solution to the following problem:*

$$\begin{cases} \left(\tau + W_s''(z_0)\frac{Z^2}{2}\right) W' - W_s''(z_0) Z W + i W^{(3)} = 0, \quad Z \neq 0, \\ [W]\,|_{Z=0} = -\tau, \ [W']\big|_{Z=0} = 0, \ [W'']\,|_{Z=0} = -W_s''(z_0), \\ \lim\limits_{Z\to\pm\infty} W = 0, \quad \textit{exponentially}. \end{cases} \tag{3.12}$$

Noting that $\mathrm{Im}\tau < 0$, plugging the asymptotic expansion of λ_ϵ given in (3.11) into the representation (3.9), one can get the instability of the solution to the problem (3.7) immediatcly.

Sketch of The Proof of Theorem 3.1:
Step 1: Construct an approximate solution of the linearized problem (3.3) in the form:

$$(u_\epsilon, v_\epsilon, w_\epsilon)(t,x,y,z) = e^{i\epsilon^{-1}(y-ax)}\big(U_\epsilon, V_\epsilon, W_\epsilon\big)(t,z) \tag{3.13}$$

such that

$$C_0^{-1} e^{\frac{\sigma_0 t}{\sqrt{\epsilon}}} \le \|(U_\epsilon, V_\epsilon)(t,\cdot)\|_{L^2_\alpha} \le C_0 e^{\frac{\sigma_0 t}{\sqrt{\epsilon}}} \tag{3.14}$$

and

$$\begin{cases} \partial_t u_\epsilon + (u^s\partial_x + v^s\partial_y)u_\epsilon + w_\epsilon u_z^s - \partial_z^2 u_\epsilon = r_\epsilon^1, \\ \partial_t v_\epsilon + (u^s\partial_x + v^s\partial_y)v_\epsilon + w_\epsilon v_z^s - \partial_z^2 v_\epsilon = r_\epsilon^2, \\ \partial_x u_\epsilon + \partial_y v_\epsilon + \partial_z w_\epsilon = 0, \\ (u_\epsilon, v_\epsilon, w_\epsilon)|_{z=0} = 0, \qquad \lim\limits_{z\to+\infty}(u_\epsilon, v_\epsilon) = 0, \end{cases} \tag{3.15}$$

where $(r_\epsilon^1, r_\epsilon^2)(t,x,y,z) := e^{i\epsilon^{-1}(y-ax)}(R_\epsilon^1, R_\epsilon^2)(t,z)$ satisfies

$$\|(R_\epsilon^1, R_\epsilon^2)(t,\cdot)\|_{L^2_\alpha} \le C_1\epsilon^{-\frac{1}{4}} e^{\frac{\sigma_0 t}{\sqrt{\epsilon}}}. \tag{3.16}$$

Step 2: Suppose the assertion (3.5) is not true, i.e., for all $\sigma > 0$, there exists $\delta > 0$, $m \ge 0$ and $\mu \in [0, \frac{1}{4})$, and

$$\sup_{0\le s\le t\le\delta} \|e^{-\sigma(t-s)\sqrt{|\partial_x|}} T(t,x)\|_{L(H^m_\alpha, H^{m-\mu}_\alpha)} < +\infty. \tag{3.17}$$

Denoted by

$$T_\epsilon(t,s)\big((U_0,V_0)\big) \;:=\; e^{-i\epsilon^{-1}(y-ax)}T(t,s)\Big(e^{i\epsilon^{-1}(y-ax)}(U_0,V_0)\Big).$$

Then, from the assumption (3.17), one has

$$\|T_\epsilon(t,s)\|_{L(L^2_\alpha)} \le C\epsilon^{-\mu}e^{\frac{\sqrt{a}\sigma(t-s)}{\sqrt{\epsilon}}}, \quad \forall\, 0\le s\le t\le\delta \tag{3.18}$$

which implies the estimate for the exact solution of (3.3):

$$\|(U,V)(t,\cdot)\|_{L^2_\alpha} \le C\epsilon^{-\mu}e^{\frac{\sqrt{a}\sigma t}{\sqrt{\epsilon}}}. \tag{3.19}$$

On the other hand, by comparing the problem (3.3) with (3.15), we know that the error $(\tilde U,\tilde V) := (U,V)-(U_\epsilon,V_\epsilon)$ satisfies

$$(\tilde U,\tilde V)(t,\cdot) \;=\; \int_0^t T_\epsilon(t,s)\big((R^1_\epsilon,R^2_\epsilon)(s,\cdot)\big)ds, \quad \forall\, t\le\delta,$$

which implies that by choosing $\sqrt{a}\sigma < \sigma_0$,

$$\|(\tilde U,\tilde V)(t,\cdot)\|_{L^2_\alpha} \le C\epsilon^{-\mu-\frac14}\int_0^t e^{\frac{\sqrt{a}\sigma(t-s)}{\sqrt{\epsilon}}}e^{\frac{\sigma_0 s}{\sqrt{\epsilon}}}ds \le C\epsilon^{\frac14-\mu}e^{\frac{\sigma_0 t}{\sqrt{\epsilon}}}, \tag{3.20}$$

by using (3.18) and (3.16).

Thus, from (3.14) we obtain that for $t<\delta$ and sufficiently small ϵ,

$$\|(U,V)(t,\cdot)\|_{L^2_\alpha} \ge C_0^{-1}e^{\frac{\sigma_0 t}{\sqrt{\epsilon}}}$$

which gives a contradiction to (3.19) as $\sqrt{a}\sigma < \sigma_0$.

Remark 3.1 (1) When $U_s' > 0$, the condition (3.4) given in Theorem 3.1 is equivalent to

$$\frac{d}{dz}\Big(\frac{V_s'}{U_s'}\Big) \not\equiv 0.$$

which is equivalent to

$$\frac{d}{dz}\Big(\frac{V_s}{U_s}\Big) \not\equiv 0,$$

by using $U_s(0)=V_s(0)=0$.

(2) It is easy to know that when $\frac{d}{dz}\big(\frac{V_s}{U_s}\big)\equiv 0$ holds, one also has $\frac{\partial}{\partial z}\big(\frac{v^s(t,z)}{u^s(t,z)}\big)\equiv 0$ for the shear flow given in (3.2), i.e., the tangential flow direction of the shear flow is invariant with respect to the normal variable z. In contrast to the instability result

proved in Theorem 3.1, it is very interesting to study whether the flow is stable when the tangential flow direction in the boundary layer is invariant with respect to the normal variable z. It is the goal of the next subsection!

3.2 A Well-Posedness Result

Assume that the $x-$component of the velocity in the Euler flow (1.3) is positive on the boundary, $U(t,x,y) > 0$ in (3.1), then without loss of generality, we can take the trace of the Euler flow being the form of

$$(U(t,x,y), k(t,x,y)U(t,x,y), 0; p(t,x,y)), \quad U(t,x,y) > 0, \tag{3.21}$$

on the boundary $\{z=0\}$.

As mentioned in Remark 3.1(2), we wish to study the stability of the following solution pattern for the problem (3.1) of the Prandtl equations in three space variables,

$$(u(t,x,y,z), k(t,x,y)u(t,x,y,z), w(t,x,y,z)), \tag{3.22}$$

then, from (3.1) we know that the above flow satisfies the following problem in $\{t>0, (x,y)\in D, z>0\}$,

$$\begin{cases} \partial_t u + (u\partial_x + ku\partial_y + w\partial_z)u + \partial_x p - \partial_z^2 u = 0 \\ \partial_t(ku) + (u\partial_x + ku\partial_y + w\partial_Z)(ku) + \partial_y p - k\partial_z^2 u = 0 \\ \partial_x u + \partial_y(ku) + \partial_z w = 0 \\ u|_{z=0} = w|_{z=0} = 0, \quad \lim_{Z\to+\infty} u(t,x,y,z) = U(t,x,y) \end{cases} \tag{3.23}$$

which implies

$$u\left[\partial_t k + u(\partial_x + k\partial_y)k\right] - k\partial_x p + \partial_y p = 0. \tag{3.24}$$

If we assume that

$$\partial_z u > 0,$$

then from (3.24) we get that $\partial_t k = 0$ and

$$(\partial_x + k\partial_y)k = 0, \quad \partial_y p - k\partial_x p = 0. \tag{3.25}$$

From now on, we give the following assumptions:

(A1) k is a smooth function of (x,y), and satisfies the constraint

$$\partial_x k + k\partial_y k = 0. \tag{3.26}$$

(A2) The velocity $U(t, x, y)$ and pressure $p(t, x, y)$ of the outflow satisfy

$$U(t, x, y) > 0,$$

and

$$\partial_y p - k\partial_x p = 0 \qquad \text{i.e. } \nabla p \parallel (U, kU) \tag{3.27}$$

for all $t > 0$ and $(x, y) \in D$.

Denote by $\Omega_T = (0, T] \times \Omega$, with $\Omega = \{(x, y, z)|\ (x, y) \in D, z \in \mathbb{R}_+\}$, and D being a bounded domain of $\mathbb{R}^2$. Let us consider the following problem in Ω_T:

$$\begin{cases} \partial_t u + (u\partial_x + v\partial_y + w\partial_z)u + \partial_x p - \partial_z^2 u = 0 \\ \partial_t v + (u\partial_x + v\partial_y + w\partial_z)v + \partial_y p - \partial_z^2 v = 0 \\ \partial_x u + \partial_y v + \partial_z w = 0 \\ (u, v, w)|_{z=0} = 0, \quad \lim\limits_{z\to+\infty}(u, v) = (U(t, x, y), k(x, y)U(t, x, y)) \\ (u, v)|_{t=0} = (u_0(x, y, z), k(x, y)u_0(x, y, z)) \\ (u, v)|_{\Gamma_-} = (u_1(t, x, y, z), k(x, y)u_1(t, x, y, z)) \end{cases} \tag{3.28}$$

where $\Gamma_- = (0, T] \times \gamma_- \times [0, +\infty)$, with $\gamma_- = \{(x, y) \in \partial D|\ (1, k(x, y)) \cdot \vec{n}(x, y) < 0\}$, and $\vec{n}(x, y)$ being the unit outward normal vector of D on $(x, y) \in \partial D$, under the assumption $\partial_z u_0(x, y, z) > 0$.

For the above problem, first we have:

Lemma 3.2 *If $(u, v, w)(t, x, y, z)$ is the classical solution to the problem* (3.28), *then $v(t, x, y, z) = k(x, y)u(t, x, y, z)$.*

This lemma can be easily obtained by observing from (3.28) that $ku - v$ satisfies a scalar degenerate parabolic equation. One can find the detail in [11].

By using Lemma 3.2, we only need to study the solution being the form

$$(u(t, x, y, z), k(x, y)u(t, x, y, z), w(t, x, y, z)) \tag{3.29}$$

of the problem (3.28), from which we know that (u, w) satisfy the following reduced problem in Ω_T:

$$\begin{cases} \partial_t u + (u\partial_x + ku\partial_y + w\partial_z)u - \partial_z^2 u = -\partial_x p, \\ \partial_x u + \partial_y(ku) + \partial_z w = 0, \\ u|_{z=0} = w|_{z=0} = 0, \quad \lim\limits_{z\to+\infty} u = U(t, x, y), \\ u|_{\Gamma_-} = u_1(t, x, y, z), \quad u|_{t=0} = u_0(x, y, z), \end{cases} \tag{3.30}$$

It is obvious that the form of the problem (3.30) is quite similar to that of the problem (2.1) which was studied in Sect. 2 for the two-dimensional Prandtl equations, so by using either the Crocco transformation as in [21] or the energy method [1], we can conclude the local existence of a classical solution to the problem (3.30) in the class of $u(t, x, y, z)$ being strictly monotonic in the normal variable z. Moreover, as in [30], under an additional favorite assumption on the pressure, we can obtain a global weak solution to this problem. These results are summarized in the following theorem.

Theorem 3.2 ([10, 11]) *(1) There is a unique local classical solution to the problem* (3.30) *with the $x-$directional tangential velocity $u(t, x, y, z)$ being strictly monotonic in $z > 0$, under the assumption $\partial_z u_0 > 0,\ \partial_z u_1 > 0$ for $z \geq 0$.*

(2) Moreover, if

$$p_x(t, x, y) \leq 0, \quad \text{for } t > 0,\ (x, y) \in D. \tag{3.31}$$

then there is a global weak solution to the problem derived from (3.30) *by using the Crocco transformation.*

Remark 3.2 The assumption (3.27) on the outer flow leads to avoid the appearance of the secondary flow in the boundary layer, a complete new unstable mechanism noted by Moore in [18] for the three-dimensional Prandtl layer.

The next goal is to study the stability of this special flow. For any given special solution

$$(u^s(t, x, y, z), k(x, y)u^s(t, x, y, z), w^s(t, x, y, z)), \tag{3.32}$$

of the problem (3.1), with

$$\partial_z u^s(t, x, y, z) > 0, \qquad k_x + kk_y = 0$$

in $\Omega_T = \{0 < t \leq T, (x, y) \in D, z > 0\}$, consider the following linearized problem of (3.1) at the background state (3.32):

$$\begin{cases} \partial_t u + (u^s\partial_x + ku^s\partial_y + w^s\partial_z)u + (u\partial_x + v\partial_y + w\partial_z)u^s - \partial_z^2 u = f_1, \\ \partial_t v + (u^s\partial_x + ku^s\partial_y + w^s\partial_z)v + (u\partial_x + v\partial_y + w\partial_z)(ku^s) - \partial_z^2 v = f_2, \\ \partial_x u + \partial_y v + \partial_z w = 0, \\ (u, v, w)|_{z=0} = 0, \quad \lim\limits_{z\to+\infty}(u, v) = 0, \quad (u, v)|_{\Gamma_-} = (u_1, v_1)(t, x, y, z), \\ (u, v)|_{t=0} = (u_0, v_0)(x, y, z), \end{cases} \tag{3.33}$$

where $\Gamma_- = (0, T] \times \gamma_- \times \mathbb{R}_z^+$, with $\gamma_- = \{(x, y) \in \partial D|\ (1, k(x, y)) \cdot \vec{n}(x, y) < 0\}$, and $\vec{n}(x, y)$ being the unit outward normal vector of D on $(x, y) \in \partial D$.

Now, we are going to study energy estimates for the solutions of the problem (3.33), which is given in the following two steps.

Step 1: Set $\tilde{v}(t,x,y,z) = k(x,y)u(t,x,y,z) - v(t,x,y,z)$. By the relation $k_x + kk_y = 0$, from (3.33) it is easy to know that $\tilde{v}(t,x,y,z)$ satisfies the following problem,

$$\begin{cases} \partial_t \tilde{v} + (u^s\partial_x + ku^s\partial_y + w^s\partial_z)\tilde{v} + k_y u^s \tilde{v} - \partial_z^2 \tilde{v} = kf_1 - f_2, \quad in\ \Omega_T, \\ \tilde{v}|_{z=0} = 0, \quad \lim\limits_{z\to+\infty} \tilde{v} = 0, \quad \tilde{v}|_{\Gamma_-} = (ku_1 - v_1)(t,x,y,z), \\ \tilde{v}|_{t=0} = (ku_0 - v_0)(x,y,z). \end{cases} \tag{3.34}$$

One can easily estimate the solution $\tilde{v}$ of (3.34) by using the classical energy method.

Step 2: Rewrite the problem (3.33) by using that $v = ku - \tilde{v}$ as follows:

$$\begin{cases} \partial_t u + (u^s\partial_x + ku^s\partial_y + w^s\partial_z)u + (u\partial_x + ku\partial_y + w\partial_z)u^s \\ \qquad\qquad -\partial_z^2 u = f_1 + \tilde{v}\partial_y u^s, \ in\ \Omega_T, \\ \partial_x u + \partial_y(ku) + \partial_z w = \partial_y \tilde{v}, \quad in\ \Omega_T, \\ (u,w)|_{z=0} = 0, \quad \lim\limits_{z\to+\infty} u = 0, \quad u|_{\Gamma_-} = u_1(t,x,y,z)|_{\Gamma_-}, \\ u|_{t=0} = u_0(x,y,z). \end{cases} \tag{3.35}$$

Noting that $\partial_z u^s(t,x,y,z) > 0$, as in [1], for the problem (3.35), we introduce the transformation:

$$h = \partial_z\left(\frac{u}{\partial_z u^s}\right), \qquad or \qquad u = \partial_z u^s \int_0^z h d\tilde{z}. \tag{3.36}$$

Then, from (3.35) we know that $h(t,x,y,z)$ satisfies the following initial boundary value problem for a scalar integro-differential equation:

$$\begin{cases} \partial_t h + [u^s\partial_x + ku^s\partial_y + w^s\partial_z]h - 2\partial_z(\eta h) + \partial_z\big[(\zeta - k_y u^s)\int_0^z h\, ds\big] \\ \qquad\qquad -\partial_z^2 h = \partial_z(\tilde{f} + \partial_y u^s \frac{\tilde{v}}{\partial_z u^s}) - \partial_y \tilde{v}, \\ (\partial_z h + 2\eta h)|_{z=0} = -\tilde{f}|_{z=0}, \quad h|_{\Gamma_-} = h_1(t,x,y,z) \triangleq \partial_z(\frac{u_1(t,x,y,z)}{\partial_z u^s(t,x,y,z)|_{\Gamma_-}}), \\ h|_{t=0} = h_0(x,y,z) \triangleq \partial_z(\frac{u_0(x,y,z)}{\partial_z u^s(0,x,y,z)}), \end{cases} \tag{3.37}$$

with

$$\eta = \frac{\partial_z^2 u^s}{\partial_z u^s}, \quad \zeta = \frac{(\partial_t + u^s\partial_x + ku^s\partial_y + w^s\partial_z - \partial_z^2)\partial_z u^s}{\partial_z u^s}, \quad \tilde{f} = \frac{f_1}{\partial_z u^s}.$$

Similar to [1], one can estimate the solution $h = \partial_z(\frac{u}{\partial_z u^s})$ to the problem (3.37) in a weighted norm, from which we get the estimate of the solution $u(t,x,y,z)$. Combining this estimate with that of $\tilde{v}$ from (3.34), we conclude the estimate of $v(t,x,y,z)$. The estimate of $w(t,x,y,z)$ follows immediately from the divergence-free constraint given in (3.33).

Therefore, we obtain the following result:

Theorem 3.3 ([11]) *The classical solution* $(u^s(t,x,y,z), k(x,y)u^s(t,x,y,z), w^s(t,x,y,z))$ *of the problem* (3.28) *constructed in Theorem 3.2 is linearly stable with respect to any three-dimensional perturbation of the initial data, boundary data, and force terms in* (3.1).

Acknowledgments This research was supported in part by National Natural Science Foundation of China (NNSFC) under Grant No. 91230102 and 91530114, and the Shanghai Committee of Science and Technology under Grant No. 15XD1502300. The author would like to thank Chengjie Liu, Chao-Jiang Xu and Tong Yang for the nice collaborations over the years on the mathematical theory of the Prandtl boundary layers.

References

1. Alexandre, R., Wang, Y.-G., Xu, C.-J., Yang, T.: Well-posedness of the Prandtl equation in Sobolev spaces. J. Am. Math. Soc. **28**, 745–784 (2015)
2. Caflisch, R.E., Sammartino, M.: Existence and singularities for the Prandtl boundary layer equations. Z. Angew. Math. Mech. **80**, 733–744 (2000)
3. Cannone, M., Lombardo, M.C., Sammartino, M.: Well-posedness of the Prandtl equation with non compatible data. Nonlinearity **26**, 3077–3100 (2013)
4. Gerard-Varet, D., Dormy, E.: On the ill-posedness of the Prandtl equation. J. Am. Math. Soc. **23**, 591–609 (2010)
5. Gong, S.-B., Guo, Y., Wang, Y.-G.: Boundary layer problems for the two dimensional compressible Naiver-Stokes equations. Anal. Appl. **14**, 1–37 (2016)
6. Grenier, E.: On the nonlinear instability of Euler and Prandtl equations. Commun. Pure Appl. Math. **53**, 1067–1091 (2000)
7. Guo, Y., Nguyen, T.: Prandtl boundary layer expansions of steady Navier-Stokes flows over a moving plate. arXiv:1411.6984
8. Guo, Y., Nguyen, T.: A note on the Prandtl boundary layers. Commun. Pure Appl. Math. **64**, 1416–1438 (2011)
9. Hörmander, L.: The boundary problems of physical geodesy. Arch. Ration. Mech. Anal. **62**, 1–52 (1982)
10. Liu, C.-J., Wang, Y.-G., Yang, T.: A global existence of weak solutions to the Prandtl equations in three space variables. arXiv:1509.03856
11. Liu, C.-J., Wang, Y.-G., Yang, T.: A well-posedness theory for the Prandtl equations in three space variables. arXiv:1405.5308
12. Liu, C.-J., Wang, Y.-G., Yang, T.: On the ill-posedness of the Prandtl equations in three space dimensions. Arch. Ration. Mech. Anal. **220**, 83–108 (2016)
13. Liu, C.-J., Wang, Y.-G.: Derivation of Prandtl boundary layer equations for the incompressible Navier-Stokes equations in a curved domain. Appl. Math. Lett. **34**, 81–85 (2014)
14. Lombardo, M.C., Cannone, M., Sammartino, M.: Well-posedness of the boundary layer equations. SIAM J. Math. Anal. **35**, 987–1004 (2003) (electronic)
15. Lopes Filho, M.C., Mazzucato, A.L., Nussenzveig Lopes, H.J., Taylor, M.: Vanishing viscosity limit and boundary layers for circularly symmetric 2D flows. Bull. Braz. Math. Soc. (N.S.) **39**, 471–513 (2008)
16. Maekawa, Y.: On the inviscid limit problem of the vorticity equations for viscous incompressible flows in the half plane. Commun. Pure Appl. Math. **67**, 1045–1128 (2014)
17. Masmoudi, N., Wong, T.-K.: Local-in-time existence and uniqueness of solutions to the Prandtl equations by energy methods. Pure Appl. Math. **68**, 1683–1741 (2015)

18. Moore, F.K.: Three-dimensional boundary layer theory. Adv. Appl. Mech. **4**, 159–228 (1956)
19. Moser, J.: A new technique for the construction of solutions of nonlinear differential equations. Proc. Nat. Acad. Sci. **47**, 1824–1831 (1961)
20. Nash, J.: The imbedding problem for Riemannian manifolds. Ann. Math. **63**, 20–63 (1956)
21. Oleinik, O.A., Samokhin, V.N.: Mathematical Models in Boundary Layer Theory. Chapman & Hall/CRC (1999)
22. Oleinik, O.A.: The Prandtl system of equations in boundary layer theory. Soviet Math. Dokl. **4**, 583–586 (1963)
23. Prandtl, L.: Über Flüssigkeitsbewegungen bei sehr kleiner Reibung. In: Verh. Int. Math. Kongr., Heidelberg 1904 Teubner 1905, 484–494
24. Sammartino, M., Caflisch, R.E.: Zero viscosity limit for analytic solutions of the Navier-Stokes equations on a half-space, I. Existence for Euler and Prandtl equations. Commun. Math. Phys. **192**, 433–461 (1998)
25. Sammartino, M., Caflisch, R.E.: Zero viscosity limit for analytic solutions of the Navier-Stokes equations on a half-space, II. Construction of the Navier-Stokes solution. Commun. Math. Phys. **192**, 463–491 (1998)
26. Wang, Y.-G., Xie, F., Yang, T.: Local well-posedness of the boundary layer equations from the compressible isentropic Navier-Stokes system in half plane. SIAM J. Math. Anal. **47**, 321–346 (2015)
27. Weinan, E., Engquist, B.: Blow up of solutions of the unsteady Prandtl equation. Commun. Pure Appl. Math. **50**, 1287–1293 (1997)
28. Weinan, E.: Boundary layer theory and the zero-viscosity limit of the Navier-Stokes equation. Acta Math. Sin. (Engl. Ser.) **16**, 207–218 (2000)
29. Xin, Z.-P., Zhang, L.: On the global existence of solutions to the Prandtl system. Adv. Math. **181**, 88–133 (2004)
30. Zhang, P., Zhang, Z.: Long time well-posedness of Prandtl system with small and analytic initial data. J. Funct. Anal. **270**, 2591–2615 (2016)

Visual Exploration of Complex Functions

Elias Wegert

Abstract The technique of domain coloring allows one to represent complex functions as images on their domain. It endows functions with an individual face and may serve as simple and efficient tool for their visual exploration. The emphasis of this paper is on *phase plots*, a special variant of domain coloring. Though these images utilize only the argument (phase) of a function and neglect its modulus, analytic (and meromorphic) functions are uniquely determined by their phase plot up to a positive constant factor. Following (Wegert in Not AMS 58:78–780, 2011 [49], Wegert in Visual Complex Functions. An Introduction with Phase Portraits, Springer Basel, 2012 [53]), we introduce phase plots and several of their modifications and explain how properties of functions can be reconstructed from these images. After a survey of related results, the main part is devoted to a number of applications which illustrate the usefulness of phase plots in teaching and research.

Keywords Complex functions · Visualization · Special functions · Analytic landscape · Phase plot · Phase portrait · Argument principle · Padé approximation · Riemann zeta function · Gravitational lenses · Filter design

Mathematics Subject Classification (2010). Primary 30-01; Secondary 30A99 · 33-01

1 Introduction

Graphical representations of functions belong to the most useful tools in mathematics and its applications. While graphs of (scalar) real-valued functions can be depicted easily, the situation is quite different for complex functions. Even the graph of a

E. Wegert (✉)
Institute of Applied Mathematics, TU Bergakademie Freiberg,
Akademiestraße 6, 09596 Freiberg, Germany
e-mail: wegert@math.tu-freiberg.de

T. Qian and L.G. Rodino (eds.), *Mathematical Analysis, Probability and Applications – Plenary Lectures*, Springer Proceedings in Mathematics & Statistics 177, DOI 10.1007/978-3-319-41945-9_10

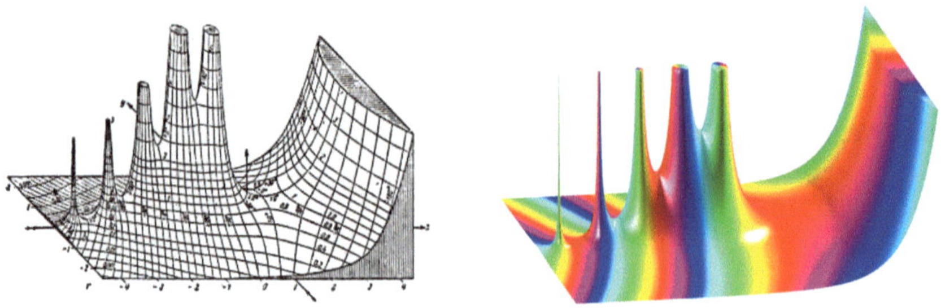

Fig. 1 Analytic landscapes of the Gamma function

Fig. 2 *Color circle*, coded phase, and domain coloring of $\mathbb{C}$

complex analytic function in one variable is a surface in four dimensional space, and hence not so easily drawn.

The first pictorial representations of complex functions in history are *analytic landscapes*, i.e., graphs of $|f|$; probably introduced by Edmond Maillet [31] in 1903. The analytic landscape of Euler's Gamma function in the famous book [19] by Jahnke and Emde achieved an almost iconic status (see Fig. 1, left). Impressive contemporary pictures of analytic landscapes can be seen on "The Wolfram Special Function Site" [55].

Analytic landscapes involve only the *modulus* $|f|$ of the function f, its *argument* arg f is lost. In the era of black and white illustrations this shortcoming was often compensated by complementing the analytic landscape with lines of constant argument. Today we can do this much better using *colors*.

Since the ambiguous argument arg z of a complex number is only determined up to an additive multiple of 2π, we prefer to work with the well-defined *phase* $z/|z|$ of z. Phase lives on the complex unit circle $\mathbb{T}$ and can easily be encoded by *colors* using the standard hsv color wheel (Fig. 2, left). The *colored analytic landscape* is the graph of $|f|$, colored according to the phase of f (Fig. 1, right).

2 Domain Coloring

In practice, it is often difficult to generate analytic landscapes which allow one to read off properties of the function easily and precisely. An alternative approach is not only simpler but even more general: Instead of drawing a graph, one can depict a function directly on its domain by color coding its values *completely*, as in the image on the right-hand side of Fig. 2.

Such coloring techniques for complex-valued functions have been in use at least since the 1980s (Larry Crone [9], see Hans Lundmark [26]), but they became popular only with Frank Farris' review [10] of Tristan Needham's book "Visual Complex Analysis" and its complement [11]. Farris also coined the name "domain coloring."

2.1 Phase Plots and Their Modifications

In contrast to "standard" domain coloring, which color codes the complete values of f by a *two dimensional* color scheme, *phase plots* display only the phase $\psi(f) := f/|f|$, thus requiring just a *one dimensional* color space with a *circular topology*. To also admit zeros and poles, we extend this definition by $\psi(0) := 0$ and $\psi(\infty) := \infty$, and associate black to 0 and white to ∞, respectively.

At the first glance it seems to be of no advantage to depict the phase of a function instead of its modulus. But indeed there is some subtle asymmetry between these two entities. In fact there are at least three reasons why phase plots outperform analytic landscapes, as can be seen in Fig. 3. First, phase has a small range (the unit circle), while the range of the modulus of an analytic function is usually quite large. As a consequence, the visual resolution is much higher for the phase than for the absolute value.

Second, reconstruction of (missing) information is simpler and more accurate for phase plots, as will be shown in the following section. In particular zeros, poles, and essential singularities can be clearly identified.

Last but not least, the analytic landscape is a three-dimensional object which usually must be projected for visualization, while the phase plot is a flat color image on the domain of the function, which allows one to read off information more precisely.

Since phase occupies only one dimension of the color space (which is usually the three-dimensional RGB space), additional information can be easily incorporated. If, for example, the modulus of f is encoded by a gray scale, we get standard domain

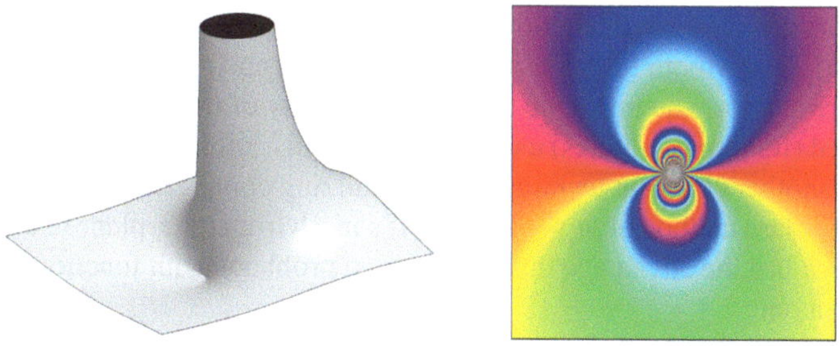

Fig. 3 Analytic landscape versus phase plot of $f(z) = e^{1/z}$

Fig. 4 Color schemes and representations of $f(z) = \dfrac{z-1}{z^2+z+1}$

coloring. Figure 4 illustrates four useful color schemes and the corresponding phase plots of $f(z) = (z-1)/(z^2+z+1)$ in the square $|\mathrm{Re}\, z| < 2$, $|\mathrm{Im}\, z| < 2$. The upper row depicts the color scheme in the w-plane; *pulling back* the colors to the z-plane via the mapping $w = f(z)$ yields the images in the lower row.

The leftmost column corresponds to the pure (plain) phase plot, while the rightmost images show standard domain coloring, respectively. The second column involves a gray component which is a sawtooth function of $\log |f|$, like

$$g = \lceil \log |f| \rceil - \log |f|.$$

Here $x \mapsto \lceil x \rceil$ is the *ceiling function*, which determines the smallest integer not less than x. The jumps in the gray component generate *contour lines* of $|f|$, i.e., lines of constant modulus. In between two such lines darker colors correspond to smaller values of $|f|$. From one line to the next the modulus of f increases *by a constant factor*, which allows one to determine the values much more accurately than from standard domain coloring. Another advantage is that this coloring is insensitive to the range of the function. A similar modification was used in the third column, but here discontinuities of the shading enhance some *isochromatic lines* (sets of constant phase). In the fourth column we have applied both shading schemes simultaneously, which generates a (logarithmically scaled) polar *tiling* of the range plane. The frequencies of the sawtooth functions encoding modulus and phase are chosen such that the tiles are "almost square." Due to the conformality of the mapping, this property is preserved (for almost all tiles) under pull back. This color scheme resembles a *grid mapping*, another common technique for visualizing complex functions. Compared with the standard method of *pushing forward* a mesh from the z-plane to the w-plane, *pulling back* has the advantage that there are no problems with functions of valence greater than one.

It is worth noticing that the shading method works with almost no additional computational costs, is absolutely stable, and does not require sophisticated numerical algorithms for computing contour lines.

2.2 *How to Read Phase Plots*

Which properties of an analytic function are reflected in its phase plot and how can we extract the information?

First of all it is important to note that *meromorphic* functions are almost uniquely determined by their phase plot: if two such functions (in a connected domain D) have the same phase plot (in an open subset of D), then one is a positive scalar multiple of the other (see [53]).

2.2.1 Zeros, Poles, and Saddle Points

Many features of a complex function can be read off from the local structure of its phase plot: not only zeros and poles of f, but also zeros of f' (saddle points). A simple criterion can be derived from the local normal form $f(z) = a + (z - z_0)^m g(z)$ with $g(z_0) \neq 0$, $a \in \mathbb{C}$ and $m \in \mathbb{Z}$. If z_0 is a zero or a pole of f, we have $a = 0$ (with $m > 0$ or $m < 0$, respectively), otherwise $a \neq 0$ and $m - 1 \geq 0$ is the order of the zero of f'. The following definition is needed in a more precise local classification of phase plots given in [48].

Definition 2.1 A phase plot $P := \psi \circ f$ is said to be (locally) conformally equivalent at a point z_0 to the phase plot $Q = \psi \circ g$ at w_0, if there exists a neighborhood U of z_0, a neighborhood V of w_0, and a bijective conformal mapping φ of U onto V such that $Q\big(\varphi(z)\big) = P(z)$ for all $z \in U \setminus \{z_0\}$.

In this definition, we admit that P and Q are defined only in punctured neighborhoods of z_0 and w_0, respectively.

Theorem 2.2 *Let $f : D \to \widehat{\mathbb{C}}$ be a meromorphic function. Then, for any $z_0 \in D$, the phase plot of f at z_0 is conformally equivalent to the phase plot of the following functions g at* 0*:*

(i) *If $f(z_0) \in \mathbb{C} \setminus \{0\}$ and $f'(z_0) \neq 0$, then $g(z) = \psi(f(z_0)) \exp z$.*
(ii) *If $f(z_0) \in \mathbb{C} \setminus \{0\}$ and f' has a zero of order $m \geq 1$ at z_0, then $g(z) = \psi(f(z_0)) \exp(z^{m+1})$.*
(iii) *If f has a zero of order $m \geq 1$ at z_0, then $g(z) = z^m$.*
(iv) *If f has a pole of order $m \geq 1$ at z_0, then $g(z) = z^{-m}$.*

It follows from (iii) and (iv) that not only the location z_0 but also the multiplicity m of zeros and poles can easily read off from the phase plot of f: in the vicinity of

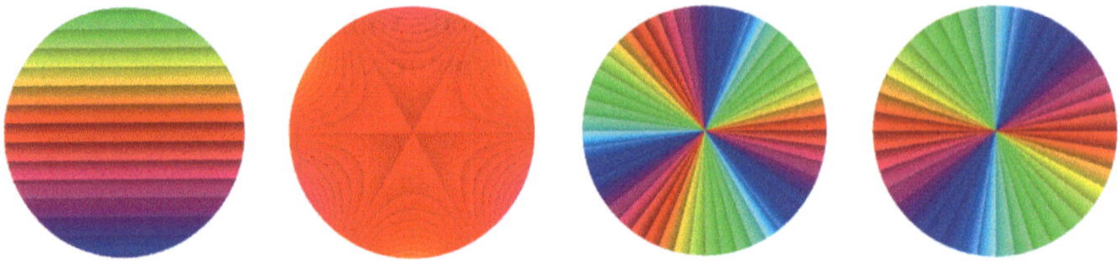

Fig. 5 Local normal forms of (enhanced) phase plots

z_0 it looks like a rotated phase plot of z^m or z^{-m}, respectively. In particular, zeros and poles can be distinguished by the different orientations of colors.

In the second case (ii), the point z_0 is said to be a *saddle* of f of order m. A saddle of order m is the common crossing point of $m+1$ isochromatic lines. In the pure phase plot saddles appear as diffuse spots and it needs some training to detect them. Using a color scheme with enhanced isochromatic lines makes this much easier.

Figure 5 shows the prototypes of (enhanced) phase plots in the four cases with $f(z_0) = 1$, a saddle of order 2 with $f(z_0) = 1$, a zero of order 3, and a pole of order 2, respectively.

2.2.2 Isolated Singularities

It is clear that removable singularities cannot be seen in the phase plot of a function, and we already know how poles look like. So what about essential singularities? Do they always manifest themselves as in Fig. 3? The answer is basically yes, but the statement of a strict result needs some terminology.

Let $f : D \to \widehat{\mathbb{C}}$ be a nonconstant meromorphic function. For any (color) $c \in \mathbb{T}$, let

$$S(c) := \{z \in D : \psi(f(z)) = c\}$$

be the subset of the domain D where the phase plot of f has color c. After removing from $S(c)$ all points z where $f'(z) = 0$, the remaining set splits into a finite or countable number of connected components. These are smooth curves which we call *isochromatic lines* in the phase plot of f.

Theorem 2.3 *An isolated singularity z_0 of f is an essential singularity if and only if for some (and then for any color) $c \in \mathbb{T}$ any neighborhood of z_0 intersects infinitely many isochromatic lines with color c.*

The result follows from Picard's Great Theorem (see [53]); an elementary proof, based on the Casorati-Weierstrass Theorem, is in [49], Theorem 4.4.6.

We point out that a corresponding result for the lines of constant *argument* does not hold, since the values of arg $f(z)$ on isochromatic lines with the same color may be different. For example, any continuous branch of arg $\exp(1/z)$ in $\mathbb{C} \setminus \{0\}$ attains different values on distinct isochromatic lines.

Fig. 6 Phase plots illustrating the growth of functions

2.2.3 Growth

The Cauchy–Riemann equations for (any continuous branch of) the logarithm $\log f = \ln|f| + \mathrm{i} \arg f$ imply that the isochromatic lines are orthogonal to the contour lines $|f| = \text{const}$. Consequently, the isochromatic lines are the lines of steepest ascent/descent of $|f|$. The direction in which $|f|$ increases can easily be determined: for example, when walking on a yellow line in ascending direction, we have red on the right and green to the left.

To go a little beyond this qualitative result, let s denote the unit vector parallel to the gradient of $|f|$ and $n := \mathrm{i}s$. Using the Cauchy–Riemann equations for $\log f$ we get

$$|f'|/|f| = \partial_s \ln|f| = \partial_n \arg f.$$

The left-hand side is the modulus of the logarithmic derivative f'/f; it measures the *growth* of $|f|$ (in the direction of its gradient) relative to the absolute value of f. The right-hand side is the *density of the isochromatic lines*. So, at least in principle, we can read the growth of a function from its (pure) phase plot. In practice it is more convenient to use the enhanced variant with contour lines of $|f|$.

Note that (almost) parallel stripes (with constant density of isochromatic lines) indicate *exponential growth*. The phase plots in Fig. 6 show an exponential function (left), a function growing faster than exponentially from left to right (middle), and the sum of three exponential functions e^{az} with different complex values of a. Knowing the size of the depicted domain, an experienced observer can read off the three values of a.

2.2.4 Periodic Functions

Clearly, the phase of a (doubly) periodic function is (doubly) periodic, but what about the converse? If, for example, a phase plot is doubly periodic, can we then be sure that it represents an elliptic function ?

Fig. 7 Three prototypes of periodic phase plots

Though there are only *two* classes (simply and doubly periodic) of nonconstant periodic meromorphic functions on $\mathbb{C}$, we can observe *three* different types of periodic phase plots as shown in Fig. 7 (from left to right: an exponential function, the cosine function and a Weierstrass $\wp$-function).

Motivated by these pictures we say that a (nonconstant) phase plot P is

(i) *striped* if there exists $p_0 \neq 0$ such that for all $p = \alpha p_0$ with $\alpha \in \mathbb{R}$

$$P(z+p) = P(z) \text{ for all } z, \tag{2.1}$$

(ii) *simply periodic* if there exists $p_0 \neq 0$ such that (2.1) holds if and only if $p = k\,p_0$ for all $k \in \mathbb{Z}$,

(iii) *doubly periodic* if there exist $p_1, p_2 \neq 0$ with $p_1/p_2 \notin \mathbb{R}$ such that (2.1) holds if and only if $p = k_1 p_1 + k_2 p_2$ for all $k_1, k_2 \in \mathbb{Z}$.

While it is easy to characterize striped and simply periodic phase plots, the doubly periodic case is more subtle. In the following theorem σ denotes the Weierstrass Sigma function:

$$\sigma(z) := z \prod_{\lambda \in \Lambda \setminus \{0\}} \exp\left(\frac{z}{\lambda} + \frac{z^2}{2\lambda^2}\right)\left(1 - \frac{z}{\lambda}\right),$$

where $\Lambda := p_1\mathbb{Z} + p_2\mathbb{Z}$ is the grid generated by the primitive periods p_1 and p_2. We further define u_1, u_2 and q_1, q_2 by

$$u_j := \sum_{\lambda \in \Lambda \setminus \{0, p_j\}} \frac{1}{\lambda(\lambda - p_j)^2}, \qquad q_j := p_j^2 u_j - 3/p_j. \tag{2.2}$$

Theorem 2.4 *The phase plot of a nonconstant meromorphic function f on $\mathbb{C}$ is*

(i) striped *if and only if there exist $a, b \in \mathbb{C}$ with $a \neq 0$ such that*

$$f(z) = \mathrm{e}^{az+b},$$

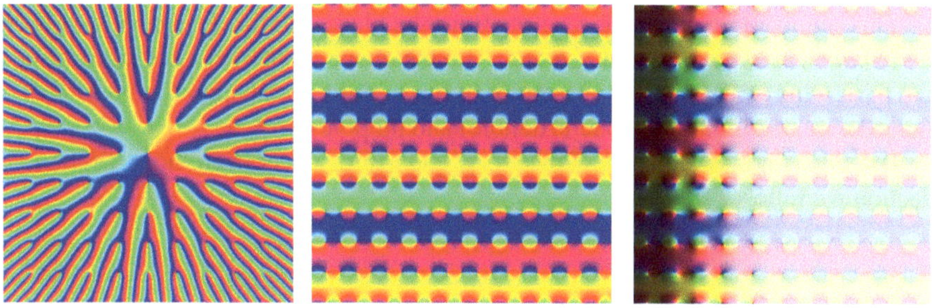

Fig. 8 A Weierstrass Sigma function σ and $\sigma(z)/\sigma(z-b)$

(ii) simply periodic *with primitive period p if and only if there exist a simply periodic function* $g:\mathbb{C}\to\mathbb{C}$ *with period p and a real number a such that*

$$f(z) = \mathrm{e}^{az/p}\, g(z),$$

(iii) doubly periodic *with primitive periods* p_1 *and* p_2 $(p_1/p_2 \notin \mathbb{R})$ *if and only if f can be represented as*

$$f(z) = \mathrm{e}^{az}\, g(z)\frac{\sigma(z)}{\sigma(z-b)},$$

where g is elliptic with periods p_1 *and* p_2, *and* $a, b \in \mathbb{C}$ *satisfy*

$$\mathrm{Im}\,(ap_j) \equiv \mathrm{Im}\,(bq_j) \pmod{2\pi}, \qquad j = 1, 2,$$

with q_j *defined in* (2.2).

For a proof of (i) and (ii) see [53], assertion (iii) is due to Marius Stefan [40].

Figure 8 shows a Weierstrass Sigma function and a quotient $\sigma(z)/\sigma(z-b)$ which has a doubly periodic phase plot (middle), but is not an elliptic function (right).

3 Phase Diagrams

The phase plot of a function contains information which is nonlocal. As an example, we consider the *argument principle*: Let $f : D \to \widehat{\mathbb{C}}$ be meromorphic in the domain D and assume that D contains the closure of the interior of a (positively oriented) Jordan curve J. If f has neither zeros nor poles on J, then the winding number $\mathrm{wind}_J P_f$ (about the origin) of the phase $P_f := \psi \circ f$ along J is the difference of the number $n(f, J)$ of zeros and the number $p(f, J)$ of poles of f inside J (counted according to their multiplicity),

$$n(f, J) - p(f, J) = \mathrm{wind}_J P_f.$$

This number can easily be read off from the phase plot of f, and we call it the *chromatic winding number* of f along J. In the image on the left-hand side of Fig. 9 we have $\mathrm{wind}_J P_f = 4$. The phase plot in the middle reveals that the interior of J indeed contains four zeros (one is double) and no poles of f.

But this is not yet the end of the story, one can discover even more. In the next section we follow [48].

3.1 The Phase Flow

The isochromatic lines in the phase plot of f are the flow lines of the vector field

$$V_f : D \to \mathbb{C},\ z \mapsto -\frac{f(z)\,\overline{f'(z)}}{|f(z)|^2 + |f'(z)|^2}$$

(see Fig. 9, middle and right). With an appropriate definition at zeros and poles of f the vector field V_f is smooth and vanishes exactly at the zeros and poles of f and f', which we call *singular points* of V_f.

The flow generated by the vector field V_f is said to be the *phase flow* of f. Endowing the phase plot with the orbits of this flow yields the *phase diagram* of f. Using standard techniques from the theory of dynamical systems, one can characterize the *orbits* of V_f and describe the *basins of attraction* of zeros (for details see [48]).

3.2 The Extended Argument Principle

If J is a Jordan curve in D which does not contain singular points of V_f, the *directional winding number* $\mathrm{wind}_J V_f$ of f along J is the winding number (about the origin) of V_f along J. In the rightmost image of Fig. 9 we have

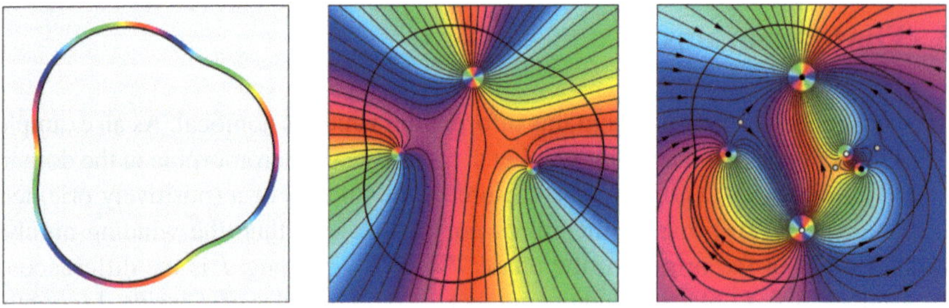

Fig. 9 Argument principle, phase flow and winding numbers

$$\mathrm{wind}_J P_f = 1, \qquad \mathrm{wind}_J V_f = 2.$$

Analyzing the phase diagram using index theory reveals a relation between the two winding numbers of f along J and the numbers $n(f', J)$ and $p(f', J)$ of zeros and poles of f' inside J, respectively.

Theorem 3.1 ([48]) *Let f be meromorphic in D and assume that the positively oriented Jordan curve J and its interior are contained in D. If neither f nor f' have zeros or poles on J, then*

$$n(f', J) - p(f', J) = \mathrm{wind}_J P_f - \mathrm{wind}_J V_f$$

Note that (at least in principle, but not always in practice) both winding numbers can be read off from the phase plot of f in an arbitrarily small neighborhood of J.

If f is holomorphic, the argument principle and Theorem 3.1 allow one to determine the number of zeros of f and f' inside J from the phase plot of f near J. In Fig. 9 (left) we have $\mathrm{wind}_J P_f = 4$ and $\mathrm{wind}_J V_f = 1$, so that $n(f, J) = 4$ and $n(f', J) = 3$.

An important special case pertains to the situation when f is holomorphic and the isochromatic lines of f are nowhere tangent to J. Since the latter implies $\mathrm{wind}_J V_f{=}1$, Theorem 3.1 tells us that then $n(f', J) = n(f, J) - 1$. This yields a short proof of Walsh's theorem on the location of critical points of Blaschke products [46–48].

4 Applications

In this section we discuss applications of phase plots which we believe to be useful—though in some examples the mathematical background is rather trivial.

4.1 Software Implementation

When one needs to compute special functions numerically, it is tempting to download code which is freely available on the internet. In many cases this may be an easy and efficient way to solve the problem, but one should be aware that there is no guarantee that software does what it claims to do.

An example is shown in Fig. 10. The image on the left is a phase plot of the complex Gamma function, computed with a certified Matlab routine. The picture in the middle displays the phase plot of a want-to-be Gamma function in the same domain, computed with code from a dubious source. Though the overall impression is almost the same, a closer look reveals that something must be wrong on the negative real axis near the left boundary of the domain. Zooming in a little closer, we discover a beautiful bug sitting at the end of an artificial branch cut.

Fig. 10 A bug in software for evaluating the Gamma function

Fig. 11 Implementations of the logarithmic Gamma Function

Since this is not the only incident which can be reported, one should be very careful when using software without knowing what it really computes. Though looking at phase plots can by no means ensure correctness of computations, it may help to discover some inconsistencies quite easily.

4.2 Multivalued Functions

Computations involving multivalued functions, like complex nth roots or the complex logarithm, are often challenging because usually only their main branch is implemented in standard software. In particular, composing such functions without taking care for choosing the appropriate branches may lead to fallacious results.

Let us consider the logarithmic Gamma function `GammaLog`(z) as an example. The three images of Fig. 11 show $\log\Gamma(z)$ (left), another implementation from the web which claims to be `GammaLog`(z) (middle), and a version having a branch cut along the negative real line (which is the standard definition). After some training, phase plots make it quite easy to understand the structure of spurious branch cuts, but removing them can be very tedious.

4.3 Riemann Surfaces

Phase plots may serve as convenient tool for constructing Riemann surfaces. We demonstrate this for the Riemann surface of the inverse of the sine function $f(z) = \sin z$. Basically this procedure involves three steps:
Step 1. Look at the phase plot of f in the z-plane and determine the basins of attraction of the zeros (first row of Fig. 12). In the case at hand the basins are vertical stripes $k\pi < \operatorname{Re} z < (k+1)\pi$, $k \in \mathbb{Z}$. Every such basin is mapped onto a copy of the complex w-plane, slit along the rays $[-\infty, -1]$ and $[1, +\infty]$ (second row).
Step 2. Change the coloring of the z-plane to the standard color scheme (phase plot of the identity, see first row of Fig. 13).
Step 3. Push the colors forward from the fundamental domains to the w-plane by $w = f(z)$. This generates phase plots of f^{-1} on the different sheets of its Riemann surface (second row of Fig. 13).

Gluing the rims of branch cuts according to their neighboring relations (which usually, but not always, can be seen from the phase plots), yields the Riemann surface on which the phase plot of g can be displayed (see Fig. 14).

Thomas Banchoff [5] and Michael Trott [43, 44] described techniques for visualizing complex functions on domain-colored Riemann surfaces. This topic was studied in more detail by Konrad Poehlke and Konstantin Polthier [33]. In two subsequent papers [34] and [32] (with M. Niesen) they propose algorithms for the automatic construction of Riemann surfaces with prescribed branch points and branch indices.

Fig. 12 The sine function mapping strips to $\mathbb{C}$

Fig. 13 Three branches of the inverse sine function

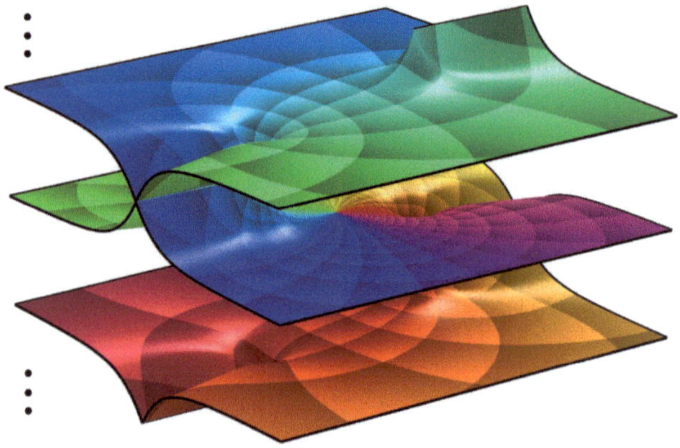

Fig. 14 The inverse sine function on its Riemann surface

The image on the left of Fig. 15 (reproduced from [32] with permission) shows such a surface composed of five sheets.

Another (more specialized) approach to automated computation of Riemann surfaces of algebraic curves is described in Stefan Kranich's PhD thesis [24]. The image on the right of Fig. 15 is the Riemann surface of the folium of Descartes, defined implicitly by the equation $z^3 + w^3 - 3zw = 0$ (reprinted in scaled form with permission).

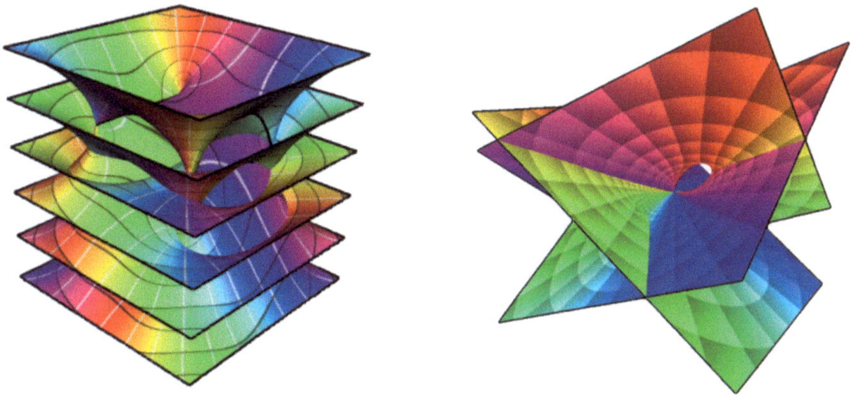

Fig. 15 Automatically generated Riemann surfaces

Fig. 16 Canonical embedding of planar graphs via Belyi functions

A general point-based algorithm for rendering implicit surfaces in $\mathbb{R}^4$ using interval arithmetic for topological robustness and (phase) coloring as substitute for the fourth dimension is described by Bordignon et al. [7].

4.4 Belyi Functions

A planar graph can be embedded in the (complex) plane in many different ways. Given one such embedding (like the one on the left-hand side of Fig. 16), one may ask whether it can be continuously deformed into a "canonical shape" without changing the vertex–edge relation and with no crossings throughout this whole process. Surprisingly, such a representation exists (depicted on the right of Fig. 16). A theorem due to Gennadii Belyi [6] tells us that every planar graph G can be represented by a rational function R (in the special class of so-called Belyi functions) with the following properties: The zeroes of R are exactly the vertices of G (red points) and the edges of G (black lines) are the preimages of the interval [0, 1] under R. In every face of G there is exactly one pole of R (white points) and the preimages of $[1, +\infty]$ (white lines) connect the poles to one point (gray) on each of the edges bounding the face containing that pole. (The area outside the graph is considered to be a face with its pole at infinity.) Moreover, all edges run into a vertex with equal angles between neighboring edges and the (white) lines originating at the poles intersect the edges perpendicularly. Last but not least, every face is the basin of attraction (see Sect. 3 and [48]) of the associated pole (with respect to the reverse phase flow).

The actual computation of the Belyi function associated with a given graph is a challenging problem. Donald Marshall developed an approach via conformal welding [27, 28] and implemented it using his software ZIPPER [29, 30]. I am grateful to him for providing the coefficients of the Belyi function shown in Fig. 16.

4.5 Filters and Controllers

In signal and control theory (linear, causal, time invariant, and stable) systems are described by transfer functions, which are analytic in the right half plane. In practice, most transfer functions are rational functions with poles in the left half plane. In the frequency domain the system acts on an input as multiplication operator with its transfer function T. In particular, the frequency response $T(\mathrm{i}\omega)$ tells one what the system does with harmonic input signals $e^{\mathrm{i}\omega t}$: The values $|T(\mathrm{i}\omega)|$ and $\arg T(\mathrm{i}\omega)$ are the *gain* and the *phase shift* induced by the system operating at frequency ω.

The phase plot on the left of Fig. 17 is the transfer function of a Butterworth filter—a low pass filter, which damps high frequency signals. This can be seen from its frequency response on the imaginary axis: the white segment is the passband where $|T(\mathrm{i}\omega)| \approx 1$, in the stopband (black) $|T(\mathrm{i}\omega)|$ decays for increasing values of ω. Using the contour lines and the phase coloring one can read off the frequency response directly from the phase plot of T and, for instance, construct Bode and Nyquist plots.

Santiago Garrido and Luis Moreno [15] developed more elaborate techniques involving phase plots for designing controllers. The two images in the middle and

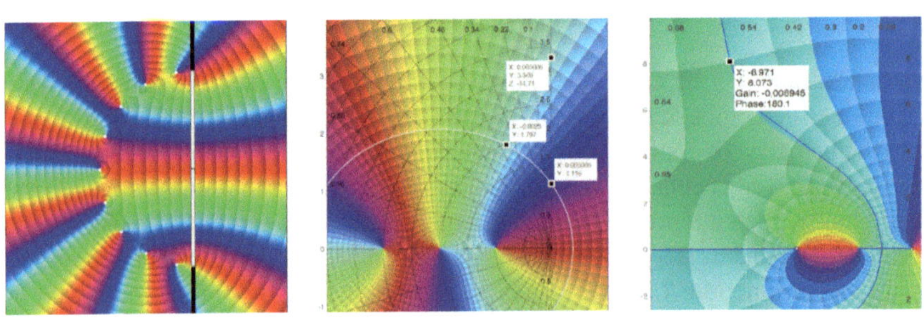

Fig. 17 Design of filters and controllers

on the right of Fig. 17 show some screenshots from their software (I am grateful to the authors for providing these images).

4.6 Numerical Algorithms

In recent years, domain coloring techniques have proven useful tools in analyzing numerical algorithms. Compared with numerical values, like the norm of the error (function) of an approximate solution, images deliver much more structural information. This may, for instance, improve the understanding of the method's global behavior. A further advantage of phase plots is the high sensitivity of the phase $\psi(z)$ for small values of z, which lets them act as a looking glass focused at the origin.

4.6.1 Iterative Methods

Numerical methods often use iterative procedures to find successively better approximations to solutions of a problem. There are many options to display relevant information about the global behavior of these methods. In order to demonstrate how coloring techniques can be used in this context, we consider zero finding for a complex function f (for a more detailed exposition see Varona [45]).

Most iterative methods start with an initial value z_0 and calculate $z_1, z_2, \ldots$ recursively by

$$z_{k+1} = z_k - \lambda_k\, f(z_k), \tag{4.1}$$

where λ_k is a parameter which may be constant (Whittaker's method) or depending on z_k. For $\lambda_k := 1/f'(z_k)$ we get the popular Newton method which converges (locally) quadratically. A skillful choice of λ_k leads to an accelerated convergence of the approximating sequence. For example, the "double convex acceleration of Whittaker's method" (DCAW method) uses the iteration formula (see [18, 45])

$$z_{k+1} = z_k - \frac{f(z_k)}{2\,f'(z_k)} \left[1 - g(z_k) + \frac{1 + g(z_k)}{1 - g(z_k)\big(1 - g(z_k)\big)} \right],$$

where $g(z) = f(z) f''(z) / \big(2 f'(z)^2\big)$. This method has convergence order 3.

Figure 18 displays the results of some experiments for solving $f(z) := z^5 - 1 = 0$ by the recursion (4.1). We see the fourth iterate z_4 (upper row) and the residue $f(z_4)$ (lower row) as functions of the initial point z_0 for different methods, namely (from left to right) Whittaker's method with $\lambda = 0.15$, Newton's method, an accelerated Whittaker method (see [45]), and the DCAW method.

The pictures in the upper row are plain phase plots, showing the emanating basins of attraction of the zeros; the five dominating colors correspond to the phase of these zeros. In the lower row we used domain coloring, encoding the modulus by a gray

Fig. 18 Iterates and residues for several zero-finding methods

scale, to get a better feeling for the magnitude of the residue. In the (almost) black regions the absolute value of $f(z_4)$ is in the range of 10^{-15}, in the bright domains the iterates converge to the point at infinity.

4.6.2 Numerical Differentiation

As another simple example, we consider approximations of the first derivative of a function f by the difference quotients

$$f_1(z) := \frac{f(z+h) - f(z)}{h}, \qquad f_2(z) := \frac{f(z+h) - f(z-h)}{2h},$$
$$f_3(z) := \frac{f(z+\mathrm{i}h) - f(z-\mathrm{i}h)}{2\mathrm{i}h}, \qquad f_4(z) := \frac{f_2(z) + f_3(z)}{2}.$$

Figure 19 shows phase plots of the error functions $f' - f_k$ $(k = 1, 2, 3, 4)$ for $f(z) = 1/z$ with $h = 10^{-3}$ (first row) and $h = 10^{-5}$ (second row). What do we see here?

The difference quotients f_k approximate f' with order h^n, where $n = 1, 2, 2, 4$ for $k = 1, 2, 3, 4$, respectively. A straightforward computation using Taylor series then shows that the error function satisfies

$$e_k(z) := f_k(z) - f'(z) = c_k\, h^n\, f^{(n+1)}(z) + O(h^{n+1})$$

so that we basically should see a phase plot of $c_k\, f^{(n+1)}(z) = c_k(-z)^{-n-2}$. This is indeed the case in the first three pictures for $h = 10^{-3}$. For f_4 the error is already so small, that rounding effects (cancelation of digits) manifest themselves near the boundary. For $h = 10^{-5}$, similar effects can also be observed for e_2 and e_3, while in the fourth picture the error function e_4 is completely dominated by noise (for a

Fig. 19 Error functions for numerical differentiation

Fig. 20 Evaluation of Cauchy integrals

computer expert the emerging structure may reveal information about the implemented arithmetic).

The most interesting observation is that one can read off the approximation order n directly: applying the method to $f(z) = z^{-1}$, the resulting phase plot shows a pole of order $n + 2$ at the origin. Similar types of experiments can be designed for other approximation methods.

4.6.3 Numerical Integration

Evaluation of integrals is another topic which can nicely be illustrated and studied using phase plots. In Fig. 20, we demonstrate this for a Cauchy integral of an analytic function f. The exact values of the integral are displayed in the figure on the left-hand side. Outside the contour of integration the integral vanishes, at points z surrounded by the contour its values are equal to $k\,f(z)$, where k is the winding number of the contour about z (here k is either 1 or 2).

The other two figures show approximations of the integral, evaluated by the trapezoidal rule with 200 and 1200 nodes, respectively. We see poles, sitting at the contour of integration, induced by the pole of the Cauchy kernel. The gear-like pattern in the exterior domain has almost parallel isochromatic lines, indicating rapid decay of the function (in the direction perpendicular to the contour); it is bounded by a chain of zeros aligned along the contour of integration. This chain may be related to Jentzsch's theorem and its generalizations ([8, 20, 46]).

Austin, Kravanja and Trefethen [1] used phase plots in order to compare different methods (Cauchy integrals, polynomial and rational interpolation) for computing values $f(z)$ and $f^{(m)}(z)$ of analytic and meromorphic functions in a disk from samples at the boundary of that disk.

4.6.4 Padé Approximation

Phase plots of rational functions can "visually approximate" any image, drawn solely with saturated colors from the hsv color wheel (for a precise statement see [50]). Particularly nice images arise from rational functions with zeros and poles forming special patterns, as it happens, for instance, in Padé approximation. In turn, these images may help to understand special aspects of these approximations.

The first row of Fig. 21 shows the function $f(z) = \tan z^4$ to be approximated (left), and two Padé approximants of order [100, 100]. The function depicted in the middle is computed by a standard method, the function on the right is the output of a stabilized ("robust") algorithm developed by Gonnet et al. [17]. Though the pictures can barely be distinguished, the structural differences become obvious in the next two rows, depicting phase plots of the numerator polynomial p (left) and the denominator polynomial q (middle), as well as the error function $f - f_{[100,100]}$ (right). The upper row corresponds to the standard algorithm, while the lower row visualizes the output of the stabilized algorithm. Apparently the first one produces a lot of spurious zeros in both polynomials p and q, which are (almost) canceled in the quotient p/q.

The black line in the error plots on the right-hand side is the unit circle. The almost(!) unstructured part in the middle (the influence of the zero-pole-cancelation is seen here) is due to small fluctuations about zero.

The computations are performed with the Matlab routine `padeapprox` of the Chebfun toolbox (for details see [17]).

4.6.5 Differential Equations

Numerous classes of special functions (Bessel, Airy, hypergeometric, etc.) arise as solutions of second order ordinary differential equations (ODEs). Computing these functions often requires elaborate numerical methods. A particularly hard case is given by the six Painlevé equations, which are prototypes of equations

Fig. 21 Padé approximation of $f(z) = \tan z^4$

$$u'' = F\left(z, u, u'\right),$$

where F is a rational function of its arguments, and have single-valued solutions u for all choices of their two initial conditions.

Solutions of Painlevé equations often have widely scattered poles, which were for a long time perceived as "numerical mine fields." Only in 2011, the first effective numerical algorithm for calculating their solutions was described by Bengt Fornberg and Andre Weideman [14].

Figure 22 shows special solutions of the Painlevé I (left) and Painlevé II equation (right), computed by Fornberg and Weideman (I am grateful to the authors for providing the data). The phase plot does not only deliver much more information than the plain zero-pole-pattern usually displayed in texts about Painlevé equations—one does not even need *compute* the zeros and poles, they show up automatically.

Fig. 22 Solutions of the Painlevé I and II equations

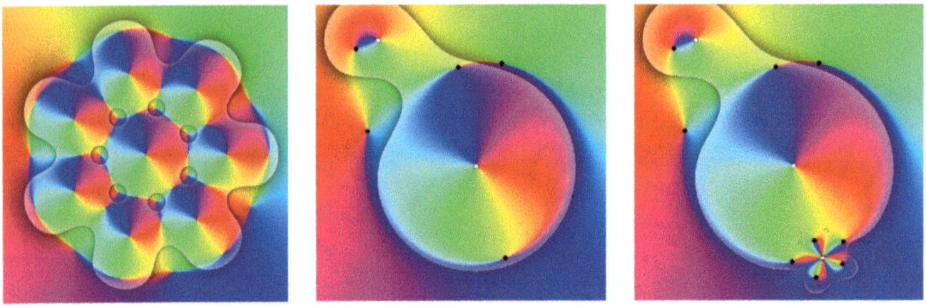

Fig. 23 Rational harmonic functions in gravitational lensing

4.7 Gravitational Lensing

Though phase plots allow one only to reconstruct meromorphic functions (almost) uniquely, they may nevertheless help to explore more general classes of functions. As an example we consider a problem involving *rational harmonic* functions which arises in gravitational lensing.

In 2006, Dmitri Khavinson and Genevra Neumann [22] proved that functions of the form $f(z) = r(z) - \overline{z}$, where r is a rational function of degree $n \geq 2$, can have at most $5n - 5$ zeros. That this bound is sharp follows from an example given by the astrophysicist Sun Hong Rhie in 2003. In the context of her paper [35], the zeros of f represent the images produced from a single light source by a gravitational lens formed by n point masses, located at the n poles of $r(z)$.

The picture on the left of Fig. 23 is a phase plot of Rhie's example for $n = 8$, having 35 zeros. Due to the term $\overline{z}$, the function f is not meromorphic, and hence a pure phase plot does not depict all properties one is interested in. The modified color scheme of the image allows one to read off where the mapping $z \mapsto f(z)$ is orientation preserving (brighter colors) or orientation reserving (darker colors). This is important to distinguish between zeros and poles: in the brighter regions the orientation of colors near a zero of f is the same as in the color wheel; in the darker regions the orientation is reversed.

Rhie's example was highly symmetric, and it is very unlikely that heavy cosmic objects (galaxies) form such a pattern. So it was greatly appreciated by the community of astrophysicists when Robert Luce, Olivier Sète and Jörg Liesen [37, 38] found a more general recursive construction of maximal gravitational lenses without symmetry. Phase plots played a prominent role in their investigations see [25]. The two images on the right of Fig. 23 (provided by the authors) illustrate how five zeros emerge from introducing an additional pole near a former zero located in the orientation preserving region of f.

4.8 The Riemann Zeta Function

Without doubt the Riemann Zeta function is one of the most fascinating mathematical objects. A reformulation of Bagchi's general universality theorem ([2, 21]) implies that its phase plot in the right half $1/2 < \operatorname{Re} z < 1$ of the critical strip is incredibly colorful (see [53]). Figure 24 displays a collection of phase plots of Zeta in the critical strip. Each rectangle has width 1 and covers about 20 units in the direction of the imaginary axis, with some overlap between neighboring rectangles. Since our visual system is trained in pattern detection, it usually does not take long until one discovers a diagonal structure. This observation inspired Jörn Steuding and me to study mean values of the Zeta function on (vertical) arithmetic progressions. Sampling ζ at points with fixed distance d, we expected that the asymptotic behavior of the mean values should be nontrivial if d is in resonance with the observed stochastic period. This could be confirmed by the following result from our paper [41].

Theorem 4.1 *Fix* $s \in \mathbb{C} \setminus \{1\}$ *with* $0 < \sigma := \operatorname{Re} s \leq 1$, $t := \operatorname{Im} s \geq 0$, *and let* $d = 2\pi/\log m$, *where* $m \geq 2$ *is an integer. Then, for* $M \to +\infty$,

$$\frac{1}{M} \sum_{0 \leq k < M} \zeta(s + \mathrm{i}k\,d) = \frac{1}{1 - m^{-s}} + O(M^{-\sigma} \log M).$$

Fig. 24 The Riemann Zeta function in the critical strip

Fig. 25 A color scheme for generating stream lines

We point out that this result does not really explain the observed stripes (which correspond to $d = 2$). To do this, one should consider mean values of the *phase* of Zeta instead of Zeta itself—another challenging problem ....

5 Concluding Remarks

Visualization of complex functions may facilitate new views on known results, raise interesting questions at all levels of difficulty, and, as the last example has shown, may inspire research.

Besides phase plots and standard domain coloring many other color schemes may be useful to illustrate and investigate special features of a function. So the striped patterns in Fig. 25 are convenient to display flow lines, while the chess-board-like structures in Fig. 26 are more appropriate to visualize conformal mappings. In this figure, the domain of the mapping is displayed in the upper row, while the lower row shows the corresponding image domains.

Cristina Ballantine and Dorin Ghisa [3, 4] used very beautiful color schemes to visualize Blaschke products, and Ghisa [16] analyzes several special functions (including the Gamma function and Riemann's Zeta function) using their coloring techniques.

Going a step further, one can put any image in the range plane of a complex function and pull it back to the domain, which may have fascinating and appealing results. For some masterpieces (and the theoretical background) we refer to Frank Farris work [12, 13].

Applications of phase plots in teaching comprise the visualization of converging power series, Weierstrass' disk chain method, Riemann surfaces, and other topics of standard lectures on complex functions. With *dynamic phase plots* one can interactively study the dependence of functions on parameters—such hands-on approaches allow students to become familiar with abstract concepts by doing their own experiments.

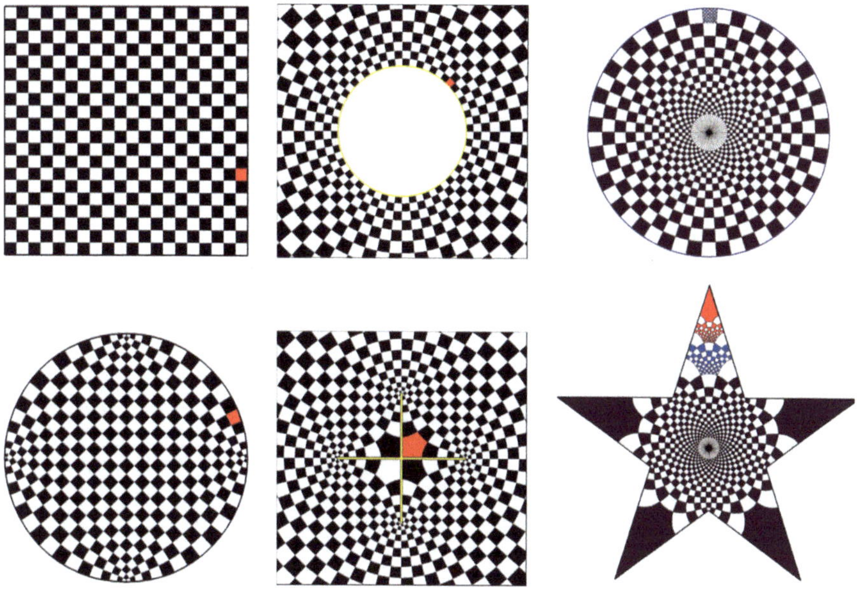

Fig. 26 Color schemes for visualizing conformal mappings

A comprehensive teaching-oriented introduction to complex functions and phase plots is given in the author's textbook [49]. A mathematical calendar featuring this theme can be downloaded at [54].

Matlab software for generating phase plots and colored analytic landscapes on plain domains and the Riemann sphere with various color schemes is available at the Matlab exchange platform [51, 52]. For implementations in Mathematica, we refer to Thaller [42], Trott [44], Sandoval-Romero and Hernández-Garduño [36], and Shaw [39]. Visual Basic code can be downloaded from Larry Crone's website [9]. A stand-alone Java implementation of phase plots of elementary functions is available as part of the Cinderella project by Ulrich Kortenkamp and Jürgen Richter-Gebert [23].

References

1. Austin, A.P., Kravanja, P., Trefethen, L.N.: Numerical Algorithms based on analytic function values at roots of unity. SIAM J. Numer. Anal. **52**, 1795–1821 (2014)
2. Bagchi, B.: A universality theorem for Dirichlet L-functions. Mat. Z. **181**, 319–334 (1982)
3. Ballantine, C., Ghisa, D.: Color visualization of Blaschke self-mappings of the real projective plane. Rev. Roum. Math. Pures Appl. **54**, 375–394 (2009)
4. Ballantine, C., Ghisa, D.: Colour visualization of Blaschke product mappings. Complex Var. Elliptic Equat. **55**, 201–217 (2010)
5. Banchoff, T.: Complex function graphs. http://www.math.brown.edu/~banchoff/gc/script/CFGInd.html

6. Belyi, G.V.: Galois extensions of a maximal cyclotomic field. Izvestiya Akademii Nauk SSSR **14**, 269–276 (1979). (Russian). English translation: Mathematics USSR Izvestija, **14**, 247–256 (1980)
7. Bordignon, A.L., Sa, L., Lopes, H., Pesco, S., de Figueiredo, L.H.: Point-based rendering of implicit surfaces in R4. Comput. Graph. **37**, 873–884 (2013)
8. Blatt, H.P., Blatt, S., Luh, W.: On a generalization of Jentzsch's theorem. J. Approx. Theory **159**, 26–38 (2009)
9. Crone, L.: Color graphs of complex functions. http://www1.math.american.edu/People/lcrone/ComplexPlot.html Accessed 17 Mar 2016
10. Farris, F.A.: Review of Visual Complex Analysis. By Tristan Needham. Am. Math. Monthly **105**, 570–576 (1998)
11. Farris, F.A.: Visualizing complex-valued functions in the plane. http://www.maa.org/visualizing-complex-valued-functions-in-the-plane (2016). Accessed 17 Mar 2016
12. Farris, F.A.: Symmetric yet organic: Fourier series as an artist's tool. J. Math. Arts **7**, 64–82 (2013)
13. Farris, F.A.: Creating Symmetry: The Artful Mathematics of Wallpaper Patterns, 230p. Princeton University Press (2015)
14. Fornberg, B., Weideman, J.A.C.: A Numerical methodology for the Painlevé equations. J. Comput. Phys. **230**, 5957–5973 (2011)
15. Garrido, S., Moreno, L.: PM diagram of the transfer function and its use in the design of controllers. J. Math. Syst. Sci. **5**, 138–149 (2015)
16. Ghisa, D.: Fundamental Domains and the Riemann Hypothesis, 148p. Lap Lambert Academic Publishing (2012)
17. Gonnet, P., Güttel, S., Trefethen, L.N.: Robust Padé approximation via SVD. Siam Rev. **55**, 101–117 (2013)
18. Hernández, M.A.: An acceleration procedure of the Whittaker method by means of convexity. Zb. Rad. Prirod.-Mat. Fak. **20**, 27–38 (1990)
19. Jahnke, E., Emde, F.: Funktionentafeln mit Formeln und Kurven. Teubner (1933)
20. Jentzsch, R.: Untersuchungen zur Theorie der Folgen analytischer Funktionen. Acta. Math. **41**, 219–251 (1918)
21. Karatsuba, A.A., Voronin, S.M.: The Riemann Zeta-Function. Walter de Gruyter (1992)
22. Khavinson, D., Neumann, G.: From the fundamental theorem of algebra to astrophysics: a "harmonious" path. Not. AMS **55**, 666–675 (2008)
23. Kortenkamp, U., Richter-Gebert, J.: Phase Diagrams of Complex Functions. http://science-to-touch.com/CJS/CindyJS/complexFunctions/ (2016). Accessed 17 Mar 2016
24. Kranich, S.: Continuity in dynamic geometry. An algorithmic approach. Ph.D. thesis, TU Munich (2016)
25. Luce, R., Sète, O., Liesen, J.: A note on the maximum number of zeros of $r(z) - \bar{z}$. Comput. Methods Funct. Theory **15**, 439–448 (2015)
26. Lundmark, H.: Visualizing complex analytic functions using domain coloring. http://users.mai.liu.se/hanlu09/complex/domain_coloring.html (2016). Accessed 18 Mar 2016
27. Marshall, D.E.: Conformal welding for finitely connected regions. Comput. Methods Funct. Theory **11**, 655–669 (2011)
28. Marshall, D.E.: Conformal welding and planar graphs. http://www.birs.ca/events/2015/5-day-workshops/15w5052/videos/watch/201501120953-Marshall.html (2016). Accessed 15 Mar 2016
29. Marshall, D.E.: Numerical conformal mapping software: zipper. https://www.math.washington.edu/~marshall/zipper.html (2016). Accessed 15 Mar 2016
30. Marshall, D.E., Rohde, S.: Convergence of a variant of the Zipper algorithm for conformal mapping. SIAM J. Numer. Anal. **45**, 2577–2609 (2007)
31. Maillet, E.: Sur les lignes de décroissance maxima des modules et les équations algébraiques ou transcendantes. J. de l'Éc. Pol. **8**, 75–95 (1903)
32. Nieser, M., Poelke, K., Polthier, K.: Automatic generation of Riemann surface meshes. In: Advances in Geometric Modeling and Processing. Lecture Notes in Computer Science, vol. 6130, pp. 161–178. Springer (2010)

33. Poelke, K., Polthier, K.: Lifted domain coloring. Comput. Graph. Forum **28**, 735–742 (2009)
34. Poelke, K., Polthier, K.: Domain coloring of complex functions: an implementation-oriented introduction. IEEE Comput. Graphics Appl. **32**, 90–97 (2012)
35. Rhie, S.H.: n-point gravitational lenses with 5(n-1) images. ArXiv Astrophysics arXiv:astro-ph/0305166 (2003)
36. Sandoval-Romero, Á., Hernández-Garduño, A.: Domain coloring on the riemann sphere. Math. J. **17** (2015)
37. Sète, O., Luce, R., Liesen, J.: Perturbing rational harmonic functions by poles. Comput. Methods Funct. Theory **15**, 9–35 (2015)
38. Sète, O., Luce, R., Liesen, J.: Creating images by adding masses to gravitational point lenses. Gen. Relativ. Gravit. **47**, 42 (2015)
39. Shaw, W.T.: Complex Analysis with Mathematica. Cambridge University Press (2006)
40. Stefan, M.B.: On doubly periodic phases. Proc. Am. Math. Soc. **142**, 3149–3152 (2011)
41. Steuding, J., Wegert, E.: The Riemann zeta function on arithmetic progressions. Exp. Math. **21**, 235–240 (2012)
42. Thaller, B.: Visualization of complex functions. Math. J. **7**, 163–180 (1999)
43. Trott, M.: Visualization of Riemann surfaces of algebraic functions. Math. Educ. Res. **6**, 15–36 (1997)
44. Trott, M.: Visualization of Riemann surfaces. http://library.wolfram.com/infocenter/Demos/15/ (2016). Accessed 17 Mar 2016
45. Varona, J.L.: Graphic and numerical comparison between iterative methods. Math. Intell. **24**, 37–46 (2002)
46. Walsh, J.L.: The location of critical points of analytic and harmonic functions. American Mathematical Society Colloquium Publications **34**, 386p, New York (1950)
47. Walsh, J.L.: Note on the location of zeros of extremal polynomials in the non-Euclidean plane. Acad. Serbe Sci. Publ. Inst. Math. **4**, 157–160 (1952)
48. Wegert, E.: Phase diagrams of meromorphic functions. Comput. Methods Funct. Theory **10**, 639–661 (2010)
49. Wegert, E.: Visual Complex Functions. An Introduction with Phase Portraits. Springer Basel (2012)
50. Wegert, E.: Complex functions and images. Computational Methods and Function Theory **13**, 3–10 (2013)
51. Wegert, E.: Phase plots of complex functions. http://www.mathworks.com/matlabcentral/fileexchange/44375 (2016). Accessed 15 Mar 2016
52. Wegert, E.: The complex function explorer. http://www.mathworks.com/matlabcentral/fileexchange/45464 (2016). Accessed 15 Mar 2016
53. Wegert, E., Semmler, G.: Phase plots of complex functions: a journey in illustration. Not. AMS **58**, 768–780 (2011)
54. Wegert, E., Semmler, G., Gorkin, P., Daepp, U.: Complex Beauties. Mathematical calendars featuring phase plots. http://www.mathcalendar.net (2016). Accessed 15 Mar 2016
55. Wolfram Research, The Wolfram Functions Site. http://www.functions.wolfram.com (2016). Accessed 15 Mar 2016

Integral Transform Approach to Time-Dependent Partial Differential Equations

Karen Yagdjian

Abstract In this review, we present an integral transform that maps solutions of some class of the partial differential equations with time independent coefficients to solutions of more complicated equations, which have time-dependent coefficients. We illustrate this transform by applications to several model equations. In particular, we give applications to the generalized Tricomi equation, the Klein–Gordon and wave equations in the curved spacetimes such as Einstein-de Sitter, de Sitter, anti-de Sitter, and the spacetimes of the black hole embedded into de Sitter universe.

Keywords Generalized Tricomi equation · de Sitter spacetime · Einstein-de Sitter spacetime · Global solutions · Strauss exponent

Mathematics Subject Classication. 35C15 · 35Q75 · 35Q05 · 83F05 · 76H05

1 Introduction

In this review, we present an integral transform that maps solutions of some class of the partial differential equations with time independent coefficients to solutions of more complicated equations, which have coefficients depending on time in some specific way. That integral transform leads to representation formulae, which for many equations exhaust all solutions. We also give survey of some results which were obtained by means of those representation formulas. That transform was used in a series of papers [28–31, 33, 84–95] to investigate in a unified way several equations such as the linear and semilinear Tricomi equations, Gellerstedt equation, the wave equation in Einstein-de Sitter spacetime, the wave and the Klein–Gordon equations in the de Sitter and anti-de Sitter spacetimes. The listed equations play

---

K. Yagdjian (✉)
School of Mathematical and Statistical Science, University of Texas Rio Grande Valley, Edinburg, TX 78539, USA
e-mail: karen.yagdjian@utrgv.edu

T. Qian and L.G. Rodino (eds.), *Mathematical Analysis, Probability and Applications – Plenary Lectures*, Springer Proceedings in Mathematics & Statistics 177, DOI 10.1007/978-3-319-41945-9_11

an important role in the gas dynamics, elementary particle physics, quantum field theory in curved spaces, and cosmology.

Consider for the smooth function $f = f(x,t)$ the solution $w = w(x,t;b)$ to the problem

$$w_{tt} - A(x,\partial_x)w = 0, \;\; w(x,0;b) = f(x,b), \;\; w_t(x,0;b) = 0, \;\; t \in [0,T_1] \subseteq \mathbb{R}, \;\; x \in \Omega \subseteq \mathbb{R}^n, \tag{1.1}$$

with the parameter $b \in I = [t_0, T] \subseteq \mathbb{R}$, $t_0 < T \leq \infty$, and with $0 < T_1 \leq \infty$. Here Ω is a domain in $\mathbb{R}^n$, while $A(x,\partial_x)$ is the partial differential operator $A(x,\partial_x) = \sum_{|\alpha|\leq m} a_\alpha(x) D_x^\alpha$. For $M \in \mathbb{C}$, we are going to present the integral operator

$$\mathcal{K}[w](x,t) = 2\int_{t_0}^{t} db \int_0^{|\phi(t)-\phi(b)|} K(t;r,b;M) w(x,r;b)dr, \quad x \in \Omega, \;\; t \in I, \tag{1.2}$$

which maps the function $w = w(x,r;b)$ into solution $u = u(x,t)$ of the equation

$$u_{tt} - a^2(t)A(x,\partial_x)u - M^2 u = f, \qquad x \in \Omega\,, \;\; t \in I. \tag{1.3}$$

In fact, the function $u = u(x,t)$ takes initial values as follows

$$u(x,t_0) = 0, \;\; u_t(x,t_0) = 0, \quad x \in \Omega\,.$$

Here $\phi = \phi(t)$ is a distance function produced by $a = a(t)$, that is $\phi(t) = \int_{t_0}^t a(\tau)\,d\tau$, while $M \in \mathbb{C}$ is a constant. Moreover, we also give the corresponding operators, which generate solutions of the source-free equation and takes nonvanishing initial values. These operators are constructed in [87, 88] in the case of $A(x,\partial_x) = \Delta$, where Δ is the Laplace operator on $\mathbb{R}^n$, and, consequently, the Eq. (1.1) is the wave equation. More general equations with x-dependent coefficients are treated in [33]. In the present review, we restrict ourselves to the smooth functions, but it is evident that similar formulas, with the corresponding interpretations, are applicable to the distributions as well. (For details see, e.g., [87].)

In order to motivate our approach, we consider the solution $w = w(x,t;b)$ to the Cauchy problem

$$w_{tt} - \Delta w = 0, \;\; (t,x) \in \mathbb{R}^{1+n}, \;\; w(x,0;b) = \varphi(x,b), \;\; w_t(x,0;b) = 0, \;\; x \in \mathbb{R}^n, \tag{1.4}$$

with the parameter $b \in I \subseteq \mathbb{R}$. We denote that solution by $w_\varphi = w_\varphi(x,t;b)$; if φ is independent of the second time variable b, then we write simply $w_\varphi(x,t)$. There are well-known explicit representation formulas for the solution of the problem (1.4). (See, e.g., [75].)

The starting point of the integral transform approach suggested in [84] is the Duhamel's principle (see, e.g., [75]), which has been revised in order to prepare the ground for generalization. Our *first observation* is that the function

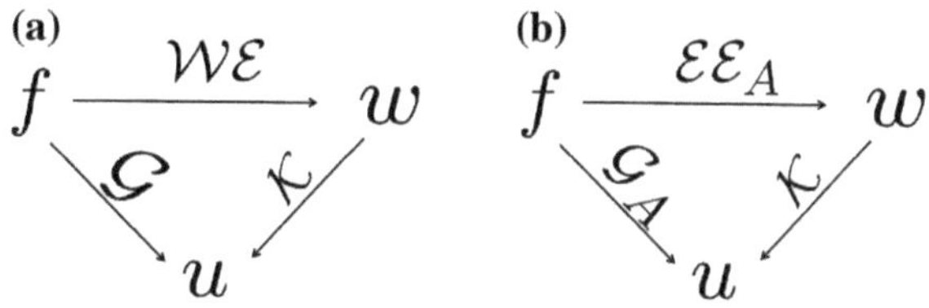

Fig. 1 **a** Case of wave equation $A(x, \partial_x) = \Delta$ **b** Case of general $A(x, \partial_x)$

$$u(x, t) = \int_{t_0}^{t} db \int_{0}^{t-b} w_f(x, r; b)\, dr\,, \tag{1.5}$$

is the solution of the Cauchy problem $u_{tt} - \Delta u = f(x, t)$ in $\mathbb{R}^{n+1}$, and $u(x, t_0) = 0$, $u_t(x, t_0) = 0$ in $\mathbb{R}^n$, if the function $w_f = w_f(x; t; b)$ is a solution of the problem (1.4), where $\varphi = f$. The *second observation* is that in (1.5) the upper limit $t - b$ of the inner integral is generated by the propagation phenomena with the speed which equals to one. In fact, that is a distance function. Our *third observation* is that the solution operator $\mathcal{G} : f \longmapsto u$ can be regarded as a composition of two operators. The first one

$$\mathcal{WE} : f \longmapsto w$$

is a Fourier Integral Operator, which is a solution operator of the Cauchy problem for wave equation. The second operator

$$\mathcal{K} : \ w \longmapsto u$$

is the integral operator given by (1.5). We regard the variable b in (1.5) as a "subsidiary time". Thus, $\mathcal{G} = \mathcal{K} \circ \mathcal{WE}$ and we arrive at the diagram of Fig. 1.

Based on the first diagram, we have generated in [89] a class of operators for which we have obtained explicit representation formulas for the solutions, and, in particular, the representations for the fundamental solutions of the partial differential operator. In fact, this diagram brings into a single hierarchy several different partial differential operators. Indeed, if we take into account the propagation cone by introducing the distance function $\phi(t)$, and if we provide the integral operator (1.5) with the kernel $K(t; r, b; M)$, as in (1.2), then we actually generate new representations for the solutions of different well-known equations with x-independent coefficients. (See, for details, [89].)

Our *fourth observation* is that if we plug into (1.5) the solution $w = w(x; t; b)$ of the Dirichlet problem for the elliptic equation $w_{tt} + \Delta w = 0$, $(t, x) \in \mathbb{R}^{1+n}$, $w(x, 0; b) = f(x, b)$, $x \in \mathbb{R}^n$, then the integral (1.5) defines the solution u of the equation $u_{tt} + \Delta u = f(x, t) + \int_{t_{in}}^{t} w_t(x; 0; b)\, db$, such that $u(x, t_{in}) = 0$, $u_t(x, t_{in}) = 0$. Thus, the integral (1.5), regarded as an *integral transform*, can be used for elliptic equations as well. That integral transform is interesting in its own right. If we want to remove the integral of the right-hand side of the equation, then we have to restrict ourselves to some particular functions f since the Cauchy problem $w_{tt} + \Delta w = 0$, $w(x, 0; b) = f(x, b)$, $w_t(x, 0; b) = 0$, is solvable, even locally, not for every, even smooth, function f.

Moreover, in [93, 95] we extended the class of the equations for which we can obtain explicit representation formulas for the solutions, by varying the first mapping. More precisely, consider the diagram (b) of Fig. 1, where $w = w_{A,\varphi}(x, t; b)$ is a solution to the problem (1.1) with the parameter $b \in I \subseteq \mathbb{R}$. If we have a resolving operator of the problem (1.1), then, by applying (1.2), we can generate solutions of another equation. Thus, $\mathcal{G}_A = \mathcal{K} \circ \mathcal{E}\mathcal{E}_A$. The new class of equations contains operators with x-depending coefficients, and those equations are not necessarily hyperbolic.

We believe that the integral transform and the representation formulas for the solutions that we describe in this article fill up the gap in the literature on that topic.

2 Linear Equations in the de Sitter Spacetime

Now let us restrict ourselves to the Klein–Gordon equation in the de Sitter spacetime, that is $a(t) = e^{-t}$ in (1.3). Recently the equations in the de Sitter and anti-de Sitter spacetimes became the focus of interest for an increasing number of authors (see, e.g., [1, 5–8, 11–20, 23, 32, 43–49, 59–61, 67, 79, 80, 92] and the bibliography therein) which investigate those equations from a wide spectrum of perspectives. The creation of a tool for the investigation of the local and global solvability in the problems for these linear and nonlinear equations appears to be a worthwhile undertaking.

To formulate the main result of this paper we need the following notations. First, we define a *chronological future* $D_+(x_0, t_0)$ and a *chronological past* $D_-(x_0, t_0)$ of the point (x_0, t_0), $x_0 \in \mathbb{R}^n$, $t_0 \in \mathbb{R}$, as follows: $D_\pm(x_0, t_0) := \{(x, t) \in \mathbb{R}^{n+1}\,;\ |x - x_0| \le \pm(e^{-t_0} - e^{-t})\}$. Then, for $(x_0, t_0) \in \mathbb{R}^n \times \mathbb{R}$, $M \in \mathbb{C}$, we define the function

$$
\begin{aligned}
E(x, t; x_0, t_0; M) := {} & 4^{-M} e^{M(t_0+t)}\Big((e^{-t_0} + e^{-t})^2 - (x - x_0)^2\Big)^{M-\frac{1}{2}} \\
& \times F\Big(\frac{1}{2} - M, \frac{1}{2} - M; 1; \frac{(e^{-t_0} - e^{-t})^2 - (x - x_0)^2}{(e^{-t_0} + e^{-t})^2 - (x - x_0)^2}\Big),
\end{aligned} \tag{2.6}
$$

where $(x, t) \in D_+(x_0, t_0) \cup D_-(x_0, t_0)$ and $F\big(a, b; c; \zeta\big)$ is the hypergeometric function. (For definition of the hypergeometric function, see, e.g., [9].) When no ambiguity arises, like in (2.6), we use the notation $x^2 := |x|^2$ for $x \in \mathbb{R}^n$. Thus, the function E depends on $r^2 = (x - x_0)^2$, that is $E(x, t; x_0, t_0; M) = E(r, t; 0, t_0; M)$. According to Theorem 2.12 [93], the function $E(r, t; 0, t_0; M)$ solves the following one-dimensional Klein–Gordon equation in the de Sitter spacetime:

$$
E_{tt}(r, t; 0, t_0; M) - e^{-2t} E_{rr}(r, t; 0, t_0; M) - M^2 E(r, t; 0, t_0; M) = 0\,.
$$

The kernels $K_0(z, t; M)$ and $K_1(z, t; M)$ are defined by

$$K_0(z, t; M) := -\left[\frac{\partial}{\partial b} E(z, t; 0, b; M)\right]_{b=0}, \tag{2.7}$$

$$K_1(z, t; M) := E(z, t; 0, 0, M). \tag{2.8}$$

The Eq. (1.3) is said to be an *equation with imaginary (real) mass* if $M^2 > 0$ $(-M^2 \geq 0)$; here $M \in \mathbb{C}$. From now on we assume that $a_\alpha \in C(\Omega)$. For the Klein–Gordon equation (1.3) we have the following result.

Theorem 2.1 ([93]) *For $f \in C(\Omega \times I)$, $I = [0, T]$, $0 < T \leq \infty$, and φ_0, $\varphi_1 \in C(\Omega)$, let the function $v_f(x, t; b) \in C^{m,2,0}_{x,t,b}(\Omega \times [0, 1 - e^{-T}] \times I)$ be a solution to the problem*

$$\begin{cases} v_{tt} - A(x, \partial_x)v = 0\,, & x \in \Omega\,, \quad t \in [0, 1 - e^{-T}]\,, \\ v(x, 0; b) = f(x, b)\,, & v_t(x, 0; b) = 0\,, \quad b \in I, \quad x \in \Omega\,, \end{cases} \tag{2.9}$$

and the function $v_\varphi(x, t) \in C^{m,2}_{x,t}(\Omega \times [0, 1 - e^{-T}])$ be a solution of the problem

$$\begin{cases} v_{tt} - A(x, \partial_x)v = 0, & x \in \Omega\,, \quad t \in [0, 1 - e^{-T}]\,, \\ v(x, 0) = \varphi(x), & v_t(x, 0) = 0\,, \quad x \in \Omega\,. \end{cases} \tag{2.10}$$

Then the function $u = u(x, t)$ defined by

$$\begin{aligned} u(x, t) = {} & 2\int_0^t db \int_0^{\phi(t)-\phi(b)} v_f(x, r; b)E(r, t; 0, b; M)\, dr + e^{\frac{t}{2}} v_{\varphi_0}(x, \phi(t)) \\ & + 2\int_0^{\phi(t)} v_{\varphi_0}(x, s)K_0(s, t; M)ds + 2\int_0^{\phi(t)} v_{\varphi_1}(x, s)K_1(s, t; M)ds, \quad x \in \Omega,\ t \in I, \end{aligned} \tag{2.11}$$

where $\phi(t) := 1 - e^{-t}$, solves the problem

$$\begin{cases} u_{tt} - e^{-2t}A(x, \partial_x)u - M^2u = f, & x \in \Omega\,, \ t \in I, \\ u(x, 0) = \varphi_0(x)\,, \quad u_t(x, 0) = \varphi_1(x), & x \in \Omega\,. \end{cases} \tag{2.12}$$

Here the kernels E, K_0 and K_1 have been defined in (2.6), (2.7) and (2.8), respectively.

We note that the operator $A(x, \partial_x)$ is of arbitrary order, that is, the equation of (2.12) can be an evolution equation, not necessarily hyperbolic. Then, the problems in (2.9) and (2.12) can be a mixed initial-boundary value problem involving the boundary condition. Indeed, assume that $\Omega \subset \mathbb{R}^n$ is domain with smooth boundary $\partial\Omega$, and that $\nu = \nu(x)$ is a unit normal vector. Let $\alpha = \alpha(x)$ and $\beta = \beta(x)$ be continuous functions, $\alpha, \beta \in C(\partial\Omega)$. If $v = v(t, x)$ satisfies the boundary condition

$$\alpha(x)v(x, t) + \beta(x)\partial_\nu v(x, t) = 0 \quad \text{for all} \quad t \in [0, 1 - e^T], \quad x \in \partial\Omega\,,$$

then the function $u = u(x, t)$ fulfills the same boundary condition

$$\alpha(x)u(x,t) + \beta(x)\partial_\nu u(x,t) = 0 \quad \text{for all} \quad t \in I, \quad x \in \partial\Omega\,.$$

Next, we stress that interval $[0, 1 - e^{-T}] \subseteq [0, 1]$, which appears in (2.9), reflects the fact that de Sitter model possesses the horizon [38]; existence of the horizon in the de Sitter model is widely used to define an asymptotically de Sitter space (see, e.g., [7, 79]) and to involve geometry into the analysis of the operators on the de Sitter space (see, e.g., [10, 59, 64, 78]).

The special case of Theorem 2.1, when $A(x, \partial_x) = \Delta$, one can find in [87, 88]. The case of the anti-de Sitter spacetime is discussed in [82]. The proof given in those papers is based on the well-known explicit representation formulas for the wave equation, the Riemann function, the spherical means, and the Asgeirsson's mean value theorem. The main outcome, resulting from the application of all those tools, is the derivation of the final representation formula and the kernels E, K_0, and K_1. Having in the hand the integral transform and the final formulas, the straightforward proof by substitution, which works also for the equations with coefficients depending on x, is given in [93].

Among possible applications of the integral transform method are the $L^p - L^q$ estimates, Strichartz estimates, Huygens' principle, global and local existence theorem for semilinear and quasilinear equations. Below, we give examples of the equations with the variable coefficients those are amenable to the integral transform method.

Example 1 The metric g in the de Sitter type spacetime, that is, $g_{00} = g^{00} = -1, g_{0j} = g^{0j} = 0$, $g_{ij}(x,t) = e^{2t}\sigma_{ij}(x)$, $|g(x,t)| = e^{2nt}|\det\sigma(x)|$, $g^{ij}(x,t) = e^{-2t}\sigma^{ij}(x)$, i, $j = 1, 2, \ldots, n$, where $\sum_{j=1}^{n}\sigma^{ij}(x)\sigma_{jk}(x) = \delta_{ik}$, and δ_{ij} is Kronecker's delta. The linear covariant Klein–Gordon equation in the coordinates is

$$\psi_{tt} - \frac{e^{-2t}}{\sqrt{|\det\sigma(x)|}}\sum_{i,j=1}^{n}\frac{\partial}{\partial x^i}\left(\sqrt{|\det\sigma(x)|}\sigma^{ij}(x)\frac{\partial}{\partial x^j}\psi\right) + n\psi_t + m^2\psi = f\,.$$

Here m is a physical mass of the particle. If we introduce the new unknown function $u = e^{nt/2}\psi$, then the equation takes the form of the Klein–Gordon equation

$$u_{tt} - \frac{e^{-2t}}{\sqrt{|\det\sigma(x)|}}\sum_{i,j=1}^{n}\frac{\partial}{\partial x^i}\left(\sqrt{|\det\sigma(x)|}\sigma^{ij}(x)\frac{\partial}{\partial x^j}u\right) + M^2 u = f\,,$$

where $M^2 = m^2 - \frac{n^2}{4}$ is the square of the so-called curved (or effective) mass. For the last equation we set

$$A(x,\partial_x)u = \frac{1}{\sqrt{|\det\sigma(x)|}}\sum_{i,j=1}^{n}\frac{\partial}{\partial x^i}\left(\sqrt{|\det\sigma(x)|}\sigma^{ij}(x)\frac{\partial}{\partial x^j}u\right)\,.$$

If $\Omega = \Pi$ is a non-Euclidean space of constant negative curvature and the equation of the problems (2.9) and (2.10) is a non-Euclidean wave equation, then the explicit representation formulas are known (see, e.g., [40, 51]) and the Huygens' principle is a consequence of those formulas. Thus, for a non-Euclidean wave equation, due to Theorem 2.1, the functions $v_f(x,t;b)$ and $v_\varphi(x,t)$ have explicit representations, and the arguments of [87, 94] allow us to derive for the solution $u(x,t)$ of the problem (2.12) in the de Sitter type metric with hyperbolic spatial geometry the explicit representation, the $L^p - L^q$ estimates, and to examine the Huygens' principle. Precise statements will be published in the forthcoming paper.

Example 2 This example we introduce as a toy model, which helps to understand the properties of *the black hole formally embedded in the de Sitter universe*. The metric tensor $g_{\mu\nu}$ is generated by line element $ds^2 = -\left(1 - \dfrac{2GM_{bh}}{c^2 r}\right) c^2 dt^2 + e^{\frac{2ct}{R}}\left(1 - \dfrac{2GM_{bh}}{c^2 r}\right)^{-1} dr^2 + e^{\frac{2ct}{R}} r^2 (d\theta^2 + \sin^2\theta\, d\phi^2)$. The Ricci tensor of that background is

$$\mathcal{R}_{\mu\nu} = \frac{3}{\left(1 - \frac{2GM}{c^2 r}\right) R^2} \begin{pmatrix} -c^2\left(1 - \frac{2GM}{c^2 r}\right) & \frac{2RGM}{3cr^2} & 0 & 0 \\ \frac{2RGM}{3cr^2} & e^{\frac{2ct}{R}} \frac{1}{\left(1 - \frac{2GM}{c^2 r}\right)} & 0 & 0 \\ 0 & 0 & e^{\frac{2ct}{R}} r^2 & 0 \\ 0 & 0 & 0 & e^{\frac{2ct}{R}} r^2 \sin^2(\theta) \end{pmatrix},$$

while the the Riemannian curvature is $\mathcal{R} = \frac{12}{R^2\left(1 - \frac{2GM}{c^2 r}\right)}$. Hence,

$$\mathcal{R}_{\mu\nu} = \frac{\mathcal{R}}{4} \begin{pmatrix} -c^2\left(1 - \frac{2GM}{c^2 r}\right) & \frac{2RGM}{3cr^2} & 0 & 0 \\ \frac{2RGM}{3cr^2} & e^{\frac{2ct}{R}} \frac{1}{\left(1 - \frac{2GM}{c^2 r}\right)} & 0 & 0 \\ 0 & 0 & e^{\frac{2ct}{R}} r^2 & 0 \\ 0 & 0 & 0 & e^{\frac{2ct}{R}} r^2 \sin^2(\theta) \end{pmatrix}.$$

The Einstein tensor $G_{\mu\nu} := R_{\mu\nu} - \frac{1}{2}\mathcal{R} g_{\mu\nu}$ for this metric is or

$$G_{\mu\nu} = -\frac{\mathcal{R}}{4} g_{\mu\nu} - \frac{\mathcal{R}}{4} \begin{pmatrix} 0 & R\frac{2GM}{3cr^2} & 0 & 0 \\ R\frac{2GM}{3cr^2} & 0 & 0 & 0 \\ 0 & 0 & 0 & 0 \\ 0 & 0 & 0 & 0 \end{pmatrix}.$$

The metric g is *an asymptotically Einstein metric*, in the sense that

$$\mathcal{R}_{\mu\nu} = (k + O(r^{-1}))g_{\mu\nu} + O(r^{-2}).$$

At the same time, the metric g is *an asymptotically hyperbolic (de Sitter) metric*, in the sense that

$$\mathcal{R}_{\mu\nu} = k(r) g_{\mu\nu} + O(r^{-2}) , \quad \text{as} \quad r \to \infty .$$

The stress–energy tensor T can be calculated as follows

$$\frac{8\pi G}{c^4} T = G_{\mu\nu} + \Lambda g_{\mu\nu} = \mathcal{R}_{\mu\nu} - \frac{1}{2}\mathcal{R} g_{\mu\nu} + \Lambda g_{\mu\nu} , \qquad \Lambda = \frac{3}{R^2} .$$

Hence,

$$\frac{8\pi G}{c^4} T = -\frac{\mathcal{R}}{4}\frac{2GM}{c^2 r} g_{\mu\nu} - \frac{\mathcal{R}}{4}\frac{2GM}{c^2 r}\begin{pmatrix} 0 & \frac{cR}{3r} & 0 & 0 \\ \frac{cR}{3r} & 0 & 0 & 0 \\ 0 & 0 & 0 & 0 \\ 0 & 0 & 0 & 0 \end{pmatrix} .$$

Consequently, T is of Type II (see, [38, p. 89]). It is evident that the *weak energy condition*, that is

$$T_{\mu\nu} u^\mu u^\nu \geq 0 \qquad \text{for all time-like vectors } u ,$$

is not satisfied unless $u^0 u^1 \leq 0$. But according to the next lemma it is satisfied on some conic set consisting of the time-like vectors with $u^0 u^1 \geq 0$. Thus, the following lemma addresses the weak energy condition (see, e.g., [38, p. 89], [17, p. 51]).

Lemma 2.2 *The set of all vectors u such that $T_{\mu\nu} u^\mu u^\nu \geq 0$ contains all time-like vectors with $u^0 u^1 < 0$ as well as a conic set of vectors with $u^0 u^1 > 0$ and satisfying*

$$c^2 \left(1 - \frac{2GM}{c^2 r}\right) (u^0)^2 - e^{\frac{2ct}{R}} \left(1 - \frac{2GM}{c^2 r}\right)^{-1} (u^1)^2$$
$$\times \left(\frac{R}{3r} e^{\frac{-ct}{R}} + \sqrt{\left(\frac{R^2}{9r^2} e^{\frac{-2ct}{R}} + 1\right) + \left(1 - \frac{2GM}{c^2 r}\right) r^2 \left((u^2/u^1)^2 + (u^3/u^1)^2 \sin^2(\theta)\right)} \right)^2 > 0 .$$

Proof The lemma can be proved by straightforward calculations. □

The stress–energy tensor is

$$\frac{8\pi G}{c^4} T_{\mu\nu} = \frac{\mathcal{R}}{4}\frac{2GM}{c^2 r}\begin{pmatrix} c^2(1 - \frac{2GM}{c^2 r}) & -\frac{cR}{3r} & 0 & 0 \\ -\frac{cR}{3r} & -e^{\frac{2ct}{R}} \frac{1}{(1 - \frac{2GM}{c^2 r})} & 0 & 0 \\ 0 & 0 & -e^{\frac{2ct}{R}} r^2 & 0 \\ 0 & 0 & 0 & -e^{\frac{2ct}{R}} r^2 \sin^2(\theta) \end{pmatrix} .$$

The eigenvalues of the stress–energy tensor are as follows

$$\lambda_1 = \frac{\mathcal{R}}{4}\frac{2GM}{c^2 r}\left(-\frac{1}{2\alpha}\left(e^{\frac{2ct}{R}} - c^2\alpha^2\right) + \sqrt{\frac{1}{4\alpha^2}\left(e^{\frac{2ct}{R}} + c^2\alpha^2\right)^2 + \frac{c^2R^2}{9r^2}}\right),$$

$$\lambda_2 = \frac{\mathcal{R}}{4}\frac{2GM}{c^2 r}\left(-\frac{1}{2\alpha}\left(e^{\frac{2ct}{R}} - c^2\alpha^2\right) - \sqrt{\frac{1}{4\alpha^2}\left(e^{\frac{2ct}{R}} + c^2\alpha^2\right)^2 + \frac{c^2R^2}{9r^2}}\right),$$

$$\lambda_3 = \frac{\mathcal{R}}{4}\frac{2GM}{c^2 r}\left(-e^{\frac{2ct}{R}} r^2\right), \quad \lambda_4 = \frac{\mathcal{R}}{4}\frac{2GM}{c^2 r}\left(-e^{\frac{2ct}{R}} r^2 \sin^2(\theta)\right),$$

where we denoted $\alpha := 1 - \frac{2GM}{c^2 r}$.

The following lemma addresses the *dominant energy condition* (see, e.g., [38, p. 91], [17, p. 51]).

Lemma 2.3 *The vector* $-T^{\nu}_{\ \mu}u^{\mu}$ *is time-like and future directed for all time-like future directed vectors u, such that* $u^0u^1 \leq 0$. *The conic set of all future directed vectors u, such that* $u^0u^1 > 0$ *and* $-T^{\nu}_{\ \mu}u^{\mu}$ *is time-like and future directed, contains all vectors satisfying*

$$c^2\left(1 - \frac{2GM}{c^2 r}\right)(u^0)^2 - e^{\frac{2ct}{R}}\left(1 - \frac{2GM}{c^2 r}\right)^{-1}(u^1)^2$$
$$\times\left(\frac{2e^{-\frac{ct}{R}}R}{3\left(1 - e^{-\frac{2ct}{R}}\frac{R^2}{9r^2}\right)r} + \sqrt{\frac{4e^{-\frac{2ct}{R}}R^2}{9\left(1 - e^{-\frac{2ct}{R}}\frac{R^2}{9r^2}\right)^2 r^2} + 1}\right)^2 \geq 0.$$

Proof The lemma can be proved by straightforward calculations. □

Hence, we have an *asymptotically dominant energy condition*: for every $\varepsilon > 0$ there is sufficiently large $M(\varepsilon) > 0$ such that, for every "ε-time-like vector" u such that $u^0u^1 > 0$ and

$$-\left(1 - \frac{2GM}{c^2 r}\right)c^2(u^0)^2 + (1+\varepsilon)e^{\frac{2ct}{R}}\left(1 - \frac{2GM}{c^2 r}\right)^{-1}(u^1)^2 < 0$$

the following inequality $T_{\mu\nu}u^{\mu}u^{\nu} \geq 0$ holds for all t, r with sufficiently large $t + r \geq M(\varepsilon)$, more precisely,

$$\frac{1}{9\left(1 - e^{-\frac{2ct}{R}}\frac{R^2}{9r^2}\right)^2 r^2}\left(2e^{-\frac{ct}{R}}R + 3\left(1 + e^{-\frac{2ct}{R}}\frac{R^2}{9r^2}\right)r\right)^2 - 1 \leq \varepsilon.$$

In order to discuss the *strong energy condition* we calculate $\frac{8\pi G}{c^4}T^{\mu}_{\mu} = -4\frac{\mathcal{R}}{4}\frac{2GM}{c^2r}$. Then

$$\rho_{\alpha\beta} = \frac{\mathcal{R}}{4}\frac{2GM}{c^2r}g_{\alpha\beta} - \frac{\mathcal{R}}{4}\frac{2GM}{c^2r}\begin{pmatrix} 0 & \frac{cR}{3r} & 0 & 0 \\ \frac{cR}{3r} & 0 & 0 & 0 \\ 0 & 0 & 0 & 0 \\ 0 & 0 & 0 & 0 \end{pmatrix}.$$

Consider the radial case; we need

$$\rho_{\mu\nu}u^{\mu}u^{\nu} = \frac{\mathcal{R}}{4}\frac{2GM}{c^2r}\left(-(u^0)^2c^2(1-\frac{2GM}{c^2r}) - 2\frac{cR}{3r}u^0u^1 + (u^1)^2e^{\frac{2ct}{R}}\frac{1}{(1-\frac{2GM}{c^2r})}\right) \geq 0$$

and the time-like vector u:

$$g_{\mu\nu}u^{\mu}u^{\nu} = -(u^0)^2c^2(1-\frac{2GM}{c^2r}) + (u^1)^2e^{\frac{2ct}{R}}\frac{1}{(1-\frac{2GM}{c^2r})} < 0$$

Thus, even *an asymptotically strong energy condition* is violated. There are many matter configurations which violate the strong energy condition. In particular, the violation of the strong energy condition takes place for a scalar field in the de Sitter expansion. (See, [25, Sect. 9.7.3])

The covariant wave equation in the black hole embedded in de Sitter universe background is

$$\begin{aligned} &-\left(1-\frac{2GM_{bh}}{c^2r}\right)^{-1}\frac{1}{c^2}\frac{\partial^2\psi}{\partial t^2} - \frac{3}{cR}\left(1-\frac{2GM_{bh}}{c^2r}\right)^{-1}\frac{\partial\psi}{\partial t} \\ &+e^{-\frac{2ct}{R}}\left\{\left(1-\frac{2GM_{bh}}{c^2r}\right)\frac{\partial^2\psi}{\partial r^2} + \frac{2}{r}\left(1-\frac{GM_{bh}}{c^2r}\right)\frac{\partial\psi}{\partial r}\right. \\ &\left.+\frac{1}{r^2\sin\theta}\frac{\partial}{\partial\theta}\left(\sin\theta\frac{\partial\psi}{\partial\theta}\right) + \frac{1}{r^2\sin^2\theta}\frac{\partial}{\partial\phi}\left(\frac{\partial\psi}{\partial\phi}\right)\right\} = 0. \end{aligned}$$

For $\psi = \psi(r,t,\theta,\phi)$ we obtain the covariant wave equation ($m = 0$)

$$\begin{aligned} &\frac{\partial^2\psi}{\partial t^2} + \frac{3c}{R}\frac{\partial\psi}{\partial t} - c^2e^{-\frac{2ct}{R}}\left\{\left(1-\frac{2GM_{bh}}{c^2r}\right)^2\frac{\partial^2\psi}{\partial r^2} + \frac{2}{r}\left(1-\frac{GM_{bh}}{c^2r}\right)\left(1-\frac{2GM_{bh}}{c^2r}\right)\frac{\partial\psi}{\partial r}\right. \\ &\left.+\left(1-\frac{2GM_{bh}}{c^2r}\right)\frac{1}{r^2}\Delta_{\mathbb{S}^2}\psi\right\} = f. \end{aligned}$$

The covariant Klein–Gordon equation is ($m \neq 0$)

$$\begin{aligned} &\frac{\partial^2\psi}{\partial t^2} + \frac{3c}{R}\frac{\partial\psi}{\partial t} - c^2e^{-\frac{2ct}{R}}\left\{\left(1-\frac{2GM_{bh}}{c^2r}\right)^2\frac{\partial^2\psi}{\partial r^2} + \frac{2}{r}\left(1-\frac{GM_{bh}}{c^2r}\right)\left(1-\frac{2GM_{bh}}{c^2r}\right)\frac{\partial\psi}{\partial r}\right. \\ &\left.+\left(1-\frac{2GM_{bh}}{c^2r}\right)\frac{1}{r^2}\Delta_{\mathbb{S}^2}\psi\right\} + \frac{m^2c^4}{h^2}\psi - m^2\frac{2c^2GM_{bh}}{h^2r}\psi = 0. \end{aligned}$$

For the large r (the far field) the equation is the wave equation in FLRW spacetime, while the near field limit for small time is Schwarzschild.

We make change $u = e^{\frac{3c}{2R}t}\psi$ ($\psi = e^{-\frac{3c}{2R}t}u$) in the wave equation, then the covariant wave equation became non-covariant wave equation

$$\frac{\partial^2 u}{\partial t^2} - c^2 e^{-\frac{2ct}{R}}\left\{\left(1-\frac{2GM_{bh}}{c^2 r}\right)^2 \frac{\partial^2 u}{\partial r^2} + \frac{2}{r}\left(1-\frac{GM_{bh}}{c^2 r}\right)\left(1-\frac{2GM_{bh}}{c^2 r}\right)\frac{\partial u}{\partial r}\right.$$
$$\left. + \left(1-\frac{2GM_{bh}}{c^2 r}\right)\frac{1}{r^2}\Delta_{\mathbb{S}^2} u\right\} - \frac{9c^2}{4R^2}u = g\,,$$

where $g = e^{\frac{3c}{2R}t}f$. This is the non-covariant Klein–Gordon equation with the imaginary mass

$$u_{tt} - e^{-\frac{2ct}{R}}A(x,\partial_x)u - M^2 u = 0\,,$$

were the curved mass is $M = \frac{3c}{2R}$ and

$$A(x,\partial_x)u := c^2\left\{\left(1-\frac{2GM_{bh}}{c^2 r}\right)^2 \frac{\partial^2 u}{\partial r^2}\right.$$
$$\left. + \frac{2}{r}\left(1-\frac{GM_{bh}}{c^2 r}\right)\left(1-\frac{2GM_{bh}}{c^2 r}\right)\frac{\partial u}{\partial r} + \left(1-\frac{2GM_{bh}}{c^2 r}\right)\frac{1}{r^2}\Delta_{\mathbb{S}^2} u\right\}.$$

One can choose the unites, such that $c/R = 1$, then the non-covariant Klein–Gordon equation became

$$u_{tt} - e^{-2t}A(x,\partial_x)u - M^2 u = 0\,. \tag{2.13}$$

Thus, we are in position to apply Theorem 2.1 and to reveal the properties of the black hole in the de Sitter background.

Example 3 The Euler-Bernoulli beam equation with the variable coefficients is the equation

$$\psi_{tt} + e^{-2t}\sum_{i,j=1}^{n}\partial_{x_i}^2\left(a^{ij}(x)\partial_{x_j}^2\psi\right) = f\,.$$

Here $A(x,\partial_x) = \sum_{i,j=1}^n \partial_{x_i}^2 a^{ij}(x)\partial_{x_j}^2$ and we assume that $\sum_{i,j=1}^n a^{ij}(x)\xi_i\xi_j \geq 0$.

3 The Generalized Tricomi Equation

In this subsection, we consider the generalized Tricomi equation. For a smooth function $f = f(x,t)$ consider the solution $w = w_{A,f}(x,t;b)$ to the problem

$$w_{tt} - A(x,\partial_x)w = 0, \quad w(x,0;b) = f(x,b), \quad w_t(x,0;b) = 0, \quad t \in [0,T_1] \subseteq \mathbb{R}, \quad x \in \widetilde{\Omega} \subseteq \mathbb{R}^n, \tag{3.14}$$

with the parameter $b \in I = [t_{in}, T] \subseteq \mathbb{R}$, $0 \le t_{in} < T \le \infty$, and with $0 < T_1 \le \infty$. Here $\widetilde{\Omega}$ is a domain in $\mathbb{R}^n$, while $A(x,\partial_x)$ is the partial differential operator $A(x,\partial_x) = \sum_{|\alpha|\le m} a_\alpha(x)\partial_x^\alpha$ with smooth coefficients, $a_\alpha \in C^\infty(\widetilde{\Omega})$. We are going to present the integral operator

$$\mathcal{K}[w](x,t) = \int_{t_{in}}^{t} db \int_0^{|\phi(t)-\phi(b)|} K(t;r,b)w(x,r;b)\,dr, \quad x \in \widetilde{\Omega}, \ t \in I, \tag{3.15}$$

which maps the function $w = w(x,r;b)$ into the solution $u = u(x,t)$ of the generalized Tricomi equation

$$u_{tt} - a^2(t)A(x,\partial_x)u = f, \qquad x \in \widetilde{\Omega}\,, \ t \in I\,, \tag{3.16}$$

where $a^2(t) = t^\ell, \ell \in \mathbb{C}$. In fact, the function $u = u(x,t)$ takes initial values as follows

$$u(x,t_{in}) = 0, \quad u_t(x,t_{in}) = 0, \quad x \in \widetilde{\Omega}\,.$$

In (3.15), $\phi = \phi(t)$ is a distance function produced by $a = a(t)$, that is $\phi(t) = \int_{t_{in}}^t a(\tau)\,d\tau$. Moreover, we also introduce the corresponding operators, which generate solutions of the source-free equation and take nonvanishing initial values. These operators are constructed in [84] in the case of $\ell > 0$, $A(x,\partial_x) = \Delta$, $\widetilde{\Omega} = \mathbb{R}^n$, where Δ is the Laplace operator on $\mathbb{R}^n$, and, consequently, the Eq. (3.14) is the wave equation. In the present paper, we restrict ourselves to smooth functions, but it is easily seen that similar formulas, with the corresponding interpretations, are applicable to distributions as well.

Here we emphasize that the integral transform $\mathcal{K}$ is less singular than fundamental solution (Green function) given by $\mathcal{G}$ since the operator $\mathcal{WE}$ takes an essential part of singularities.

In this subsection we restrict ourselves to the generalized Tricomi equation, that is $a(t) = t^\ell, \ell \in \mathbb{C}$. This class includes, among others, equations of a wave propagating in the so-called Einstein-de Sitter (EdeS) universe and in the radiation dominated universe with spatial slices of the constant curvature.

The transform linking to the generalized Tricomi operator is generated by the kernel

$$K(t;r,b) = E(r,t;b;\gamma) := c_\ell \left((\phi(t)+\phi(b))^2 - r^2\right)^{-\gamma} F\left(\gamma,\gamma;1;\frac{(\phi(t)-\phi(b))^2 - r^2}{(\phi(t)+\phi(b))^2 - r^2}\right), \tag{3.17}$$

with the distance function $\phi = \phi(t)$ and the numbers γ, c_ℓ defined as follows

$$\phi(t) = \frac{2}{\ell+2}t^{\frac{\ell+2}{2}}, \quad \gamma := \frac{\ell}{2(\ell+2)}, \quad \ell \in \mathbb{C}\setminus\{-2\}, \quad c_\ell = \left(\frac{\ell+2}{4}\right)^{-\frac{\ell}{\ell+2}}, \tag{3.18}$$

while $F\big(a, b; c; \zeta\big)$ is the Gauss's hypergeometric function. Here $t_{in} = 0$.

According to Theorem 2.1 [95], the function $E(r, t; b; \gamma)$ solves the following Tricomi-type equation:

$$E_{tt}(r, t; b; \gamma) - t^{\ell} E_{rr}(r, t; b; \gamma) = 0\,, \qquad 0 < b < t\,. \tag{3.19}$$

The proof of Theorem 2.1 [95], which is given in [95], is straightforward. This theorem generalizes corresponding statement from [22]. In fact, that proof is applicable to the different distance functions $\phi = \phi(t)$, see, for instance, [93], where the case of $a(t) = e^{-t}$ is discussed.

There are four important examples of equations which are amenable to the integral transform approach, when $\ell = 3, 1, -1, -4/3$; those are the small disturbance equations for the perturbation velocity potential of a two-dimensional near sonic uniform flow of dense gases in a physical plane (see, e.g., [48, 76]), the Tricomi equation (see, e.g., [4, 13, 21, 27, 35, 36, 50, 52–54, 57, 58, 62, 65, 74, 77] and bibliography therein), the equation of waves in the radiation dominated universe (see, e.g., [25, 37] and bibliography therein) and in the EdeS spacetime (see, e.g., [25, 37, 38, 63] and bibliography therein), respectively.

To introduce the integral transform we need some special geometric structure of the domains of functions.

Definition 3.1 The set $\Omega \subseteq \overline{\mathbb{R}^{n+1}_+}$ is said to be backward time line-connected to $t = 0$, if for every point $(x, t) \in \Omega$ the line segment $\{(x, s)\,|\, s \in (0, t]\,\}$ is also in Ω; that is $\{(x, s)\,|\, s \in (0, t]\} \subseteq \Omega$.

Henceforth we just write "backward time connected" for such sets. Similarly, if $\Omega \subseteq [0, T] \times \mathbb{R}^n$, $T > 0$, then one can define a forward time line-connected to $t = T$ set. The union and the intersection of the backward time connected sets are also backward time connected. The interior and the closure of the backward time connected set are also the backward time connected sets. For every set, there exists its minimal backward time connected covering. The domain of the dependence for the wave equation is backward time connected, while domain of influence is forward time connected.

Definition 3.2 Let $\phi \in C(\overline{\mathbb{R}_+})$ be nonnegative increasing function and Ω be a backward time connected set. The backward time connected set $\Omega_\phi \subseteq \overline{\mathbb{R}^{n+1}_+}$ defined by

$$\Omega_\phi := \bigcup_{(x,t)\in\Omega} \{(x, \tau)\,|\, \tau \in (0, \phi(t)]\,\}$$

is said to be a ϕ-image of Ω.

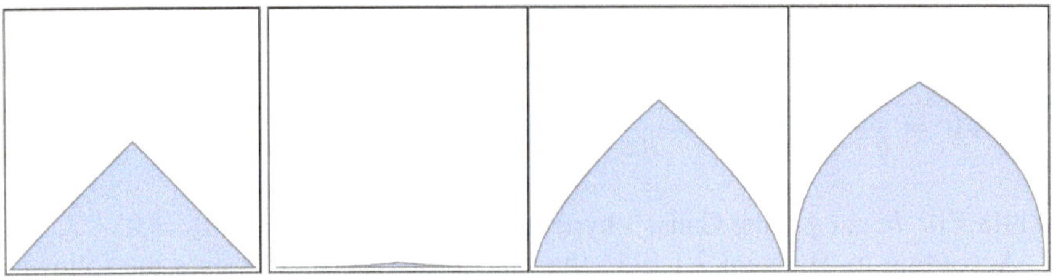

Fig. 2 $\phi(t) = (2/(\ell+2))t^{(\ell+2)/2},\ \phi(t) + |x| \le 1, x \in [-1,1],\ t \in [0,2],\ \ell = 0, -\frac{4}{3}, 1, 3$

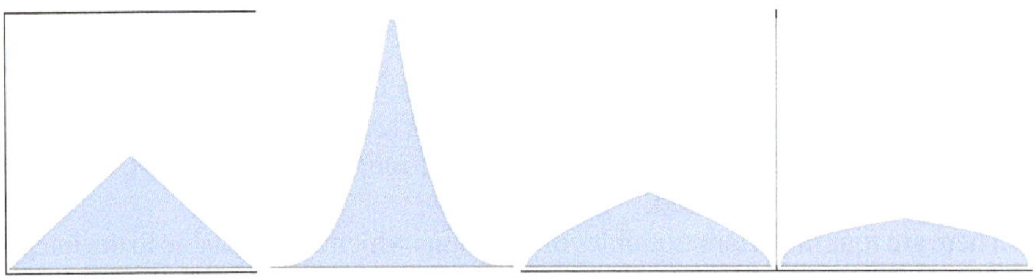

Fig. 3 $\phi(t) = (2/(\ell+2))t^{(\ell+2)/2},\ \phi(t) + |x| \le 8, x \in [-8,8],\ t \in [0,18],\ \ell = 0, -\frac{4}{3}, 1, 3$

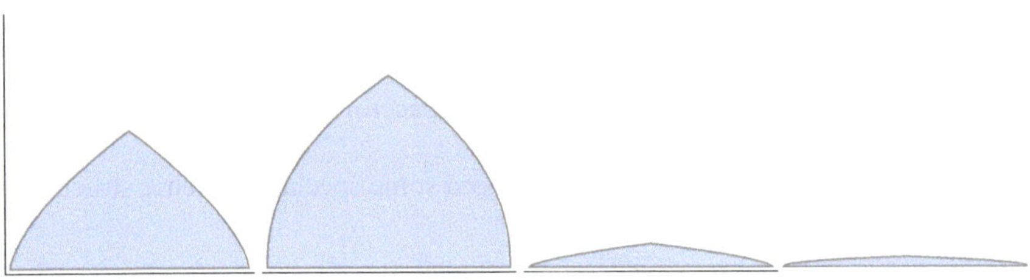

Fig. 4 Ω and Ω_ϕ with $x_0 = 1/2$; Ω and Ω_ϕ with $x_0 = 100$

On Figs. 2, 3 we illustrate the dependence domains for hyperbolic equations with $\ell = 0, -\frac{4}{3}, 1, 3$ and $A(x, \partial_x) = \Delta$.

Figure 4 illustrates the part of the domain in the hyperbolic region of the Tricomi problem (see, e.g., [52] and references therein) that has the form $\Omega := \{(x,t) \mid |x| < x_0 - \frac{2}{3}t^{\frac{3}{2}},\ -x_0 \le x \le x_0,\ t > 0\}$ for $x_0 = 1/2$ and $x_0 = 100$. Corresponding ϕ-images of Ω are $\Omega_\phi := \{(x,t) \mid |x| < x_0 - \left(\frac{2}{3}\right)^{5/2} t^{\frac{9}{4}},\ -x_0 \le x \le x_0\}$ with $x_0 = 1/2$ and $x_0 = 100$, respectively.

The next theorem describes the main property of the integral transform (3.15). We denote U_{x_0} a neighborhood of the point $x_0 \in \mathbb{R}^n$.

Theorem 3.3 ([95]) *Let $f = f(x,t)$ be a function defined in the backward time connected domain Ω. Suppose that for a given $(x_0, t_0) \in \Omega$ the function $w(x, r; b) \in C^{m,2,0}_{x,r,t}(U_{x_0} \times [0, \phi(t_0)] \times [0, t_0])$ solves the problem*

$$w_{rr} - A(x, \partial_x)w = 0 \quad \textit{in} \quad U_{x_0} \quad \textit{for all} \quad r \in (0, |\phi(t_0) - \phi(b)|)\,, \tag{3.20}$$

$$w(x, 0; b) = f(x, b) \quad \textit{in} \quad U_{x_0} \quad \textit{for all} \quad b \in (0, t_0)\,. \tag{3.21}$$

Then for $\ell > -2$ the function

$$u(x, t) = c_\ell \int_0^t db \int_0^{|\phi(t)-\phi(b)|} \left((\phi(t) + \phi(b))^2 - r^2\right)^{-\gamma} \tag{3.22}$$

$$\times F\left(\gamma, \gamma; 1; \frac{(\phi(t) - \phi(b))^2 - r^2}{(\phi(t) + \phi(b))^2 - r^2}\right) w(x, r; b)\, dr,$$

defined in the past $U_{x_0} \times [0, t_0)$ of $U_{x_0} \times \{t_0\}$, is continuous in $U_{x_0} \times [0, t_0)$ and it satisfies the equation

$$u_{tt} - t^\ell A(x, \partial_x)u = f(x, t) \tag{3.23}$$

$$+ c_\ell(\phi'(t))^2 \int_0^t (\phi(t) + \phi(b))^{-2\gamma} F\left(\gamma, \gamma; 1; \frac{(\phi(t) - \phi(b))^2}{(\phi(t) + \phi(b))^2}\right) w_r(x, 0; b)\, db\,,$$

in the sense of distributions $\mathcal{D}'(U_{x_0} \times (0, t_0))$. The function $u(x, t)$ takes the vanishing initial value $u(x, 0) = 0$ for all $x \in U_{x_0}$. Moreover, if, additionally, $\ell < 4$, then u_t is continuous in $U_{x_0} \times [0, t_0)$ and $u_t(x, 0) = 0$ for all $x \in U_{x_0}$.

We stress here that the integral transform $w \longmapsto u$ is point-wise in x and nonlocal in time. Let π_x be a projection $\pi_x : \Omega \longrightarrow \mathbb{R}^n$ of the backward time connected domain Ω, and denote $\widetilde{\Omega} := \pi_x(\Omega)$.

Corollary 3.4 *Let $f = f(x, t)$ be a function defined in the backward time connected domain Ω. Suppose that the function $w(x, r; b) \in C^{m,2,0}_{x,r,t}$ satisfies*

$$w_{rr} - A(x, \partial_x)w = 0 \quad \textit{for all} \quad (x, r) \in \Omega_\phi \quad \textit{and} \quad (x, b) \in \Omega\,,$$

$$w(x, 0; b) = f(x, b) \quad \textit{for all} \quad (x, b) \in \Omega\,.$$

Then for $\ell > -2$ the function (3.22) defined on Ω, is continuous and satisfies the Eq. (3.23) in the sense of distributions $\mathcal{D}'(\Omega)$. The function u takes the vanishing initial value $u(x, 0) = 0$ for all $x \in \widetilde{\Omega}$. If, additionally, $\ell < 4$, then u_t is continuous in Ω and $u_t(x, 0) = 0$ for all $x \in \widetilde{\Omega}$.

If the initial value problem for the operator $\partial_t^2 - A(x, \partial_x)$ admits two initial conditions, then we can eliminate the function w_r from the right-hand side of Eq. (3.23).

Theorem 3.5 ([95]) *Let $f = f(x, t)$ be a function defined in the backward time connected domain Ω. Suppose that the function $w(x, r; b) \in C^{m,2,0}_{x,r,t}$ satisfies*

$$w_{rr} - A(x, \partial_x)w = 0 \quad \textit{for all} \quad (x, r) \in \Omega_\phi \quad \textit{and for all} \quad (x, b) \in \Omega\,,$$

$$w(x, 0; b) = f(x, b), \quad w_r(x, 0; b) = 0 \quad \textit{for all} \quad (x, b) \in \Omega\,.$$

Then for $\ell > -2$ the function

$$u(x,t) = c_\ell \int_0^t db \int_0^{|\phi(t)-\phi(b)|} \left((\phi(t)+\phi(b))^2 - r^2\right)^{-\gamma} \times F\left(\gamma, \gamma; 1; \frac{(\phi(t)-\phi(b))^2 - r^2}{(\phi(t)+\phi(b))^2 - r^2}\right) w(x,r;b)\, dr,$$

defined on Ω, is continuous and satisfies the equation

$$u_{tt} - t^\ell A(x,\partial_x)u = f(x,t)\,, \tag{3.24}$$

in the sense of distributions $\mathcal{D}'(\Omega)$. The function u takes the vanishing initial value $u(x,0) = 0$ for all $x \in \widetilde{\Omega}$. If, additionally, $\ell < 4$, then u_t is continuous in Ω and $u_t(x,0) = 0$ for all $x \in \widetilde{\Omega}$.

For instance, the Cauchy problem for the second order strictly hyperbolic equation admits two initial conditions. We recall here that for the weakly hyperbolic operators $\partial_t^2 - \sum_{|\alpha|\le 2} a_\alpha(x)\partial_x^\alpha$, which satisfy the Levi conditions (see, e.g., [83]), the Cauchy problem can be solved for smooth initial data. If $m = 1$, then the problem with two initial conditions can be solved in Gevery spaces. (See, e.g., [83].) The case of $m > 2$ covers the beam equation and hyperbolic in the sense of Petrowski (p-evolution) equations. On the other hand, the Cauchy-Kowalewski theorem guarantees solvability of the problem in the real analytic functions category for the partial differential equation (3.24) with any positive ℓ and $m = 2$. Furthermore, the operator $A(x,\partial_x) = \sum_{|\alpha|\le 2} a_\alpha(x)\partial_x^\alpha$ can be replaced with an abstract operator A acting on some linear topological space of functions.

Example 1 Consider equations of the gas dynamics. (a) For the Tricomi equation in the *hyperbolic domain*,

$$u_{tt} - t\Delta u = f(x,t), \tag{3.25}$$

$\phi(t) = \frac{2}{3}t^{\frac{3}{2}}$ and $A(x,\partial_x) = \Delta$. Then for every $f \in C(\mathbb{R}^n \times [0,T])$ we can solve the Cauchy problem for the wave equation

$$w_{tt} - \Delta w = 0, \quad w(x,0;b) = f(x,b)\,, \quad w_t(x,0;b) = 0\,, \quad x \in \mathbb{R}^n, \quad t \in \left[0, \frac{2}{3}T^{\frac{3}{2}}\right]$$

in $\mathbb{R}^n \times \left[0, \frac{2}{3}T^{\frac{3}{2}}\right] \times [0,T]$. (For the explicit formula see, e.g., (3.33), (3.34).) The solution to the Cauchy problem for (3.25) with vanishing initial data is given as follows

$$u(x,t) = 3^{-1/3}2^{2/3}\int_0^t db \int_0^{\frac{2}{3}t^{\frac{3}{2}}-\frac{2}{3}b^{\frac{3}{2}}} \left(\left(\frac{2}{3}t^{\frac{3}{2}}+\frac{2}{3}b^{\frac{3}{2}}\right)^2 - r^2\right)^{-\frac{1}{6}}$$
$$\times F\left(\frac{1}{6},\frac{1}{6};1;\frac{\left(\frac{2}{3}t^{\frac{3}{2}}-\frac{2}{3}b^{\frac{3}{2}}\right)^2-r^2}{\left(\frac{2}{3}t^{\frac{3}{2}}+\frac{2}{3}b^{\frac{3}{2}}\right)^2-r^2}\right) w(x,r;b)dr, \quad t\in[0,T].$$

For the Tricomi equation in the *elliptic domain*,

$$u_{tt} + t\Delta u = f(x,t), \quad t>0, \tag{3.26}$$

we have $A(x,\partial_x) = -\Delta$ and, since the Cauchy problem for (3.26) is not well posed, Theorem 3.3 gives representation of the solutions only for some specific functions f.
(b) The small disturbance equation for the perturbation velocity potential of a two-dimensional near sonic uniform flow of dense gases in a physical plane, has been derived by Kluwick [48], Tarkenton and Cramer [76]. It leads to the equation

$$u_{tt} - t^3\Delta u = f(x,t), \tag{3.27}$$

with $\ell = 3$ and $\phi(t) = \frac{2}{5}t^{\frac{5}{2}}$, and $A(x,\partial_x) = \Delta$. The solution to the Cauchy problem for (3.27) with vanishing initial data is given as follows

$$u(x,t) = \frac{3}{10}\int_0^t db \int_0^{\frac{2}{5}t^{\frac{5}{2}}-\frac{2}{5}b^{\frac{5}{2}}} \left(\left(\frac{2}{5}t^{\frac{5}{2}}+\frac{2}{5}b^{\frac{5}{2}}\right)^2 - r^2\right)^{-\frac{3}{10}}$$
$$\times F\left(\frac{3}{10},\frac{3}{10};1;\frac{\left(\frac{2}{5}t^{\frac{5}{2}}-\frac{2}{5}b^{\frac{5}{2}}\right)^2-r^2}{\left(\frac{2}{5}t^{\frac{5}{2}}+\frac{2}{5}b^{\frac{5}{2}}\right)^2-r^2}\right) w(x,r;b)dr, \quad t>0.$$

Example 2 Consider the wave equation in the Einstein-de Sitter (EdeS) spacetime with hyperbolic spatial slices. The metric of the Einstein & de Sitter universe (EdeS universe) is a particular member of the Friedmann-Robertson-Walker metrics

$$ds^2 = -dt^2 + a_{sc}^2(t)\left[\frac{dr^2}{1-Kr^2} + r^2 d\Omega^2\right], \tag{3.28}$$

where $K = -1, 0,$ or $+1$, for a hyperbolic, flat or spherical spatial geometry, respectively. For the EdeS the scale factor is $a_{sc}(t) = t^{2/3}$. The covariant d'Alambert's operator (the Laplace-Beltrami operator),

$$\square_g \psi = \frac{1}{\sqrt{|g|}}\frac{\partial}{\partial x^i}\left(\sqrt{|g|}g^{ik}\frac{\partial \psi}{\partial x^k}\right),$$

in the spherical coordinates is

$$\square_{EdeS}\psi = -\left(\frac{\partial}{\partial x^0}\right)^2\psi - \frac{2}{t}\left(\frac{\partial\psi}{\partial x^0}\right)\psi + t^{-\frac{4}{3}}\frac{\sqrt{1-Kr^2}}{r^2}\frac{\partial}{\partial r}\left(r^2\sqrt{1-Kr^2}\frac{\partial\psi}{\partial r}\right)$$
$$+t^{-\frac{4}{3}}\frac{1}{r^2\sin\theta}\frac{\partial}{\partial\theta}\left(\sin\theta\frac{\partial\psi}{\partial\theta}\right) + t^{-\frac{4}{3}}\frac{1}{r^2\sin^2\theta}\left(\frac{\partial}{\partial\phi}\right)^2\psi\,.$$

The change $\psi = t^{-1}u$ of the unknown function leads the equation $\square_{EdeS}\psi = g$ to the equation

$$u_{tt} - t^{-4/3}A(x,\partial_x)u = f\,,$$

where

$$A(x,\partial_x)u = \frac{\sqrt{1-Kr^2}}{r^2}\frac{\partial}{\partial r}\left(r^2\sqrt{1-Kr^2}\frac{\partial u}{\partial r}\right) + \frac{1}{r^2\sin\theta}\frac{\partial}{\partial\theta}\left(\sin\theta\frac{\partial u}{\partial\theta}\right) + \frac{1}{r^2\sin^2\theta}\left(\frac{\partial}{\partial\phi}\right)^2 u\,. \tag{3.29}$$

The spatial part X of the spacetime (3.28) has a constant curvature $6K$. The operator $A(x,\partial_x)$ (3.29) is the Laplace-Beltrami operator on X. The explicit formulas for the solutions of the Cauchy problem for the wave operator on spaces with constant negative curvature are known, see, for instance, [41, 51]. Thus, Theorem 3.5 gives an explicit representation for the solution of the Cauchy problem with vanishing initial data for the wave equation in the EdeS spacetime with negative constant curvature $K < 0$. We note here that $\gamma = -1$ for the metric (3.28) (see [29]) that makes the hypergeometric function polynomial.

The next theorem represents the integral transforms for the case of the equation without a source term. In that theorem, the transformed function has nonvanishing initial values. For $\gamma \in \mathbb{C}$, $Re\,\gamma > 0$, that is for $\ell \in \mathbb{C}\setminus\overline{D_1(-1,0)} = \{z \in \mathbb{C}\,|\,|z+1| > 1\}$, and, in particular, for $\ell \in (-\infty,-2)\cup(0,\infty)$, we define the integral operator

$$(K_0v)(x,t) := 2^{2-2\gamma}\frac{\Gamma(2\gamma)}{\Gamma^2(\gamma)}\int_0^1 v(x,\phi(t)s)(1-s^2)^{\gamma-1}ds$$
$$= \phi(t)^{1-2\gamma}2^{2-2\gamma}\frac{\Gamma(2\gamma)}{\Gamma^2(\gamma)}\int_0^{\phi(t)} v(x,\tau)(\phi^2(t)-\tau^2)^{\gamma-1}d\tau\,.$$

For $\gamma \in \mathbb{C}$, $Re\,\gamma < 1$, that is for $\ell \in \mathbb{C}\setminus\overline{D_1(-3,0)} = \{z \in \mathbb{C}\,|\,|z+3| > 1\}$, and, in particular, for $\ell \in (-\infty,-4)\cup(-2,\infty)$, we define the integral operator

$$(K_1v)(x,t) := t2^{2\gamma}\frac{\Gamma(2-2\gamma)}{\Gamma^2(1-\gamma)}\int_0^1 v(x,\phi(t)s)(1-s^2)^{-\gamma}ds$$
$$= t\phi(t)^{2\gamma-1}2^{2\gamma}\frac{\Gamma(2-2\gamma)}{\Gamma^2(1-\gamma)}\int_0^{\phi(t)} v(x,\tau)(\phi^2(t)-\tau^2)^{-\gamma}d\tau\,.$$

Thus, both operators are defined simultaneously for $\gamma \in \mathbb{C}$, $0 < Re\,\gamma < 1$, and, in particular, for $\ell \in (-\infty, -4) \cup (0, \infty)$. Denote

$$a_\ell := 2^{1-2\gamma} \frac{\ell\,\Gamma(2\gamma)}{2\gamma\Gamma^2(\gamma)}, \qquad b_\ell := (\ell+2)2^{2\gamma-1} \frac{\Gamma(2-2\gamma)}{\Gamma^2(1-\gamma)}.$$

The next theorem describes the properties of the integral transforms K_0 and K_1 in the case when ℓ is a positive number.

Theorem 3.6 ([95]) *Let ℓ be a positive number and let $\Omega \subset \mathbb{R}^{n+1}_+$ be a backward time connected domain. Suppose that the function $v \in C^{m,2}_{x,t}(\overline{\Omega})$ for given $(x_0, t_0) \in \Omega$ solves the equation*

$$\partial_t^2 v - A(x, \partial_x)v = 0 \quad \textit{at} \quad x = x_0 \quad \textit{and all} \quad t \in (0, \phi(t_0)). \tag{3.30}$$

Then, the functions $K_0 v \in C^{m,2}_{x,t}(\Omega)$ and $K_1 v \in C^{m,2}_{x,t}(\Omega)$ satisfy the equations

$$\left(\partial_t^2 - t^\ell A(x, \partial_x)\right) K_0 v = a_\ell t^{\frac{\ell}{2}-1} \partial_t v(x, 0) \quad \textit{at} \quad x = x_0 \quad \textit{for all} \quad t \in (0, t_0), \tag{3.31}$$

and

$$\left(\partial_t^2 - t^\ell A(x, \partial_x)\right) K_1 v = b_\ell t^{\frac{\ell}{2}} \partial_t v(x, 0) \quad \textit{at} \quad x = x_0 \quad \textit{for all} \quad t \in (0, t_0), \tag{3.32}$$

respectively. They have at $x = x_0$ the following initial values

$$(K_0 v)(x_0, 0) = v(x_0, 0), \qquad (K_0 v)_t(x_0, 0) = 0,$$

and

$$(K_1 v)(x_0, 0) = 0, \qquad (K_1 v)_t(x_0, 0) = v(x_0, 0).$$

Thus, the value $v(x_0, 0)$ of the solutions of (3.30) is invariant under operation K_0, while the operator K_1 acts similarly to the Dirichlet-to-Neumann map.

Corollary 3.7 *Let ℓ be a positive number and $\Omega \subset \mathbb{R}^{n+1}_+$ be a backward time connected domain. Suppose that the function $v \in C^{m,2}_{x,t}(\overline{\Omega_\phi})$ solves the equation*

$$\partial_t^2 v - A(x, \partial_x)v = 0 \quad \textit{for all} \quad (x, t) \in \Omega_\phi .$$

Then, the functions $K_0 v \in C^{m,2}_{x,t}(\Omega)$ and $K_1 v \in C^{m,2}_{x,t}(\Omega)$ satisfy the equations

$$\left(\partial_t^2 - t^\ell A(x, \partial_x)\right) K_0 v = a_\ell t^{\frac{\ell}{2}-1} \partial_t v(x, 0) \quad \textit{for all} \quad (x, t) \in \Omega,$$

and

$$\left(\partial_t^2 - t^\ell A(x,\partial_x)\right) K_1 v = b_\ell t^{\frac{\ell}{2}} \partial_t v(x,0) \quad \textit{for all} \quad (x,t) \in \Omega\,,$$

respectively. They have the following initial values

$$(K_0 v)(x,0) = v(x,0)\,, \qquad (K_0 v)_t(x,0) = 0 \quad \textit{for all} \quad x \in \widetilde{\Omega}\,,$$

and

$$(K_1 v)(x,0) = 0\,, \qquad (K_1 v)_t(x,0) = v(x,0) \quad \textit{for all} \quad x \in \widetilde{\Omega}\,.$$

For the Cauchy problem with full initial data, we have the following result for the generalized Tricomi equation in the hyperbolic domain.

Theorem 3.8 ([95]) *Let ℓ be a positive number and $\Omega \subset \mathbb{R}^{n+1}_+$ be a backward time connected domain. Suppose that the functions $v_0, v_1 \in C^{m,2}_{x,t}(\overline{\Omega_\phi})$ solve the problem*

$$\begin{aligned}
&\partial_t^2 v_i - A(x,\partial_x) v_i = 0 \quad \textit{for all} \quad (x,t) \in \Omega_\phi\,,\\
&v_i(x,0) = \sum_{k=0,1} \delta_{ik}\varphi_k(x)\,, \qquad \partial_t v_i(x,0) = 0\,, \qquad i = 0,1, \quad \textit{for all} \quad (x,t) \in \widetilde{\Omega_\phi}\,.
\end{aligned}$$

Then the function $u = K_0 v_0 + K_1 v_1 \in C^{m,2}_{x,t}(\Omega)$ solves the problem

$$\begin{aligned}
&\left(\partial_t^2 - t^\ell A(x,\partial_x)\right) u = 0 \quad \textit{for all} \quad (x,t) \in \Omega\,,\\
&u(x,0) = \varphi_0(x)\,, \qquad \partial_t u(x,0) = \varphi_1(x) \quad \textit{for all} \quad x \in \widetilde{\Omega}\,.
\end{aligned}$$

In order to make this paper more self-contained, we remind here that if $A(x,\partial_x) = \Delta$, then the function $v_\varphi(x,t)$ is given by the following formulas (see, e.g., [73]): for $\varphi \in C_0^\infty(\mathbb{R}^n)$ and for $x \in \mathbb{R}^n$, $n = 2m+1$, $m \in \mathbb{N}$,

$$v_\varphi(x,t) := \frac{\partial}{\partial t}\Big(\frac{1}{t}\frac{\partial}{\partial t}\Big)^{\frac{n-3}{2}} \frac{t^{n-2}}{\omega_{n-1} c_0^{(n)}} \int_{S^{n-1}} \varphi(x+ty)\, dS_y\,, \tag{3.33}$$

while for $x \in \mathbb{R}^n$, $n = 2m$, $m \in \mathbb{N}$,

$$v_\varphi(x,t) := \frac{\partial}{\partial t}\Big(\frac{1}{t}\frac{\partial}{\partial t}\Big)^{\frac{n-2}{2}} \frac{2t^{n-1}}{\omega_{n-1} c_0^{(n)}} \int_{B_1^n(0)} \frac{1}{\sqrt{1-|y|^2}} \varphi(x+ty)\, dV_y\,. \tag{3.34}$$

The last formulas can be also written in terms of the Radon transform; for details, see [41, 51].

The case of negative ℓ requires some modifications in the setting of the initial conditions at $t = 0$. For the EdeS spacetime ($\ell = -4/3$) these modifications are suggested in [29]; they are the so-called weighted initial conditions.

One can consider the Cauchy problem for the equations with negative ℓ and with the initial conditions prescribed at $t = t_{in} > 0$. For $\ell < -2$, the hyperbolic equations in such spacetime have permanently bounded domain of influence. Nonlinear equations with a permanently bounded domain of influence were studied in [16, 18]. In particular, Choquet-Bruhat [16, 18] proved for small initial data the global existence and uniqueness of wave maps on the FLRW expanding universe with the metric $\mathbf{g} = -dt^2 + R^2(t)\sigma$ and a smooth Riemannian manifold (S, σ) of dimension $n \leq 3$, which has a time independent metric σ and a nonzero injectivity radius, and with $R(t)$ being a positive increasing function such that $1/R(t)$ is integrable on $[t_{in}, \infty)$. If the target manifold is flat, then the wave map equation reduces to a linear system.

It will be interesting to apply the integral transform approach to the maximum principle (see, e.g., [52, 66]) for the generalized Tricomi equation, to the derivation of the $L_p - L_q$ estimates (see, e.g., [47, 87]), to the mixed problem for the Friedlander model (see, e.g., [45] and references therein), to the global existence problem for the semilinear generalized Tricomi equations on the hyperbolic space (for the wave equation see, e.g., [3, 15, 55]), and to the derivation of the Price's law for the corresponding cosmological models (see, e.g., [56] and references therein).

4 The Semilinear Equations on the Curved Spacetime

In this section, we present some results obtained in [33] on the existence of a global in time solutions of the semilinear Klein–Gordon equation in the de Sitter spacetime with the time slices being Riemannian manifolds. In the spatially flat de Sitter model, this can be $\mathbb{R}^3$ and in the spatially closed and spatially open cases it can be the three-sphere $\mathbb{S}^3$ and the three-hyperboloid $\mathbb{H}^3$, respectively (see [17, p. 113]).

The metric g in the de Sitter spacetime, is defined as follows, $g_{00} = g^{00} = -1$, $g_{0j} = g^{0j} = 0$, $g_{ij}(x,t) = e^{2t}\sigma_{ij}(x)$, $i, j = 1, 2, \ldots, n$, where $\sum_{j=1}^{n} \sigma^{ij}(x)\,\sigma_{jk}(x) = \delta_{ik}$, and δ_{ij} is Kronecker's delta. The metric $\sigma^{ij}(x)$ describes the time slices. In quantum field theory the matter fields are described by a function ψ that must satisfy equations of motion. In the case of a massive scalar field, the equation of motion is the semilinear Klein–Gordon equation generated by the metric g:

$$\square_g \psi = m^2\psi + V'_\psi(x, \psi)\,.$$

Here m is a physical mass of the particle. In physical terms this equation describes a local self-interaction for a scalar particle. A typical example of a potential function would be $V(\phi) = \phi^4$. The semilinear equations are also commonly used models for general nonlinear problems.

The covariant Klein–Gordon equation in the de Sitter spacetime in the coordinates is

$$\psi_{tt} - \frac{e^{-2t}}{\sqrt{|\det \sigma(x)|}} \sum_{i,j=1}^{n} \frac{\partial}{\partial x^i}\left(\sqrt{|\det \sigma(x)|}\sigma^{ij}(x)\frac{\partial}{\partial x^j}\psi\right) + n\psi_t + m^2\psi = F(\psi)\,.$$

This is a special case of the equation

$$\psi_{tt} + n\psi_t - e^{-2t}A(x, \partial_x)\psi + m^2\psi = F(\psi)\,,$$

where $A(x, \partial_x) = \sum_{|\alpha|\le 2} a_\alpha(x)\partial_x^\alpha$ is a second order partial differential operator. More precisely, in this section, we assume that $a_\alpha(x)$, $|\alpha| = 2$, is positive definite (and symmetric).

In [90–92] a global existence of small data solutions of the Cauchy problem for the semilinear Klein–Gordon equation and systems of equations in the de Sitter spacetime with flat time slices, that is, $\sigma^{ij}(x) = \delta^{ij}$, is proved. The nonlinearity F was assumed Lipschitz continuous with exponent $\alpha \ge 0$ (see definition below). It was discovered that unlike the same problem in the Minkowski spacetime, no restriction on the order of nonlinearity is required, provided that a physical mass of the field belongs to some set, $m \in (0, \sqrt{n^2-1}/2] \cup [n/2, \infty)$. The following conjecture was made in [90].

Conjecture *The interval $(\sqrt{n^2-1}/2, n/2)$ is a forbidden mass interval for the small data global solvability of the Cauchy problem for all $\alpha \in (0, \infty)$.*

For $n = 3$ the mass m interval $(0, \sqrt{2})$ is called the Higuchi bound in quantum field theory [42]. The proof of the global existence in [90–92] is based on the special integral presentations (see Sect. 2) and $L^p - L^q$ estimates.

In this section, the small data global existence result to the case of the de Sitter spacetime with the time slices being Riemannian manifolds, is presented. To formulate the theorem we need the following description of the nonlinear term. Let $B_p^{s,q}$ be the Besov space.

Condition ($\mathcal{L}$). *The function F is said to be Lipschitz continuous with exponent $\alpha \ge 0$ in the space $B_p^{s,q}$ if there is a constant $C \ge 0$ such that*

$$\|F(x, \psi_1) - F(x, \psi_2)\|_{B_p^{s,q}} \le C\|\psi_1 - \psi_2\|_{B_{p'}^{s,q}}\left(\|\psi_1\|^\alpha_{B_{p'}^{s,q}} + \|\psi_2\|^\alpha_{B_{p'}^{s,q}}\right) \tag{4.1}$$

for all $\psi_1, \psi_2 \in B_{p'}^{s,q}$, where $1/p + 1/p' = 1$.

For the important case of $B_2^{s,2} = H_{(s)}(\mathbb{R}^n)$, define the complete metric space

$$X(R, s, \gamma) := \{\psi \in C([0,\infty); H_{(s)}(\mathbb{R}^n)) \mid \; \| \psi \|_X := \sup_{t\in[0,\infty)} e^{\gamma t} \| \psi(x,t) \|_{H_{(s)}(\mathbb{R}^n)} \le R\}$$

with the metric

$$d(\psi_1, \psi_2) := \sup_{t\in[0,\infty)} e^{\gamma t} \| \psi_1(x,t) - \psi_2(x,t) \|_{H_{(s)}(\mathbb{R}^n)}\;.$$

We denote $\mathcal{B}^\infty$ the space of all $C^\infty(\mathbb{R}^n)$ functions with uniformly bounded derivatives of all orders.

Theorem 4.1 ([33]) *Let* $A(x, \partial_x) = \sum_{|\alpha|\le 2} a_\alpha(x)\partial_x^\alpha$ *be a second-order negative elliptic differential operator with real coefficients* $a_\alpha \in \mathcal{B}^\infty$. *Assume that the nonlinear term* $F(u)$ *is a Lipschitz continuous with exponent* $\alpha > 0$ *in the space* $H_{(s)}(\mathbb{R}^n)$, $s > n/2 \ge 1$, *and* $F(0) = 0$. *Assume also that* $m \in (0, \sqrt{n^2-1}/2] \cup [n/2, \infty)$. *Then, there exists* $\varepsilon_0 > 0$ *such that, for every given functions* $\psi_0, \psi_1 \in H_{(s)}(\mathbb{R}^n)$, *such that*

$$\|\psi_0\|_{H_{(s)}(\mathbb{R}^n)} + \|\psi_1\|_{H_{(s)}(\mathbb{R}^n)} \le \varepsilon, \qquad \varepsilon < \varepsilon_0 ,$$

there exists a global solution $\psi \in C^1([0,\infty); H_{(s)}(\mathbb{R}^n))$ *of the Cauchy problem*

$$\psi_{tt} + n\psi_t - e^{-2t}A(x,\partial_x)\psi + m^2\psi = F(\psi) , \tag{4.2}$$

$$\psi(x,0) = \psi_0(x) , \quad \psi_t(x,0) = \psi_1(x) . \tag{4.3}$$

That solution $\psi(x,t)$ *belongs to the space* $X(2\varepsilon, s, \gamma)$, *that is,*

$$\sup_{t\in[0,\infty)} e^{\gamma t}\|\psi(\cdot,t)\|_{H_{(s)}(\mathbb{R}^n)} < 2\varepsilon ,$$

with γ *such that either* $0 < \gamma \le \frac{1}{\alpha+1}\left(\frac{n}{2} - \sqrt{\frac{n^2}{4} - m^2}\right)$ *if* $\sqrt{n^2-1}/2 \ge m > 0$, *or we choose* $0 \le \gamma_0 < \frac{n-1}{2}$ *if* $m = n/2$ *and* $0 \le \gamma_0 \le \frac{n-1}{2}$ *if* $m > n/2$, *then* $\gamma \le \min\left\{\gamma_0\, ,\, \frac{n}{2(\alpha+1)}\right\}$.

If $m \in (\sqrt{n^2-1}/2, n/2)$, *then for the problem with* $\psi_0 = 0$ *the global solution exists and belongs to* $X(2\varepsilon, s, \gamma)$, *where* $\gamma \in (0, \frac{1}{\alpha+1}(\frac{n}{2} - \sqrt{\frac{n^2}{4} - m^2}))$.

The range $m \in (\sqrt{n^2-1}/2, n/2)$, which seems to be a forbidden mass interval for the problem with general initial data, can be allowed if we change the setting of the problem. Indeed, if we consider the initial value problem with vanishing Cauchy data and with the source term f, then we have the following result for all $m > 0$.

Theorem 4.2 ([33]) *Let* $A(x, \partial_x) = \sum_{|\alpha|\le 2} a_\alpha(x)\partial_x^\alpha$ *be a second-order negative elliptic differential operator with real coefficients* $a_\alpha \in \mathcal{B}^\infty$. *Assume that the nonlinear term* $F(u)$ *is a Lipschitz continuous with exponent* $\alpha > 0$ *in the space* $H_{(s)}(\mathbb{R}^n)$, $s > n/2 \ge 1$, *and* $F(0) = 0$. *Assume also that* $m > 0$. *Then, there exists* $\varepsilon_0 > 0$ *such that, for every given function* $f \in X(\varepsilon, s, \gamma_{rhs})$, *such that*

$$\sup_{t\in[0,\infty)} e^{\gamma_{rhs}t}\|f(x,t)\|_{H_{(s)}(\mathbb{R}^n)} \le \varepsilon < \varepsilon_0 ,$$

there exists a global solution $\psi \in C^1([0,\infty); H_{(s)}(\mathbb{R}^n))$ *of the Cauchy problem*

$$\psi_{tt} + n\psi_t - e^{-2t}A(x,\partial_x)\psi + m^2\psi = f + F(\psi) , \tag{4.4}$$

$$\psi(x,0) = 0 , \quad \psi_t(x,0) = 0 . \tag{4.5}$$

That solution $\psi(x,t)$ *belongs to the space* $X(2\varepsilon, s, \gamma)$, *that is,*

$$\sup_{t\in[0,\infty)} e^{\gamma t}\|\psi(\cdot,t)\|_{H_{(s)}(\mathbb{R}^n)} < 2\varepsilon\,,$$

with γ *such that*

$$\begin{cases} \gamma < \dfrac{1}{\alpha+1}\gamma_{rhs} & \text{if } \ m<\dfrac{n}{2} \ \text{ and } \ \gamma_{rhs}\le \dfrac{n}{2}-\sqrt{\dfrac{n^2}{4}-m^2}\,,\\ \gamma < \dfrac{1}{\alpha+1}\left(\dfrac{n}{2}-\sqrt{\dfrac{n^2}{4}-m^2}\right) & \text{if } \ m<\dfrac{n}{2} \ \text{ and } \ \gamma_{rhs} > \dfrac{n}{2}-\sqrt{\dfrac{n^2}{4}-m^2}\,,\\ \gamma \le \min\left\{\gamma_{rhs}, \dfrac{n}{2(\alpha+1)}\right\} & \text{if } \ m\ge \dfrac{n}{2} \ \text{ and } \ \dfrac{n}{2} > \gamma_{rhs}\,,\\ \gamma \le \min\left\{\gamma_0, \dfrac{n}{2(\alpha+1)}\right\} & \text{where } \ \gamma_0<\gamma_{rhs} \ \text{ if } \ m=\dfrac{n}{2} \ \text{ and } \ \dfrac{n}{2}=\gamma_{rhs}\,,\\ \gamma \le \dfrac{n}{2(\alpha+1)} & \text{if } \ m>\dfrac{n}{2} \ \text{ and } \ \dfrac{n}{2}\le \gamma_{rhs}\,,\\ \gamma < \dfrac{n}{2(\alpha+1)} & \text{if } \ m=\dfrac{n}{2} \ \text{ and } \ \dfrac{n}{2}<\gamma_{rhs}\,. \end{cases}$$

The main tools to prove Theorems 4.1 and 4.2 are the following: (1) integral transform, which produces representations of the solutions of the linear equation, (2) decay estimates in the Besov spaces, which generate weighted Stricharz estimates, and (3) the fixed point theorem.

The values of the physical mass m which leads to the values of $M=-k+\frac{1}{2}$, $k=0,1,2,\ldots$, are called in [94] *the knot points*. One of these knot points, $m=\sqrt{n^2-1}/2$, presents the only field that obeys the Huygens principle [94]. For these values of the curved mass M the functions $F(-k,-k;1;z)$, $k=0,1,2,\ldots$, are polynomials.

It is known that the Klein–Gordon quantum fields whose squared physical masses are negative (imaginary mass) represent tachyons. (See, e.g., [12].) In [12] the Klein–Gordon equation with imaginary mass is considered, and it is shown that localized disturbances spread with at most the speed of light, but grow exponentially. The conclusion is made that free tachyons have to be rejected on stability grounds.

The Klein–Gordon quantum fields on the de Sitter manifold with imaginary mass, which take an infinite set of discrete values as follows

$$m^2=-k(k+n)\,,\quad k=0,1,2,\ldots, \tag{4.6}$$

present a family of scalar tachyonic quantum fields. Epstein and Moschella [26] give a complete study of a family of scalar tachyonic quantum fields which are linear Klein–Gordon quantum fields on the de Sitter manifold whose squared masses are negative and take an infinite set of discrete values (4.6). The corresponding linear equation is

$$\psi_{tt}+n\psi_t-e^{-2t}\Delta\psi+m^2\psi=0\,,$$

for which the kernel is $E(x, t; x_0, t_0; M)$, where $M = \sqrt{\frac{n^2}{4} + k(k+n)} = k + \frac{n}{2}$, $k = 0, 1, 2, \ldots$. If n is an odd number, then m takes value at knot points set. The nonexistence of a global in time solution of the semilinear Klein–Gordon massive tachyonic (quantum fields) equation in the de Sitter spacetime is proved in [88]. The conclusion is that the self-interacting tachyons in the de Sitter spacetime have finite lifespan. More precisely, consider the semilinear equation

$$\psi_{tt} + n\psi_t - e^{-2t}\Delta\psi - m^2\psi = c|\psi|^{1+\alpha},$$

which is commonly used model for general nonlinear problems. Then, according to Theorem 1.1 [88], if $c \neq 0, \alpha > 0$, and $m \neq 0$, then for every positive numbers ε and s there exist functions $\psi_0,\ \psi_1 \in C_0^\infty(\mathbb{R}^n)$ such that $\|\psi_0\|_{H_{(s)}(\mathbb{R}^n)} + \|\psi_1\|_{H_{(s)}(\mathbb{R}^n)} \leq \varepsilon$ but the solution $\psi = \psi(x, t)$ with the initial values (4.3) blows up in finite time.

The equation that is considered in Theorems 4.1 and 4.2 is more general than the covariant Klein–Gordon equation. Then, these theorems, after evident modification, can be applied to the smooth pseudo Riemannian manifold $(\mathcal{V}, g)$ of dimension $n+1$ and $\mathcal{V} = \mathbb{R} \times \mathcal{S}$ with $\mathcal{S}$ an n-dimensional orientable smooth manifold and g is the de Sitter metric. One important example of the equation on the smooth pseudo Riemannian manifold that is amenable to Theorem 4.1 is if $\mathcal{S}$ is a non-Euclidean space of constant negative curvature and the equation of the problem (4.2) and (4.3) is a non-Euclidean Klein–Gordon equation.

5 The Semilinear Equations in the Energy Spaces

Although the integral transform approach is not used to derive the results in the energy spaces, we review those results having in mind that the comparison with the ones obtained by the integral transform approach is very instructive and interesting.

Galstian and Yagdjian [32] proved the existence of global solutions in the energy class in the case of $n = 3, 4$ and the nonlinear term is of power type. They considered equation in the Friedmann–Lemaître–Robertson–Walker spacetimes (FLRW spacetimes) with the time slices being Riemannian manifolds. The Klein–Gordon equation in the Einstein-de Sitter and de Sitter spacetimes are important particular cases discussed in [32].

We are interesed in the waves, which obey the semilinear Klein–Gordon equation, propagating in the FLRW spacetimes. The equations in the de Sitter and Einstein-de Sitter spacetimes are the important particular cases. In this section, we review some results on the global in time existence in the energy class of solutions of the Cauchy problem.

Consider the Klein–Gordon equation in the spacetimes belonging to some family of the FLRW spacetimes. In the FLRW spacetime, one can choose coordinates so that the metric has the form $ds^2 = -dt^2 + a^2(t)d\sigma^2$. (See, e.g., [38].) This family includes, as a particular case, the metric

$$ds^2 = -\, dt^2 + t^{\ell} \sum_{i,j=1,\ldots,n} \delta_{ij} dx^i dx^j \,, \tag{5.7}$$

where δ_{ij} is the Kronecker symbol and $\ell = \frac{4}{n\gamma}$. The function $a(t)$ is the scaling factor. The time dependence of the function $a(t)$ is determined by the Einstein's field equations for gravity, which for the perfect fluid imply

$$\dot{\mu} = -3(\mu + p)\frac{\dot{a}}{a}\,, \qquad \frac{\ddot{a}}{a} = -\frac{4\pi}{3}(\mu + 3p)\,, \qquad \left(\frac{\dot{a}}{a}\right)^2 = \frac{8\pi}{3}\mu - \frac{K}{a^2}\,,$$

where μ is the proper energy density, p is pressure, and K is spatial curvature. The last equations give a differential equation for $a = a(t)$ if an equation of state (equation for the pressure) $p = p(\mu)$ is known. For pressureless, $p = 0$, matter distribution in the universe and vanishing spatial curvature, $K = 0$, the solution to that equation is

$$a(t) = a_0 t^{2/3}\,,$$

where $a_0 > 0$ is a constant. The universe expands, and its expansion decelerates since $\ddot{a} < 0$. In the radiation dominated universe, the equation of state is $p = \mu/3$ and, consequently, $a(t) = a_0 t^{1/2}$. The equation of state $p = (\gamma - 1)\mu$, which includes those two cases of the matter- and radiation- dominated universe, implies that in order to have a nonnegative pressure for a positive density, it must be assumed that $\gamma \geq 1$ for the physical space with $n = 3$ (see [17] p. 122). The spacetime with $\gamma = 1$ and $n = 3$ is called the Einstein-de Sitter universe. In [32] the significance of the restriction $\gamma \geq 1$ on the range of ℓ was revealed; in fact, it was shown that it is closely related to the nongrowth of the energy and to the existence of the global in time solution of the Cauchy problem for the Klein–Gordon equation. Another important spacetime, the so-called de Sitter spacetime, is also a member of that family and it was discussed in [32] as well.

In quantum field theory the matter fields are described by a function ψ that must satisfy equations of motion. In the case of a massive scalar field, the equation of motion is the semilinear Klein–Gordon equation generated by the metric g:

$$\frac{1}{\sqrt{|g(x)|}}\frac{\partial}{\partial x^i}\left(\sqrt{|g(x)|}g^{ik}(x)\frac{\partial \psi}{\partial x^k}\right) = m^2\psi + V'_{\psi}(x,\psi)\,. \tag{5.8}$$

In physical terms this equation describes a local self-interaction for a scalar particle. A typical example of a potential function would be $V(\phi) = \phi^4$. The semilinear equations are also commonly used models for general nonlinear problems.

To motivate the approach that was used in [32], we first consider the covariant Klein–Gordon equation in the metric (5.7), which can be written in the global coordinates as follows

$$\psi_{tt} - t^{-\ell}\Delta\psi + \frac{n\ell}{2t}\psi_t + m^2\psi + V'_\psi(x, t, \psi) = 0\,.$$

Consider the Cauchy problem with the data prescribed at some positive time t_0,

$$\psi(t_0, x) = \psi_0, \quad \psi_t(t_0, x) = \psi_1\,, \tag{5.9}$$

and look for the solution defined for all values of $t \in [t_0, \infty)$ and $x \in \mathbb{R}^n$. Let us change the unknown function $\psi = t^{-\frac{ln}{4}}u$, then for the new function $u = u(t, x)$ we obtain the equation

$$u_{tt} - t^{-\ell}\Delta u + M^2(t)u + t^{n\ell/4}V'_\psi(x, t, t^{-\frac{\ell n}{4}}u) = 0$$

with the "effective" (or "curved mass")

$$M^2_{EdS}(t) := m^2 - \frac{n\ell(n\ell - 4)}{16t^2}\,. \tag{5.10}$$

It is easily seen that for the range $(0, \frac{4}{n}]$ of the parameter ℓ the curved mass is positive while its derivative is non-positive. This is crucial for the nonincreasing property of the energy and in the derivation of the energy estimate.

Let (V, g) be smooth pseudo Riemannian manifold of dimension $n + 1$ and $V = \mathbb{R} \times S$ with S an n-dimensional orientable smooth manifold, and g be a FLRW metric. We restrict our attention to the case of $n \geq 3$ and to the spacetime with the line element $ds^2 = -dt^2 + a^2(t)\sigma$. Then we consider an *expanding universe* that means that $\dot{a}(t) > 0$. For the metric with $\dot{a}(t) > 0$ we define the norm

$$\begin{aligned} \|\,\psi\,\|_{X(t)} := {} & \|\psi_t\|_{L^\infty([t_0,t];L^2(S))} + \|a^{-1}(\cdot)\nabla_\sigma\psi\|_{L^\infty([t_0,t];L^2(S))} \\ & + \|M(\cdot)\psi\|_{L^\infty([t_0,t];L^2(S))} + \|\sqrt{\dot{a}a^{-3}}\,\nabla_\sigma\psi\|_{L^2([t_0,t]\times S)}\,, \end{aligned} \tag{5.11}$$

where $0 < t_0 < t \leq \infty$ and $M(t) \geq 0$ is a curved mass defined by:

$$M^2(t) = m^2 + \left(\frac{n}{2} - \frac{n^2}{4}\right)\left(\frac{\dot{a}(t)}{a(t)}\right)^2 - \frac{n}{2}\frac{\ddot{a}(t)}{a(t)}\,. \tag{5.12}$$

Hence, in the classification suggested in [87], mass m is large if the metric g is a de Sitter metric $-dt^2 + e^{2t}dx^2$, $x \in \mathbb{R}^n$. Here and henceforth $\dot{a}(t)$ denotes the derivative with respect to time, while the spatial variable will be denoted s in a general manifold S and x when $S = \mathbb{R}^n$. In order to describe admissible nonlinearities we make the following definition.

Condition ($\mathcal{L}$). *The smooth in s function $F = F(s,u)$, $F : S \times \mathbb{R} \longrightarrow \mathbb{R}$ is said to be Lipschitz continuous in u with exponent α, if there exist $\alpha \geq 0$ and $C > 0$ such that*

$$|F(s,u) - F(s,v)| \leq C|u-v|\,(|u|^\alpha + |v|^\alpha) \quad \text{for all } u, v \in \mathbb{R},\ x \in S\,.$$

For the continuous function $\Gamma \in C([t_0,\infty))$ denote by $C_{a,\Gamma,\alpha_0}(T)$ and $C^{(-1)}_{a,\Gamma,\alpha_0}(r)$ the function

$$C_{a,\Gamma,\alpha_0}(T) := \left(\int_{t_0}^{T} \left(\frac{a(t)}{\dot a(t)} \right)^{\frac{n\alpha_0}{4-n\alpha_0}} |\Gamma(t)|^{\frac{4}{4-n\alpha_0}}\, dt \right)^{\frac{4-n\alpha_0}{4}}, \qquad 0 < \alpha_0 < \frac{4}{n}\,.$$

and its inverse, respectively.

Theorem 5.1 ([32]) *Assume that $n = 3, 4$ and that the metric g is $g = -\,dt^2 + a^2(t)\sigma$. Suppose also that $m > 0$ and that there is a positive number c_0 such that the real-valued positive function $a = a(t)$ satisfies*

$$a(t) > 0, \quad \dot a(t) > 0 \quad \text{for all} \quad t \in [t_0,\infty)\,, \tag{5.13}$$

$$M(t) > c_0 > 0, \quad \dot M(t) \leq 0 \quad \text{for all} \quad t \in [t_0,\infty)\ . \tag{5.14}$$

Consider the Cauchy problem for the Eq. (5.8) with the derivative of potential function $V'_\psi(s,t,\psi) = -\Gamma(t)F(s,\psi)$ such that F is Lipschitz continuous with exponent α, $F(s,0) = 0$ for all $s \in S$, and either

$$|\Gamma(t)| \leq C_\Gamma \frac{\dot a(t)}{a(t)} \quad \text{for all} \quad t \in [t_0,\infty)\,, \tag{5.15}$$

where C_Γ is a constant independent of t, or, there is α_0 such that

$$C_{a,\Gamma,\alpha_0}(\infty) < \infty, \qquad 0 < \alpha_0 < \frac{4}{n}\,. \tag{5.16}$$

If $\frac{4}{n} \leq \alpha \leq \frac{2}{n-2}$, then for every $\psi_0 \in H_{(1)}(S)$ and $\psi_1 \in L^2(S)$, sufficiently small initial data, $\|\psi_0\|_{H_{(1)}(S)} + \|\psi_1\|_{L^2(S)}$, the problem (5.8), (5.9) has a unique solution $\psi \in C([t_0,\infty); H_{(1)}(S)) \cap C^1([t_0,\infty); L^2(S))$ and its norm $\|a^{\frac{n}{2}}\psi\|_{X(\infty)}$ is small.

Condition (5.14) for the norm of solutions of the equation implies that the energy of solution is nonincreasing. In the next theorem, the local existence is stated with the less restrictive conditions and with the estimate for the lifespan.

Theorem 5.2 ([32]) *Suppose that $m > 0$ and that there is a positive number c_0 such that the real-valued positive function $a = a(t)$ satisfies (5.13), (5.14). Consider the Cauchy problem for the Eq. (5.8) with the derivative of potential function $V'_\psi(s,t,\psi) = -\Gamma(t)F(s,\psi)$ such that F is Lipschitz continuous with exponent α, $F(s,0) = 0$ for all $s \in S$.*

If $0 \le \alpha \le \frac{2}{n-2}$*, then for every* $\psi_0 \in H_{(1)}(S)$ *and* $\psi_1 \in L^2(S)$ *there exists* $T_1 > t_0$ *such that the problem (5.8), (5.9) has a unique solution* $\psi \in C([t_0, T_1); H_{(1)}(S)) \cap C^1([t_0, T_1); L^2(S))$.

The lifespan of the solution can be estimated as follows

$$T_1 - t_0 \ge C C^{(-1)}_{a,\Gamma,\alpha_0}(\|\psi_0\|_{H_{(1)}(S)} + \|\psi_1\|_{L^2(S)}),$$

where C is a positive constant independent of T_1, ψ_0 *and* ψ_1.

If the nonlinear term has an energy conservative potential function, then in the next theorem the existence of the global solution for large initial data was established.

Theorem 5.3 ([32]) *Suppose that all conditions of Theorem 5.2 on n,* α*, and* $a = a(t)$*, are satisfied, and additionally,*

$$\frac{2}{n}\frac{a(t)}{\dot a(t)} V'_t(t, s, a^{-n/2}(t)w) + 2V(t, s, a^{-n/2}(t)w) - a^{-n/2}(t)w V'_\psi(s, a^{-n/2}(t)w) \le 0 \tag{5.17}$$

for all $(t, s, w) \in [t_0, \infty) \times S \times \mathbb{R}$.

Then for every $\psi_0 \in H_{(1)}(\mathbb{R}^n)$ *and* $\psi_1 \in L^2(\mathbb{R}^n)$*, the problem (5.8), (5.9) has a unique solution* $\psi \in C([t_0, \infty); H_{(1)}(\mathbb{R}^n)) \cap C^1([t_0, \infty); L^2(\mathbb{R}^n))$ *and its norm* $\|a^{\frac{n}{2}}\psi\|_{X(\infty)}$ *is finite.*

The hyperbolic equations in the de Sitter spacetime have permanently bounded domain of influence. Nonlinear equations with a permanently bounded domain of influence were studied, in particular, in [85]. In that paper the example of equation, which has a blowing-up solution for arbitrarily small data, is given. Moreover, it was discovered in [85] that the time-oscillation of the metric, due to the parametric resonance, can cause blowup phenomena for wave map type nonlinearities even for the arbitrarily small data. On the other hand in the absence of oscillations in the metric, Choquet-Bruhat [16] proved for small initial data the global existence and uniqueness of wave maps on the FLRW expanding universe with the metric $\mathbf{g} = -dt^2 + R^2(t)\sigma$ and a smooth Riemannian manifold (S, σ) of dimension $n \le 3$, which has a time independent metric σ and nonzero injectivity radius, and with $R(t)$ being a positive increasing function such that $1/R(t)$ is integrable on $[t_0, \infty)$. If the target manifold is flat, then the wave map equation reduces to a linear system. On the other hand, in the Einstein-de Sitter spacetime the domain of influence is not permanently bounded.

Although recently the equations in the de Sitter and anti-de Sitter spacetimes became the focus of interest for an increasing number of authors (see, e.g., [1, 2, 5, 8, 14, 26, 32, 44, 59, 61, 69, 79] and the bibliography therein) which investigate those equations from a wide spectrum of perspectives, there are very few papers on the semilinear Klein–Gordon equation in the de Sitter spacetime. Here, we mention some of them closely related to our main result. Baskin [8] discussed small data global energy class solutions for the scalar Klein–Gordon equation on asymptotically de

Sitter spaces, which are compact manifolds with boundary. More precisely, in [8] the following Cauchy problem is considered for the semilinear equation

$$\Box_g u + m^2 u = f(u), \qquad u(x, t_0) = \varphi_0(x) \in H_{(1)}(\mathbb{R}^n), \;\; u_t(x, t_0) = \varphi_1(x) \in L^2(\mathbb{R}^n),$$

where mass is large, $m^2 > n^2/4$, f is a smooth function and satisfies conditions $|f(u)| \le c|u|^{\alpha+1}$, $|u| \cdot |f'(u)| \sim |f(u)|$, $f(u) - f'(u) \cdot u \le 0$, $\int_0^u f(v)dv \ge 0$, and $\int_0^u f(v)dv \sim |u|^{\alpha+2}$ for large $|u|$. It is also assumed that $\alpha = \frac{4}{n-1}$. In Theorem 1.3 [8] the existence of the global solution for small energy data is stated. (For more references on the asymptotically de Sitter spaces, see the bibliography in [7, 79].)

Hintz and Vasy [44] considered the semilinear wave equations of the form

$$(\Box_g - \lambda)u = f + q(u, du)$$

on a manifold M, where q is a polynomial vanishing at least quadratically at $(0, 0)$, in an *asymptotically* de Sitter and Kerr-de Sitter spaces, as well as asymptotically Minkowski spaces. The initial data for the equation are generated by the source term f. The linear framework in [44] is based on the b-analysis, in the sense of Melrose, introduced in this context by Vasy to describe the asymptotic behavior of solutions of linear equations. Hintz and Vasy have shown the small source term f solvability of suitable semilinear wave and Klein–Gordon equations. However, the microlocal, high regularity approach that was taken in [44] does not apply to low regularity nonlinearities covered in Theorem 4.1. Their result for *asymptotically* de Sitter spacetime and polynomial semilinear term with large α covers also the range $m \in (\sqrt{n^2 - 1}/2, n/2)$. On the other hand, the important case of $n = 3$ and the quadratic nonlinearity is not covered. We note here that their results as well as Theorem 4.2 of the present paper, neither prove nor disprove the conjecture from the article [90].

Nakamura [61] considered the Cauchy problem for the semilinear Klein–Gordon equations in de Sitter spacetime with $n \le 4$ and with flat time slices. The nonlinear term is of power type for $n = 3, 4$, or of exponential type for $n = 1, 2$. For the power type semilinear term with $\frac{4}{n} \le \alpha \le \frac{2}{n-2}$ Nakamura [61] proved the existence of global solutions in the energy class.

Ringström [68] considered the question of future global nonlinear stability in the case of Einstein's equations coupled to a nonlinear scalar field. The class of potential $V(\psi)$ is restricted by the condition $V(0) > 0$, $V'(0) = 0$ and $V''(0) > 0$. Ringström proved that for given initial data, there is a maximal globally hyperbolic development of the data which is unique up to isometry. The case of Einstein's equations with positive cosmological constant was not included unless the scalar field is zero.

Rodnianski and Speck [69] proved the nonlinear future stability of the FLRW family of solutions to the irrotational Euler-Einstein system with a positive cosmological constant. More precisely, they studied small perturbations of the family of FLRW cosmological background solutions to the coupled Euler-Einstein system with a positive cosmological constant in $1 + 3$ spacetime dimensions. The background solutions model an initially uniform quiet fluid of positive energy density evolving in a spacetime undergoing exponentially accelerated expansion. Their analysis shows that under the equation of state $p = (\gamma - 1)\mu$, $0 < \gamma - 1 < 1/3$, the background metric plus fluid solutions are globally future-stable under small irrotational perturbations of their initial data.

6 The Strauss Exponent for the Semilinear Equation on the Einstein-de Sitter Spacetime

In this section, we give some new results on the global in time existence of the waves propagating in the Einstein-de Sitter spacetime. We discuss only the massless fields.

In [29] Galstian, Kinoshita, and Yagdjian considered the wave propagating in the Einstein and de Sitter spacetime. The covariant d'Alembert's operator (the Laplace-Beltrami operator) in the Einstein-de Sitter spacetime belongs to the family of the non-Fuchsian partial differential operators. In [29] the authors introduced the weighted initial value problem for the covariant (if $n = 3$) wave equation and gave the explicit representation formulas for the solutions. Based on the representation formulas they also shown the $L_p - L_q$ estimates for solutions. Then, in [31] the authors gave the parametrices in the terms of Fourier integral operators also discussed the propagation and reflection of the singularities phenomena.

In fact, Galstian, Kinoshita, and Yagdjian suggested in [29] the following weighted initial value problem

$$\begin{cases} \psi_{tt} - t^{-4/3} \bigtriangleup \psi + 2t^{-1}\psi_t = 0, \qquad t > 0,\ x \in \mathbb{R}^n, \\ \lim\limits_{t\to 0+} t\psi(x,t) = \varphi_0(x), \quad \lim\limits_{t\to 0+} \left(t\psi_t(x,t) + \psi(x,t) + 3t^{-1/3} \bigtriangleup \varphi_0(x)\right) = \varphi_1(x),\ x \in \mathbb{R}^n . \end{cases}$$

We use this setting to prescribe the initial conditions for a problem for the semilinear equation

$$\begin{cases} \psi_{tt} - t^{-4/3} \bigtriangleup \psi + 2t^{-1}\psi_t = |\psi|^p, \qquad t > 0,\ x \in \mathbb{R}^n, \\ \lim\limits_{t\to 0+} t\psi(x,t) = \varphi_0(x), \quad \lim\limits_{t\to 0+} \left(t\psi_t(x,t) + \psi(x,t) + 3t^{-1/3} \bigtriangleup \varphi_0(x)\right) = \varphi_1(x),\ x \in \mathbb{R}^n . \end{cases} \tag{6.1}$$

We define $p_{cr}(n)$ as a positive root of the equation

$$(n+3)p^2 - (n+13)p - 2 = 0 . \tag{6.2}$$

that is

$$p_{cr}(n) = \frac{n + 13 + \sqrt{n^2 + 34n + 193}}{2(n+3)} .$$

Theorem 6.1 ([34]) *Assume that* $p > 1$ *and*

$$\text{either} \quad 1 < p < 1 + \frac{6}{n} \quad \text{or} \quad 1 < p \le \frac{2n+10}{n+3} \quad \text{and} \quad p < p_{cr}(n).$$

Then for every arbitrary small number $\varepsilon > 0$ *and arbitrary* s *there exist functions,* $\varphi_0, \varphi_1 \in C_0^\infty(\mathbb{R}^n)$ *with norm*

$$\|\varphi_0\|_{H_{(s)}(\mathbb{R}^n)} + \|\varphi_1\|_{H_{(s)}(\mathbb{R}^n)} < \varepsilon$$

such that solution of the problem (6.1) blows up in finite time.

Proof If we denote

$$\mathcal{L} := \partial_t^2 - t^{-4/3} \bigtriangleup + 2t^{-1}\partial_t, \qquad \mathcal{S} := \partial_t^2 - t^{-4/3} \bigtriangleup ,$$

then we can easily check for $t \ne 0$ the following operator identity

$$t^{-1} \circ \mathcal{S} \circ t = \mathcal{L} . \tag{6.3}$$

The last equation suggests a change of unknown function ψ with u such that

$$\psi = t^{-1} u.$$

Then the problem for u is as follows:

$$\begin{cases} u_{tt} - t^{-4/3} \bigtriangleup u = t^{1-p}|u|^p, & t > 0, \ x \in \mathbb{R}^n, \\ \lim\limits_{t \to 0^+} u(x,t) = \varphi_0(x), & x \in \mathbb{R}^n, \\ \lim\limits_{t \to 0^+} \big(u_t(x,t) + 3t^{-1/3} \bigtriangleup \varphi_0(x)\big) = \varphi_1(x), & x \in \mathbb{R}^n . \end{cases} \tag{6.4}$$

Denote

$$F(t) = \int_{\mathbb{R}^n} u(x,t)\, dx .$$

Then $F \in C^2(0,T)$ provided that the function u is defined for all $(x,t) \in \mathbb{R}^n \times (0,T)$, and

$$\lim_{t \to 0^+} F(t) = \int_{\mathbb{R}^n} \lim_{t \to 0^+} u(x,t)\, dx = \int_{\mathbb{R}^n} \varphi_0(x)\, dx = C_0,$$

while

$$\lim_{t\to 0^+} F'(t) = \lim_{t\to 0^+} \int_{\mathbb{R}^n} u_t(x,t)\,dx$$
$$= \lim_{t\to 0^+} \int_{\mathbb{R}^n} \left(u_t(x,t) + 3t^{-1/3} \bigtriangleup \varphi_0(x) - 3t^{-1/3} \bigtriangleup \varphi_0(x)\right) dx$$
$$= \lim_{t\to 0^+} \int_{\mathbb{R}^n} \left(u_t(x,t) + 3t^{-1/3} \bigtriangleup \varphi_0(x)\right) dx = \int_{\mathbb{R}^n} \varphi_1(x)\,dx = C_1 .$$

Thus

$$F \in C^1[0,\infty) \cap C^2(0,\infty) .$$

From the equation we have

$$F'' = t^{1-p} \int_{\mathbb{R}^n} |u(x,t)|^p\,dx \ge 0 \quad \text{for all} \quad t > 0,$$

and from the initial conditions we derive

$$F(t) = F(\varepsilon) + \int_\varepsilon^t F'(t_1)\,dt_1 = F(\varepsilon) + \int_\varepsilon^t \left(F'(\varepsilon) + \int_\varepsilon^{t_1} F''(t_2)dt_2\right) dt_1$$
$$= F(\varepsilon) + (t-\varepsilon)F'(\varepsilon) + \int_\varepsilon^t \int_\varepsilon^{t_1} F''(t_2)dt_2\,dt_1$$
$$\ge F(\varepsilon) + (t-\varepsilon)F'(\varepsilon) \quad \text{for all} \quad t \ge 0 .$$

Set $C_0 \ge 0$ and $C_1 \ge 0$. By letting $\varepsilon \to 0^+$ we obtain

$$F(t) \ge F'(0)t + F(0) = t\int_{\mathbb{R}^n} \varphi_1(x)\,dx + \int_{\mathbb{R}^n} \varphi_0(x)\,dx \ge 0 \quad \text{for all} \quad t \ge 0 .$$

On the other hand, using the compact support of $u(\cdot,t)$ and Hölder's inequality we get with τ_n the volume of the unit ball in $\mathbb{R}^n$, and $\phi(t) = 3t^{1/3}$

$$\left|\int_{\mathbb{R}^n} u(x,t)\,dx\right|^p \le \left(\int_{|x|\le R+\phi(t)} 1\,dx\right)^{p-1} \left(\int_{|x|\le R+\phi(t)} |u(x,t)|^p\,dx\right)$$
$$\lesssim (1+t)^{\frac{n(p-1)}{3}} \left(\int_{|x|\le R+\phi(t)} |u(x,t)|^p\,dx\right)$$
$$\lesssim (R+\phi(t))^{n(p-1)} \left(\int_{|x|\le R+\phi(t)} |u(x,t)|^p\,dx\right) ,$$

where the number R is chosen such that supp φ_0, supp $\varphi_1 \subseteq \{|x| \le R\}$. Here and henceforth, if A and B are two nonnegative quantities, we use $A \lesssim B$ ($A \gtrsim B$) to denote the statement that $A \le CB$ ($AC \ge B$) for some absolute constant $C > 0$.

Hence

$$
\begin{aligned}
F''(t) &= t^{1-p} \int_{\mathbb{R}^n} |u(x,t)|^p \, dx \geq (1+t)^{1-p-\frac{n(p-1)}{3}} |F(t)|^p \\
&\gtrsim (1+t)^{-\frac{(n+3)(p-1)}{3}} |F(t)|^p \\
&\gtrsim (R+\phi(t))^{-(n+3)(p-1)} |F(t)|^p \quad \text{for all} \quad t \geq 0 .
\end{aligned} \tag{6.5}
$$

If $1 < p < 1 + \frac{6}{n}$ and $C_1 > 0$, then we can apply Kato's lemma (see, e.g., [85, Lemma 2.1]) since

$$
p - 1 > \frac{(n+3)(p-1)}{3} - 2 \Longleftrightarrow p < \frac{6}{n} + 1 \Longleftrightarrow \alpha := p - 1 < \frac{6}{n}
$$

that proves blow up for such p.

Next, we consider the case of $1 < p \leq (2n+10)/(n+3)$ and $p < p_{cr}(n)$. For $\varphi_0 \in C_0^{[\frac{n}{2}]+3}(\mathbb{R}^n)$, according to Lemma 2.3 [29], the solution of the problem

$$
\begin{cases}
\mathcal{S}u = 0, \qquad x \in \mathbb{R}^n, \quad t > 0, \\
\lim\limits_{t \to 0} u(x,t) = \varphi_0(x), \qquad \lim\limits_{t \to 0} \left(u_t(x,t) + 3t^{-1/3} \,\triangle\, \varphi_0(x) \right) = 0, \qquad x \in \mathbb{R}^n ,
\end{cases} \tag{6.6}
$$

is given by the function

$$
u(x,t) = v_{\varphi_0}(x, 3t^{1/3}) - 3t^{1/3} (\partial_r v_{\varphi_0})(x, 3t^{1/3}) , \tag{6.7}
$$

where $v_\varphi(x, 3t^{1/3})$ is the value of the solution $v(x,r)$ to the Cauchy problem for the wave equation, $v_{rr} - \Delta v = 0$, $v(x,0) = \varphi(x)$, $v_t(x,0) = 0$, taken at the point $(x,r) = (x, 3t^{1/3})$. Hence, if we assume that $\Delta\varphi = \varphi$, then we obtain

$$
v_\varphi(x,t) = \cosh(t)\varphi(x)
$$

and, consequently,

$$
u(x,t) = v_{\varphi_0}(x, 3t^{1/3}) - 3t^{1/3} (\partial_r v_{\varphi_0})(x, 3t^{1/3}) = \left(\cosh(3t^{1/3}) - 3t^{1/3} \sinh(3t^{1/3}) \right) \varphi(x) .
$$

The second independent solution with separated variables is

$$
w(x,t) = \left(\sinh(3t^{1/3}) - 3t^{1/3} \cosh(3t^{1/3}) \right) \varphi(x) .
$$

The function

$$
\begin{aligned}
v(x,t) &= \left(\cosh(3t^{1/3}) - 3t^{1/3} \sinh(3t^{1/3}) \right) \varphi(x) - \left(\sinh(3t^{1/3}) - 3t^{1/3} \cosh(3t^{1/3}) \right) \varphi(x) \\
&= \left(3\sqrt[3]{t} + 1 \right) \exp\left(-3\sqrt[3]{t} \right) \varphi(x) = (\phi(t) + 1) \exp\left(-\phi(t) \right) \varphi(x)
\end{aligned}
$$

solves the problem (6.6) with $\varphi_0 = \varphi$. Moreover, v is such that

$$v(x,0) = \varphi(x)\,, \qquad \lim_{t\to\infty} v(x,t) = 0\,.$$

Now we choose the function

$$\varphi(x) = \int_{\mathbb{S}^{n-1}} e^{x\omega}\, d\omega\,.$$

Then $\Delta\varphi(x) = \Delta \int_{\mathbb{S}^{n-1}} e^{x\omega}\, d\omega = \varphi(x)$.
Hence, the function $v(x,t)$ is the low frequency solution of the linear equation

$$v_{tt} - t^{-4/3}\Delta v = 0\,.$$

Next we define the function $F_1(t)$,

$$F_1(t) := \int_{\mathbb{R}^n} u(x,t)v(x,t)\, dx\,,$$

that is the projection of the solution on the low frequency one-dimensional eigenspace of Laplace operator. Here $F_1 \in C^2(0,T)$. We estimate the function F_1 from above as follows

$$\begin{aligned} |F_1(t)|^p &\le \left(\int_{|x|\le R+\phi(t)} |v(x,t)|^{p/(p-1)}\, dx\right)^{p-1} \left(\int_{|x|\le R+\phi(t)} |u(x,t)|^p\, dx\right) \\ &\le \left(\int_{|x|\le R+\phi(t)} |v(x,t)|^{p/(p-1)}\, dx\right)^{p-1} t^{p-1} F''(t)\,. \end{aligned}$$

Hence

$$F''(t) \ge \left(\int_{|x|\le R+\phi(t)} |v(x,t)|^{p/(p-1)}\, dx\right)^{1-p} t^{1-p}\, |F_1(t)|^p\,. \tag{6.8}$$

To find out the properties of $F_1(t)$ we need the following lemma.

Lemma 6.2 ([34]) *The function*

$$\lambda(t) = \left(3\sqrt[3]{t}+1\right)\exp\left(-3\sqrt[3]{t}\right) = (\phi(t)+1)\exp\left(-\phi(t)\right)$$

solves the equation

$$\lambda''(t) - t^{-\frac{4}{3}}\lambda(t) = 0$$

and has the following properties:

$$
\begin{aligned}
&(i)\ \lambda'(t) = -\frac{3}{\sqrt[3]{t}}\exp(-3\sqrt[3]{t}) = -\frac{9}{\phi(t)}\exp\left(-\phi(t)\right) \le 0\,,\\
&(ii)\ \lim_{t\to 0}\lambda(t) = 1\,, \qquad \lim_{t\to\infty}\lambda(t) = 0\,, \qquad \lim_{t\to\infty}\lambda'(t) = 0\,,\\
&(iii)\ \frac{\lambda'(t)}{\lambda(t)} = -\frac{9}{\phi(t)\left(\phi(t)+1\right)}\,.
\end{aligned}
$$

Proof It can be verified by straightforward calculations. □

Next we turn to the function $\varphi(x)$. It is well known ([81]) that

$$
\varphi(x) \sim C_n|x|^{-(n-1)/2}e^{|x|} \quad \text{as} \quad |x| \to \infty\,.
$$

Lemma 6.3 ([81]) *Assume that $p > 1$. Then*

$$
\int_{|x|\le\tau} |\varphi(x)|^{p/(p-1)}\,dx \le c_R\tau^{\frac{n-1}{2}\frac{p-2}{p-1}}e^{\tau\frac{p}{p-1}} \quad \textit{for all} \quad \tau \ge 1. \tag{6.9}
$$

Lemma 6.4 *([34]) Assume that $\varphi_0, \varphi_1 \in C_0^\infty(\mathbb{R}^n)$, and that*

$$
\int_{\mathbb{R}^n} \varphi_1(x)\varphi(x)\,dx \ge 18\int_{\mathbb{R}^n} \varphi_0(x)\varphi(x)\,dx > 0\,,
$$

then

$$
F_1(t) \gtrsim (9\sqrt[3]{t^2}-1)\int_{\mathbb{R}^n} \varphi_1(x)\varphi(x)\,dx \quad \textit{for all} \quad t > 1\,.
$$

Proof We have

$$
F_1(0) = \lim_{t\to 0}\int_{\mathbb{R}^n} u(x,t)v(x,t)\,dx = \int_{\mathbb{R}^n} \varphi_0(x)\varphi(x)\,dx \ge c_0 > 0\,.
$$

For every $\epsilon > 0$ we have

$$
\begin{aligned}
0 &= \int_\varepsilon^t\int_{\mathbb{R}^n} (u_{tt}(x,\tau) - \tau^{-4/3}\Delta u - \tau^{1-p}|u|^p)v(x,\tau)\,dx\,d\tau\\
&= \int_\varepsilon^t\int_{\mathbb{R}^n} u_{tt}(x,\tau)v(x,\tau)\,dx\,d\tau\\
&\quad -\int_\varepsilon^t\int_{\mathbb{R}^n} \tau^{-4/3}u(x,\tau)\Delta v(x,\tau)\,dx\,d\tau - \int_\varepsilon^t\int_{\mathbb{R}^n} \tau^{1-p}|u(x,\tau)|^p v(x,\tau)\,dx\,d\tau\,.
\end{aligned}
$$

Further,

$$\int_{\varepsilon}^{t}\int_{\mathbb{R}^n} u_{tt}(x,\tau)v(x,\tau)\,dx\,d\tau$$
$$= \int_{\mathbb{R}^n} u_t(x,\tau)v(x,\tau)\,dx\Big|_{\varepsilon}^{t} - \int_{\mathbb{R}^n} u(x,\tau)v_t(x,\tau)\,dx\Big|_{\varepsilon}^{t} + \int_{\varepsilon}^{t}\int_{\mathbb{R}^n} u(x,\tau)\tau^{-4/3}\Delta v(x,\tau)\,dx\,d\tau\,.$$

Hence,

$$\int_{\mathbb{R}^n} u_t(x,\tau)v(x,\tau)\,dx\Big|_{\varepsilon}^{t} - \int_{\mathbb{R}^n} u(x,\tau)v_t(x,\tau)\,dx\Big|_{\varepsilon}^{t} = \int_{\varepsilon}^{t}\int_{\mathbb{R}^n} \tau^{1-p}|u(x,\tau)|^p v(x,\tau)\,dx\,d\tau\,.$$

The last equation implies

$$\left(\frac{d}{d\tau}\int_{\mathbb{R}^n} u(x,\tau)v(x,\tau)\,dx - 2\int_{\mathbb{R}^n} u(x,\tau)v_t(x,\tau)\,dx\right)\Big|_{\varepsilon}^{t} = \int_{\varepsilon}^{t}\int_{\mathbb{R}^n} \tau^{1-p}|u(x,\tau)|^p v(x,\tau)\,dx\,d\tau\,.$$

It follows

$$\frac{d}{dt}F_1(t) - 2\frac{\lambda_t(t)}{\lambda(t)}\int_{\mathbb{R}^n} u(x,t)\lambda(t)\varphi(x)\,dx$$
$$= \frac{d}{dt}F_1(t)\Big|_{\varepsilon} - 2\frac{\lambda_t(\varepsilon)}{\lambda(\varepsilon)}\int_{\mathbb{R}^n} u(x,\varepsilon)\lambda(\varepsilon)\varphi(x)\,dx + \int_{\varepsilon}^{t}\int_{\mathbb{R}^n} \tau^{1-p}|u(x,\tau)|^p v(x,\tau)\,dx\,d\tau\,.$$

Consequently,

$$\frac{d}{dt}F_1(t) - 2\frac{\lambda_t(t)}{\lambda(t)}F_1(t) = \frac{d}{dt}F_1(t)\Big|_{\varepsilon} - 2\frac{\lambda_t(\varepsilon)}{\lambda(\varepsilon)}F_1(\varepsilon) + \int_{\varepsilon}^{t}\int_{\mathbb{R}^n} \tau^{1-p}|u(x,\tau)|^p v(x,\tau)\,dx\,d\tau\,.$$

It follows

$$\frac{d}{dt}\left(F_1(t)\exp\left(-\int_{\varepsilon}^{t} 2\frac{\lambda_t(\tau)}{\lambda(\tau)}\,d\tau\right)\right)$$
$$= \exp\left(-\int_{\varepsilon}^{t} 2\frac{\lambda_t(\tau)}{\lambda(\tau)}\,d\tau\right)\left\{\frac{d}{dt}F_1(t)\Big|_{\varepsilon} - 2\frac{\lambda_t(\varepsilon)}{\lambda(\varepsilon)}F_1(\varepsilon) + \int_{\varepsilon}^{t}\int_{\mathbb{R}^n} \tau^{1-p}|u(x,\tau)|^p v(x,\tau)\,dx\,d\tau\right\},$$

that is

$$\frac{d}{dt}\left(F_1(t)\left(\frac{\lambda(t)}{\lambda(\varepsilon)}\right)^{-2}\right) = \left(\frac{\lambda(t)}{\lambda(\varepsilon)}\right)^{-2}\left\{\frac{d}{dt}F_1(t)\Big|_{\varepsilon} - 2\frac{\lambda_t(\varepsilon)}{\lambda(\varepsilon)}F_1(\varepsilon)\right.$$
$$\left. + \int_{\varepsilon}^{t}\int_{\mathbb{R}^n} \tau^{1-p}|u(x,\tau)|^p v(x,\tau)\,dx\,d\tau\right\}.$$

We integrate it and obtain

$$F_1(t) = \left(\frac{\lambda(t)}{\lambda(\varepsilon)}\right)^2 \Bigg[F_1(\varepsilon) + \int_\varepsilon^t \left(\frac{\lambda(s)}{\lambda(\varepsilon)}\right)^{-2} \times \left\{\frac{d}{dt}F_1(t)\Big|_\varepsilon - 2\frac{\lambda_t(\varepsilon)}{\lambda(\varepsilon)}F_1(\varepsilon) + \int_\varepsilon^s \int_{\mathbb{R}^n} \tau^{1-p}|u(x,\tau)|^p v(x,\tau)\,dx\,d\tau\right\} ds\Bigg]. \tag{6.10}$$

On the other hand, according to (iii) of Lemma 6.2 we have $\frac{\lambda_t(t)}{\lambda(t)} = -\frac{3}{\sqrt[3]{t}(3\sqrt[3]{t}+1)}$. Consider the term

$$\frac{d}{dt}F_1(t)\Big|_\varepsilon - 2\frac{\lambda_t(\varepsilon)}{\lambda(\varepsilon)}F_1(\varepsilon)$$
$$= \int_{\mathbb{R}^n} u_t(x,\varepsilon)\lambda(\varepsilon)\varphi(x)\,dx + \int_{\mathbb{R}^n} u(x,\varepsilon)\lambda_t(\varepsilon)\varphi(x)\,dx + \frac{6}{\sqrt[3]{\varepsilon}(3\sqrt[3]{\varepsilon}+1)}\int_{\mathbb{R}^n} u(x,\varepsilon)\lambda(\varepsilon)\varphi(x)\,dx.$$

We can rewrite it as follows

$$\frac{d}{dt}F_1(t)\Big|_\varepsilon + \frac{6}{\sqrt[3]{\varepsilon}(3\sqrt[3]{\varepsilon}+1)}F_1(\varepsilon)$$
$$= \int_{\mathbb{R}^n} \{u_t(x,\varepsilon) + 3\varepsilon^{-1/3}\Delta\varphi_0(x)\}v(x,\varepsilon)\,dx - \int_{\mathbb{R}^n} 3\varepsilon^{-1/3}\Delta\varphi_0(x)v(x,\varepsilon)\,dx$$
$$-\int_{\mathbb{R}^n} \frac{3}{\sqrt[3]{\varepsilon}(3\sqrt[3]{\varepsilon}+1)}u(x,\varepsilon)v(x,\varepsilon)\,dx + \frac{6}{\sqrt[3]{\varepsilon}(3\sqrt[3]{\varepsilon}+1)}\int_{\mathbb{R}^n} u(x,\varepsilon)v(x,\varepsilon)\,dx$$
$$= \int_{\mathbb{R}^n} \{u_t(x,\varepsilon) + 3\varepsilon^{-1/3}\Delta\varphi_0(x)\}v(x,\varepsilon)\,dx$$
$$+\int_{\mathbb{R}^n} 3\varepsilon^{-1/3}\Big\{-\varphi_0(x) + \frac{1}{(3\sqrt[3]{\varepsilon}+1)}u(x,\varepsilon)\Big\}v(x,\varepsilon)\,dx\,.$$

Hence, taking into account the property of u, we derive

$$\lim_{\varepsilon\to 0^+}\left(\frac{d}{dt}F_1(t)\Big|_\varepsilon - 2\frac{\lambda_t(\varepsilon)}{\lambda(\varepsilon)}F_1(\varepsilon)\right) = \int_{\mathbb{R}^n}\varphi_1(x)\varphi(x)\,dx - 9\int_{\mathbb{R}^n}\varphi_0(x)\varphi(x)\,dx\,.$$

Now

$$\lim_{\varepsilon\to 0^+}\left(\frac{\lambda(t)}{\lambda(\varepsilon)}\right)^2\left[F_1(\varepsilon) + \int_\varepsilon^t \left(\frac{\lambda(s)}{\lambda(\varepsilon)}\right)^{-2}\left\{\frac{d}{dt}F_1(t)\Big|_\varepsilon - 2\frac{\lambda_t(\varepsilon)}{\lambda(\varepsilon)}F_1(\varepsilon)\right\}ds\right]$$
$$= \lambda(t)^2F_1(0) + \lim_{\varepsilon\to 0^+}\left(\frac{\lambda(t)}{\lambda(\varepsilon)}\right)^2\int_\varepsilon^t \left(\frac{\lambda(s)}{\lambda(\varepsilon)}\right)^{-2}\left\{\frac{d}{dt}F_1(t)\Big|_\varepsilon - 2\frac{\lambda_t(\varepsilon)}{\lambda(\varepsilon)}F_1(\varepsilon)\right\}ds$$
$$= \lambda(t)^2F_1(0) + \left\{\int_{\mathbb{R}^n}\varphi_1(x)\varphi(x)\,dx - 9\int_{\mathbb{R}^n}\varphi_0(x)\varphi(x)\,dx\right\}\lambda^2(t)\int_0^t \lambda^{-2}(s)ds$$
$$= (3\sqrt[3]{t}+1)^2\exp\left(-6\sqrt[3]{t}\right)\int_{\mathbb{R}^n}\varphi_0(x)\varphi(x)\,dx + \left\{\int_{\mathbb{R}^n}\varphi_1(x)\varphi(x)\,dx - 9\int_{\mathbb{R}^n}\varphi_0(x)\varphi(x)\,dx\right\}$$
$$\times(3\sqrt[3]{t}+1)^2\exp\left(-6\sqrt[3]{t}\right)\int_0^t(3\sqrt[3]{s}+1)^{-2}\exp\left(6\sqrt[3]{s}\right)ds\,.$$

On the other hand

$$\int_0^t (3\sqrt[3]{s}+1)^{-2} \exp\left(6\sqrt[3]{s}\right) ds = \frac{1}{18}\left(\exp\left(6\sqrt[3]{t}\right)\frac{(3\sqrt[3]{t}-1)}{3\sqrt[3]{t}+1}+1\right) \tag{6.11}$$

implies

$$\begin{aligned}
&\lim_{\varepsilon\to 0^+}\left(\frac{\lambda(t)}{\lambda(\varepsilon)}\right)^2\left[F_1(\varepsilon)+\int_\varepsilon^t\left(\frac{\lambda(s)}{\lambda(\varepsilon)}\right)^{-2}\left\{\frac{d}{dt}F_1(t)\Big|_\varepsilon-2\frac{\lambda_t(\varepsilon)}{\lambda(\varepsilon)}F_1(\varepsilon)\right\}\right]ds\\
&=(3\sqrt[3]{t}+1)^2\exp\left(-6\sqrt[3]{t}\right)\int_{\mathbb{R}^n}\varphi_0(x)\varphi(x)\,dx+\left\{\int_{\mathbb{R}^n}\varphi_1(x)\varphi(x)\,dx-9\int_{\mathbb{R}^n}\varphi_0(x)\varphi(x)\,dx\right\}\\
&\qquad\times(3\sqrt[3]{t}+1)^2\frac{1}{18}\left(\frac{(3\sqrt[3]{t}-1)}{3\sqrt[3]{t}+1}+\exp\left(-6\sqrt[3]{t}\right)\right).
\end{aligned}$$

Due to conditions of the lemma,

$$\int_{\mathbb{R}^n}\varphi_1(x)\varphi(x)\,dx-9\int_{\mathbb{R}^n}\varphi_0(x)\varphi(x)\,dx\ge\frac{1}{2}\int_{\mathbb{R}^n}\varphi_1(x)\varphi(x)\,dx>0\,.$$

Then, from (6.10), by letting $\varepsilon\to 0$, we derive

$$\begin{aligned}
F_1(t)&\ge\left(\frac{\lambda(t)}{\lambda(\varepsilon)}\right)^2\left[F_1(\varepsilon)+\int_\varepsilon^t\left(\frac{\lambda(s)}{\lambda(\varepsilon)}\right)^{-2}\left\{\frac{d}{dt}F_1(t)\Big|_\varepsilon-2\frac{\lambda_t(\varepsilon)}{\lambda(\varepsilon)}F_1(\varepsilon)+\right\}ds\right]\\
&\ge(3\sqrt[3]{t}+1)^2\exp\left(-6\sqrt[3]{t}\right)\int_{\mathbb{R}^n}\varphi_0(x)\varphi(x)\,dx\\
&\quad+(3\sqrt[3]{t}+1)^2\frac{1}{18}\left(\frac{(3\sqrt[3]{t}-1)}{3\sqrt[3]{t}+1}+\exp\left(-6\sqrt[3]{t}\right)\right)\frac{1}{2}\int_{\mathbb{R}^n}\varphi_1(x)\varphi(x)\,dx\\
&\ge(3\sqrt[3]{t}+1)^2\exp\left(-6\sqrt[3]{t}\right)\int_{\mathbb{R}^n}\varphi_0(x)\varphi(x)\,dx+(9\sqrt[3]{t^2}-1)\frac{1}{36}\int_{\mathbb{R}^n}\varphi_1(x)\varphi(x)\,dx\,.
\end{aligned}$$

Lemma is proved. □

The last lemma and (6.8) imply

$$\begin{aligned}
F''(t)&\ge\left(\int_{|x|\le R+\phi(t)}|v(x,t)|^{p/(p-1)}\,dx\right)^{1-p}t^{1-p}\,|F_1(t)|^p\\
&\ge\lambda^{-p}(t)\left(\int_{|x|\le R+\phi(t)}|\varphi(x)|^{p/(p-1)}\,dx\right)^{1-p}t^{1-p}\,|F_1(t)|^p\quad\text{for all}\quad t\ge 1\,.
\end{aligned}$$

According to the last lemma

$$
\begin{aligned}
F''(t) &\geq \lambda^{-p}(t)\left(\int_{|x|\leq R+\phi(t)} |\varphi(x)|^{p/(p-1)}\,dx\right)^{1-p} t^{1-p}\,|F_1(t)|^p \\
&\geq c_R\lambda^{-p}(t)\left((R+\phi(t))^{\frac{n-1}{2}\frac{p-2}{p-1}} e^{\phi(t)\frac{p}{p-1}}\right)^{1-p} t^{1-p}\,|F_1(t)|^p \\
&\geq c_R\lambda^{-p}(t)(R+\phi(t))^{-\frac{n-1}{2}(p-2)} e^{-\phi(t)p} t^{1-p}\,|F_1(t)|^p \\
&\geq c_R(R+\phi(t))^{-p-\frac{n-1}{2}(p-2)} t^{1-p}\left|(9t^{\frac{2}{3}}-1)\int_{\mathbb{R}^n}\varphi_1(x)\varphi(x)\,dx\right|^p .
\end{aligned}
$$

Finally

$$
F''(t) \geq \quad C_R(R+\phi(t))^{-p-\frac{n-1}{2}(p-2)} t^{1-p+\frac{2}{3}p}\left|\int_{\mathbb{R}^n}\varphi_1(x)\varphi(x)\,dx\right|^p \quad \text{for all} \quad t\geq 1 . \tag{6.12}
$$

For $t > 1$ and arbitrary $\varepsilon \in (0,1)$, it follows

$$
\begin{aligned}
F(t) &= F(\varepsilon) + \int_\varepsilon^1 F'(t_1)\,dt_1 + \int_1^t F'(t_1)\,dt_1 \\
&= F(\varepsilon) + \int_\varepsilon^1 \left\{F'(\varepsilon) + \int_\varepsilon^{t_1} F''(t_2)\,dt_2\right\} dt_1 + \int_1^t \left\{F'(\varepsilon) + \int_\varepsilon^{t_1} F''(t_2)\,dt_2\right\} dt_1 \\
&\geq F(\varepsilon) + \int_\varepsilon^1 F'(\varepsilon)\,dt_1 + \int_1^t \left\{F'(\varepsilon) + \int_1^{t_1} F''(t_2)\,dt_2\right\} dt_1 \\
&\geq F(\varepsilon) + \int_\varepsilon^1 F'(\varepsilon)\,dt_1 + \int_1^t F'(\varepsilon)\,dt_1 + \int_1^t \left\{\int_1^{t_1} F''(t_2)\,dt_2\right\} dt_1 \\
&\geq F(\varepsilon) + (t-\varepsilon)F'(\varepsilon) + \int_1^t \left\{\int_1^{t_1} F''(t_2)\,dt_2\right\} dt_1 .
\end{aligned}
$$

By letting $\varepsilon \to 0$ and using (6.12) we derive

$$
F(t) \geq tF'(0) + F(0) + c_R\left|\int_{\mathbb{R}^n}\varphi_1(x)\varphi(x)\,dx\right|^p \int_1^t\int_1^{t_2} (R+\phi(t_1))^{-p-\frac{n-1}{2}(p-2)} t_1^{1-\frac{1}{3}p}\,dt_1\,dt_2 .
$$

Set (see (6.5))

$$
r = \frac{1}{6}\left[2n+16-(n+3)p\right], \qquad q = \frac{(n+3)(p-1)}{3} .
$$

We need $r \geq 1$ that is, $p \leq (2n+10)/(n+3)$. The Kato's lemma's (see, e.g., [85, Lemma 2.1]), concerning differential inequality

$$
\begin{aligned}
F(t) &\geq c_0(1+t)^r \quad \text{for large} \quad t, \\
F''(t) &\geq (1+t)^{-q}|F(t)|^p \quad \text{for large} \quad t ,
\end{aligned}
$$

conditions are $r \geq 1, p > 1$ and

$$(p-1)r > q-2 \Longleftrightarrow (n+3)p^2 - (n+13)p - 2 < 0\,.$$

Since $p_{cr}(n)$ is defined as a positive root of the Eq. (6.2) then $p < p_{cr}(n)$. The theorem is proved. □

For $n = 3$ a positive root of the Eq. (6.2) is $p_{cr}(3) = (4+\sqrt{19})/3 > 8/3$.

Corollary 6.5 *([34]) For the covariant semilinear wave equation ($n = 3$) assume that $1 < p \leq 8/3$. Then for every arbitrary small number $\varepsilon > 0$ and arbitrary s there exist functions, $\varphi_0, \varphi_1 \in C_0^\infty(\mathbb{R}^n)$, supp $\varphi_0, \varphi_1 \subseteq \{x \in \mathbb{R}^n \mid |x| \leq R\}$ with norm*

$$\|\varphi_0\|_{H_{(s)}(\mathbb{R}^n)} + \|\varphi_1\|_{H_{(s)}(\mathbb{R}^n)} < \varepsilon$$

such that solution of the problem (6.1) with support in $\{(x,t) \mid t > 0,\ x \in B_{R+\phi(t)}(0)\}$ blows up in finite time.

Now we analyze the conditions of the theorem:

$$\text{if} \quad n \geq 37 \quad \text{then for every} \quad p < \frac{n+13+\sqrt{n^2+34n+193}}{2(n+3)}\,,$$
$$\text{if} \quad n \leq 36 \quad \text{then for every} \quad p < \frac{2n+10}{n+3}$$

the theorem implies blow up of solution.

6.1 Local in Time Solution

If $p < 2$ then by invoking results of [29] one can easily prove the existence of the local in time solution. Denote by G a solution operator of the problem

$$\begin{cases} \psi_{tt} - t^{-4/3} \bigtriangleup \psi + 2t^{-1}\psi_t = f, \qquad t > 0,\ x \in \mathbb{R}^n, \\ \lim\limits_{t\to 0^+} t\psi(x,t) = \varphi_0(x), \quad \lim\limits_{t\to 0^+} \left(t\psi_t(x,t) + \psi(x,t) + 3t^{-1/3} \bigtriangleup \varphi_0(x)\right) = \varphi_1(x),\ x \in \mathbb{R}^n\,. \end{cases}$$

with $\varphi_0(x) = \varphi_1(x) = 0$, that is $\psi = G[f]$. Let ψ_0 is the solution of the last problem with $f = 0$:

$$\begin{cases} \psi_{tt} - t^{-4/3} \bigtriangleup \psi + 2t^{-1}\psi_t = 0, \qquad t > 0,\ x \in \mathbb{R}^n, \\ \lim\limits_{t\to 0^+} t\psi(x,t) = \varphi_0(x), \quad \lim\limits_{t\to 0^+} \left(t\psi_t(x,t) + \psi(x,t) + 3t^{-1/3} \bigtriangleup \varphi_0(x)\right) = \varphi_1(x),\ x \in \mathbb{R}^n\,. \end{cases}$$

Then any solution $\psi \in C(H_{(s)}(\mathbb{R}^n) \times (0,T]) \cap C^2(H_{(s)}(\mathbb{R}^n) \times (0,T])$ of the problem (6.14) solves also the linear integral equation

$$\psi(x,t) = \psi_0(x,t) + G[|\psi(\cdot,\tau)|^p](x,t), \quad t > 0\,. \tag{6.13}$$

We define a solution of (6.1) as a solution of the last integral equation.

In fact, after a change of unknown function ψ with u such that $\psi = t^{-1}u$ the problem for u is (6.4):

$$\begin{cases} u_{tt} - t^{-4/3} \,\triangle\, u = t^{1-p}|u|^p, \qquad t > 0,\ x \in \mathbb{R}^n, \\ \lim\limits_{t\to 0^+} u(x,t) = \varphi_0(x), \qquad x \in \mathbb{R}^n, \\ \lim\limits_{t\to 0^+} \left(u_t(x,t) + 3t^{-1/3} \,\triangle\, \varphi_0(x)\right) = \varphi_1(x), \quad x \in \mathbb{R}^n\,. \end{cases} \tag{6.14}$$

Theorem 6.6 ([34]) *Assume that $1 < p < 2$. For every given $\varphi_0(x), \varphi_1(x)$, there exists $T = T(\varphi_0,\varphi_1)$ such that the problem (6.1) has a solution $\psi \in C^2((0,T(\varphi_0,\varphi_1)];$ $H_{(s)}(\mathbb{R}^n))$*

Proof The following estimate has been proven in [29] (see (3.6),(3.7) and Prop. 3.3 with $p = q = 2$,):

$$\begin{aligned} \|\psi(\cdot,t)\|_{H_{(s)}(\mathbb{R}^n)} &\le Ct^{-\frac{1}{3}} \left(t^{-\frac{2}{3}} \|\varphi_0\|_{H_{(s)}(\mathbb{R}^n)} + \|\triangle\varphi_0\|_{H_{(s)}(\mathbb{R}^n)}\right) \\ &\quad + C\|\varphi_1\|_{H_{(s)}(\mathbb{R}^n)} + \int_0^t \tau \|f(\cdot,\tau)\|_{H_{(s)}(\mathbb{R}^n)}\, d\tau \quad \text{for all} \quad t > 0\,. \end{aligned}$$

In particular,

$$\begin{aligned} \|\psi_0(\cdot,t)\|_{H_{(s)}(\mathbb{R}^n)} &\le Ct^{-\frac{1}{3}} \left(t^{-\frac{2}{3}} \|\varphi_0\|_{H_{(s)}(\mathbb{R}^n)} + \|\triangle\varphi_0\|_{H_{(s)}(\mathbb{R}^n)}\right) + C\|\varphi_1\|_{H_{(s)}(\mathbb{R}^n)} \\ &\lesssim t^{-1}\|\varphi_0\|_{H_{(s)}(\mathbb{R}^n)} + t^{-\frac{1}{3}}\|\triangle\varphi_0\|_{H_{(s)}(\mathbb{R}^n)} + \|\varphi_1\|_{H_{(s)}(\mathbb{R}^n)} \quad \text{for all } t > 0\,, \end{aligned}$$

and $t\psi_0 \in C([0,T]; H_{(s)}(\mathbb{R}^n))$ for arbitrary $T > 0$.

Then, it follows

$$\begin{aligned} t\|\psi(\cdot,t)\|_{H_{(s)}(\mathbb{R}^n)} &\le Ct^{\frac{2}{3}} \left(t^{-\frac{2}{3}} \|\varphi_0\|_{H_{(s)}(\mathbb{R}^n)} + \|\triangle\varphi_0\|_{H_{(s)}(\mathbb{R}^n)}\right) \\ &\quad + Ct\|\varphi_1\|_{H_{(s)}(\mathbb{R}^n)} + t\int_0^t \tau^{1-p} \left(\tau\|\psi(\cdot,\tau)\|_{H_{(s)}(\mathbb{R}^n)}\right)^p d\tau \\ &\le C\left(\|\varphi_0\|_{H_{(s)}(\mathbb{R}^n)} + t^{\frac{2}{3}}\|\triangle\varphi_0\|_{H_{(s)}(\mathbb{R}^n)}\right) \\ &\quad + Ct\|\varphi_1\|_{H_{(s)}(\mathbb{R}^n)} + t\int_0^t \tau^{1-p} \left(\tau\|\psi(\cdot,\tau)\|_{H_{(s)}(\mathbb{R}^n)}\right)^p d\tau \quad \text{for all} \quad t > 0\,. \end{aligned}$$

Since $t\psi$ is continuous at $t = 0$, we obtain

$$t\|\psi(\cdot,t)\|_{H_{(s)}(\mathbb{R}^n)} \le C\left(\|\varphi_0\|_{H_{(s)}(\mathbb{R}^n)} + t^{\frac{2}{3}}\|\bigtriangleup\varphi_0\|_{H_{(s)}(\mathbb{R}^n)}\right) + Ct\|\varphi_1\|_{L^p(\mathbb{R}^n)} + \max_{\tau\in[0,t]}\left(\tau\|\psi(\cdot,\tau)\|_{H_{(s)}(\mathbb{R}^n)}\right)^p t\int_0^t \tau^{1-p}\,d\tau \quad \text{for all} \quad t>0\,.$$

Hence, for $1 < p < 2$ we have

$$t\|\psi(\cdot,t)\|_{H_{(s)}(\mathbb{R}^n)} \le C\left(\|\varphi_0\|_{H_{(s)}(\mathbb{R}^n)} + t^{\frac{2}{3}}\|\bigtriangleup\varphi_0\|_{H_{(s)}(\mathbb{R}^n)}\right) + Ct\|\varphi_1\|_{L^p(\mathbb{R}^n)} + \max_{\tau\in[0,t]}\left(\tau\|\psi(\cdot,\tau)\|_{H_{(s)}(\mathbb{R}^n)}\right)^p \frac{1}{2-p}t^{3-p} \quad \text{for all} \quad t>0\,.$$

If we consider the map S defined as follows

$$S[\psi](x,t) := \psi_0(x,t) + G[|\psi(\cdot,\tau)|^p](x,t), \quad t\in[0,T]\,,$$

then the last estimate implies S is a contraction for small T

$$t\|\psi(\cdot,t)\|_{H_{(s)}(\mathbb{R}^n)} \le C\left(\|\varphi_0\|_{H_{(s)}(\mathbb{R}^n)} + t^{\frac{2}{3}}\|\bigtriangleup\varphi_0\|_{H_{(s)}(\mathbb{R}^n)}\right) + Ct\|\varphi_1\|_{L^p(\mathbb{R}^n)} + \max_{\tau\in[0,t]}\left(\tau\|\psi(\cdot,\tau)\|_{H_{(s)}(\mathbb{R}^n)}\right)^p \frac{1}{2-p}t^{3-p} \quad \text{for all} \quad t>0\,.$$

This proves the theorem. □

6.2 Equation Without Singularity

The next theorem shows that the singularity of the coefficients at $t = 0$ does not cause the blow up of Theorem 6.1. In fact, it is caused by the semilinear term. Consider the following Cauchy problem

$$\begin{cases} \psi_{tt} - t^{-2k}\bigtriangleup\psi + 2t^{-1}\psi_t = |\psi|^p, & t>1,\ x\in\mathbb{R}^n, \\ \psi(x,1) = \varphi_0(x), \quad \psi_t(x,1) = \varphi_1(x), & x\in\mathbb{R}^n, \end{cases} \tag{6.15}$$

where $k \in (0, 1)$. Let $p_{cr}(n, k)$ be a positive root of the equation

$$(1-k)(n+3)p^2 - (n+5-k(n+1))p - 2 + 2k = 0\,,$$

that is

$$p_{cr}(n,k) := \frac{n+5-k(n+1)+\sqrt{k^2(n+5)^2-2k(n(n+14)+29)+n(n+18)+49}}{2(1-k)(n+3)}\,.$$

The numbers $p_{cr}(k)$ and $p_{cr}(n,k)$ can be regarded as an analog of the Strauss exponent that was defined for the semilinear wave equation in the Minkowski spacetime. (See, e.g., [39, 81, 86].)

The Eq. (6.15) is strictly hyperbolic for every bounded interval of time and it has smooth coefficients. Consequently, for every smooth initial functions φ_0 and φ_1 the problem (6.15) has a local solution. According to the next theorem for $n=3$ and $p<3$ the small data solution can blow up. Thus, for $n=3$ and $p<3$ a local in time solution, in general, cannot be prolonged to the global solution.

Theorem 6.7 ([34]) *Assume that* $p>1$ *and*

$$\text{either}\quad 1<p<1+\frac{2}{n(1-k)}\quad \text{or}\quad 1<p\le 2\frac{n-1+2/(1-k)}{n+3}\quad \text{and}\quad p<p_{cr}(n,k)\,.$$

Then for every arbitrary small number $\varepsilon>0$ *and arbitrary* s *there exist functions,* $\varphi_0,\varphi_1\in C_0^\infty(\mathbb{R}^n)$ *with norm*

$$\|\varphi_0\|_{H_{(s)}(\mathbb{R}^n)}+\|\varphi_1\|_{H_{(s)}(\mathbb{R}^n)}<\varepsilon$$

such that solution of the problem (6.15) blows up in finite time.

Proof We use operators $\mathcal{L}$ and $\mathcal{S}$ which are introduced above: $\mathcal{L}:=\partial_t^2-t^{-2k}\triangle+2t^{-1}\partial_t$, $\mathcal{S}:=\partial_t^2-t^{-2k}\triangle$, and for $t\ne 0$ the following operator identity $t^{-1}\circ\mathcal{S}\circ t=\mathcal{L}$. The last equation suggests a change of unknown function ψ with u such that $\psi=t^{-1}u$. Then the problem for u is as follows:

$$\begin{cases} u_{tt}-t^{-2k}\triangle u=t^{1-p}|u|^p, & t>1,\ x\in\mathbb{R}^n,\\ u(x,1)=u_0(x)\,,\quad u_0(x):=\varphi_0(x), & x\in\mathbb{R}^n,\\ u_t(x,1)=u_1(x)\,,\quad u_1(x):=\varphi_0(x)+\varphi_1(x), & x\in\mathbb{R}^n, \end{cases}\tag{6.16}$$

Denote

$$F(t)=\int_{\mathbb{R}^n}u(x,t)\,dx\,.$$

Then $F\in C^2[1,T]$ provided that the function u is defined for all $(x,t)\in\mathbb{R}^n\times[1,T]$, and

$$F(1)=\int_{\mathbb{R}^n}u_0(x)\,dx=C_0>0,$$

while

$$F'(1)=\int_{\mathbb{R}^n}u_1(x)\,dx=C_1\ \ge 0.$$

From the equation we have

$$F'' = t^{1-p} \int_{\mathbb{R}^n} |u(x,t)|^p \, dx \geq 0 \quad \text{for all} \quad t > 1,$$

and from the initial conditions we derive

$$\begin{aligned} F(t) &= F(1) + \int_1^t F'(t_1)\, dt_1 = F(1) + \int_1^t \left(F'(1) + \int_1^{t_1} F''(t_2) dt_2 \right) dt_1 \\ &= F(1) + (t-1)F'(1) + \int_1^t \int_1^{t_1} F''(t_2) dt_2 \, dt_1 \\ &\geq F(1) + (t-1)F'(1) \geq 0 \quad \text{for all} \quad t \geq 1 \,. \end{aligned}$$

Hence

$$F(t) \geq F'(1)(t-1) + F(1) = (t-1) \int_{\mathbb{R}^n} u_1(x)\, dx + \int_{\mathbb{R}^n} u_0(x)\, dx \geq 0 \quad \text{for all} \quad t \geq 1 \,.$$

On the other hand, using the compact support of $u(\cdot, t)$ and Hölder's inequality we get with τ_n the volume of the unit ball in $\mathbb{R}^n$, and $\phi(t) = \frac{1}{1-k} t^{1-k}$

$$\begin{aligned} \left| \int_{\mathbb{R}^n} u(x,t)\, dx \right|^p &\leq \left(\int_{|x| \leq R + \phi(t) - \phi(1)} 1 \, dx \right)^{p-1} \left(\int_{|x| \leq R + \phi(t) - \phi(1)} |u(x,t)|^p \, dx \right) \\ &\lesssim (1+t)^{n(p-1)(1-k)} \left(\int_{|x| \leq R + \phi(t) - \phi(1)} |u(x,t)|^p \, dx \right) \\ &\lesssim (R + \phi(t))^{n(p-1)} \left(\int_{|x| \leq R + \phi(t) - \phi(1)} |u(x,t)|^p \, dx \right), \end{aligned}$$

where the number R is chosen such that supp φ_0, supp $\varphi_1 \subseteq \{|x| \leq R\}$. Here and henceforth, if A and B are two nonnegative quantities, we use $A \lesssim B$ ($A \gtrsim B$) to denote the statement that $A \leq CB$ ($AC \geq B$) for some absolute constant $C > 0$. Hence

$$F''(t) = t^{1-p} \int_{\mathbb{R}^n} |u(x,t)|^p \, dx \geq (1+t)^{1-p-n(p-1)(1-k)} |F(t)|^p \quad \text{for all} \quad t \geq 1 \,. \tag{6.17}$$

We denote

$$r = 1 \,, \qquad q := (p-1) + n(p-1)(1-k) = (p-1)(1 + n(1-k)) \,. \tag{6.18}$$

Consider the first case $1 < p < 1 + \frac{2}{n(1-k)}$. If $1 < p < 1 + \frac{2}{n(1-k)}$ and $C_1 > 0$, then we can apply Kato's lemma (see, e.g., [85, Lemma 2.1]) since

$$p-1>(p-1)(1+n(1-k))-2 \Longleftrightarrow p<1+\frac{2}{n(1-k)}$$

that proves blow up of solution.

Consider the second case. For this case, we set $\phi(t) := \frac{t^{1-k}}{1-k}$ and choose

$$\begin{aligned} v(x,t) &= \widetilde{\lambda}(t)\varphi(x)\,, \\ \widetilde{\lambda}(t) &:= \frac{1}{K_{\frac{1}{2-2k}}\left(\frac{1}{1-k}\right)}\sqrt{t}K_{\frac{1}{2-2k}}\left(\frac{t^{1-k}}{1-k}\right) = \frac{1}{K_{\frac{1}{2-2k}}\left(\phi(1)\right)}\sqrt{t}K_{\frac{1}{2-2k}}\left(\phi(t)\right)\,. \end{aligned}$$

where $K_a(z)$ is the modified Bessel function of the second kind. The function $\widetilde{\lambda} = \widetilde{\lambda}(t)$ solves the following equation

$$\lambda_{tt} - t^{-2k}\lambda = 0\,.$$

It is easy to verify the following limit

$$\lim_{t\to\infty} \sqrt{t}K_{\frac{1}{2-2k}}\left(\phi(t)\right) = 0\,.$$

Hence

$$v(x,1) = \varphi(x)\,, \qquad \lim_{t\to\infty} v(x,t) = 0\,.$$

The following lemma can be easily checked.

Lemma 6.8 ([34]) *There is a number* $\Lambda_0 >$ *such that*

$$\Lambda_1(k) := -\widetilde{\lambda}_t(1) = \frac{K_{\frac{1-2k}{2-2k}}\left(\frac{1}{1-k}\right)}{K_{\frac{1}{2-2k}}\left(\frac{1}{1-k}\right)} > \Lambda_0 \quad \textit{for all} \quad k\in[0,1)\,.$$

Assume that $u_0, u_1 \in C_0^\infty$, $\operatorname{supp} u_0, u_1 \subseteq \{x\in\mathbb{R}^n \mid |x|\le R\}$. Now we turn to the function

$$F_1(t) := \int_{\mathbb{R}^n} u(x,t)v(x,t)\,dx$$

and obtain

$$\begin{aligned} |F_1(t)|^p &\le \left(\int_{|x|\le R+\phi(t)-\phi(1)} |v(x,t)|^{p/(p-1)}\,dx\right)^{p-1}\left(\int_{|x|\le R+\phi(t)-\phi(1)} |u(x,t)|^p\,dx\right) \\ &\le \left(\int_{|x|\le R+\phi(t)-\phi(1)} |v(x,t)|^{p/(p-1)}\,dx\right)^{p-1} t^{p-1}F''(t)\,. \end{aligned}$$

The equation and the last estimate imply

$$F''(t) \geq \left(\int_{|x|\leq R+\phi(t)-\phi(1)} |v(x,t)|^{p/(p-1)}\, dx \right)^{1-p} t^{1-p}\, |F_1(t)|^p \,. \tag{6.19}$$

Lemma 6.9 [34] *Assume that* $u_0, u_1 \in C_0^\infty$, supp u_0, supp $u_1 \subseteq \{x \in \mathbb{R}^n \mid |x| \leq R\}$, *and*

$$\Lambda_1(k) \int_{\mathbb{R}^n} u_0(x)\varphi(x)dx + \int_{\mathbb{R}^n} u_1(x)\varphi(x)dx \geq c_0 \int_{\mathbb{R}^n} u_0(x)\varphi(x)dx > 0\,. \tag{6.20}$$

Then, there exists sufficiently large $T > 1$ *such that for the solution* $u = u(x,t)$ *of the problem (6.16) with the support in* $\{x \in \mathbb{R}^n \mid |x| \leq R + \phi(t) - \phi(1)\}$ *one has*

$$F_1(t) \geq \frac{1}{16} t^k \left\{ \Lambda_1(k) \int_{\mathbb{R}^n} u_0(x)\varphi(x)dx + \int_{\mathbb{R}^n} u_1(x)\varphi(x)dx \right\} \quad \textit{for all} \quad t > T\,. \tag{6.21}$$

Proof We have

$$F_1(1) = \int_{\mathbb{R}^n} u(x,1)v(x,1)\, dx = \int_{\mathbb{R}^n} u_0(x)\varphi(x)\, dx \geq c_0 > 0$$

and

$$\begin{aligned} 0 &= \int_1^t \int_{\mathbb{R}^n} (u_{tt}(x,\tau) - \tau^{-2k}\Delta u - \tau^{1-p}|u|^p) v(x,\tau)\, dx\, d\tau \\ &= \int_1^t \int_{\mathbb{R}^n} u_{tt}(x,\tau)) v(x,\tau)\, dx\, d\tau - \int_1^t \int_{\mathbb{R}^n} \tau^{-2k} u \Delta v(x,\tau)\, dx\, d\tau - \int_1^t \int_{\mathbb{R}^n} \tau^{1-p}|u|^p v(x,\tau)\, dx\, d\tau\,. \end{aligned}$$

Further,

$$\begin{aligned} &\int_1^t \int_{\mathbb{R}^n} u_{tt}(x,\tau) v(x,\tau)\, dx\, d\tau \\ &= \int_{\mathbb{R}^n} u_t(x,\tau) v(x,\tau)\, dx \Big|_{\tau=1}^{\tau=t} - \int_{\mathbb{R}^n} u(x,\tau) v_t(x,\tau)\, dx \Big|_{\tau=1}^{\tau=t} + \int_1^t \int_{\mathbb{R}^n} u(x,\tau) t^{-2k} \Delta v(x,\tau)\, dx\, d\tau\,. \end{aligned}$$

Hence,

$$0 = \int_{\mathbb{R}^n} u_t(x,\tau) v(x,\tau)\, dx \Big|_{\tau=1}^{\tau=t} - \int_{\mathbb{R}^n} u(x,\tau) v_t(x,\tau)\, dx \Big|_{\tau=1}^{\tau=t} - \int_1^t \int_{\mathbb{R}^n} \tau^{1-p} |u(x,\tau)|^p v(x,\tau)\, dx\, d\tau\,.$$

That is

$$\int_{\mathbb{R}^n} u_t(x,\tau) v(x,\tau)\, dx \Big|_{\tau=1}^{\tau=t} - \int_{\mathbb{R}^n} u(x,\tau) v_t(x,\tau)\, dx \Big|_{\tau=1}^{\tau=t} = \int_1^t \int_{\mathbb{R}^n} \tau^{1-p} |u(x,\tau)|^p v(x,\tau)\, dx\, d\tau$$

implies

$$\left(\frac{d}{d\tau}\int_{\mathbb{R}^n} u(x,\tau)v(x,\tau)\,dx - 2\int_{\mathbb{R}^n} u(x,\tau)v_t(x,\tau)\,dx\right)\Big|_{\tau=1}^{\tau=t} = \int_1^t\int_{\mathbb{R}^n}\tau^{1-p}|u(x,\tau)|^p v(x,\tau)\,dx\,d\tau$$

and

$$\begin{aligned}&\frac{d}{dt}F_1(t) - 2\,\widetilde{\lambda}_t(t)\int_{\mathbb{R}^n} u(x,t)\varphi(x)\,dx\\ &= \frac{d}{dt}F_1(t)\Big|_{t=1} - 2\,\widetilde{\lambda}_t(1)\int_{\mathbb{R}^n} u(x,1)\varphi(x)\,dx + \int_1^t\int_{\mathbb{R}^n}\tau^{1-p}|u(x,\tau)|^p v(x,\tau)\,dx\,d\tau\,.\end{aligned}$$

On the other hand

$$\frac{\widetilde{\lambda}_t(t)}{\widetilde{\lambda}(t)} = -\frac{t^{-k}K_{1+\frac{1}{2(k-1)}}\left(\frac{t^{1-k}}{1-k}\right)}{K_{\frac{1}{2-2k}}\left(\frac{t^{1-k}}{1-k}\right)} = -\frac{t^{-k}K_{1+\frac{1}{2(k-1)}}(\phi(t))}{K_{\frac{1}{2-2k}}(\phi(t))} < 0 \quad \text{for all} \quad t>0\,,$$

$$\lim_{t\to\infty}\frac{\widetilde{\lambda}_t(t)}{\widetilde{\lambda}(t)} = 0\,, \quad \frac{\widetilde{\lambda}_t(1)}{\widetilde{\lambda}(1)} = \widetilde{\lambda}_t(1) = -\frac{K_{\frac{1-2k}{2-2k}}\left(\frac{1}{1-k}\right)}{K_{\frac{1}{2-2k}}\left(\frac{1}{1-k}\right)}\,.$$

Then, according to Lemma 6.8

$$\begin{aligned}&\frac{d}{dt}F_1(t) - 2\frac{\widetilde{\lambda}_t(t)}{\widetilde{\lambda}(t)}\int_{\mathbb{R}^n} u(x,t)\,\widetilde{\lambda}(t)\varphi(x)\,dx\\ &= \frac{d}{dt}F_1(t)\Big|_1 - 2\frac{\widetilde{\lambda}_t(1)}{\widetilde{\lambda}(1)}\int_{\mathbb{R}^n} u(x,1)\,\widetilde{\lambda}(1)\varphi(x)\,dx + \int_1^t\int_{\mathbb{R}^n}\tau^{1-p}|u(x,\tau)|^p v(x,\tau)\,dx\,d\tau\,.\end{aligned}$$

Consequently

$$\frac{d}{dt}F_1(t) - 2\frac{\widetilde{\lambda}_t(t)}{\widetilde{\lambda}(t)}F_1(t) = \frac{d}{dt}F_1(t)\Big|_1 - 2\frac{\widetilde{\lambda}_t(1)}{\widetilde{\lambda}(1)}F_1(1) + \int_1^t\int_{\mathbb{R}^n}\tau^{1-p}|u(x,\tau)|^p v(x,\tau)\,dx\,d\tau\,,$$

that is,

$$\begin{aligned}&\frac{d}{dt}\left(F_1(t)\left(\widetilde{\lambda}(t)\right)^{-2}\right)\\ &= \left(\widetilde{\lambda}(t)\right)^{-2}\left\{\frac{d}{dt}F_1(t)\Big|_1 + 2\Lambda_1(k)F_1(1) + \int_1^t\int_{\mathbb{R}^n}\tau^{1-p}|u(x,\tau)|^p v(x,\tau)\,dx\,d\tau\right\}\,,\end{aligned}$$

where $\Lambda_1(k) = -\widetilde{\lambda}_t(1) = \dfrac{K_{1+\frac{1}{2(k-1)}}\left(\frac{1}{1-k}\right)}{K_{\frac{1}{2-2k}}\left(\frac{1}{1-k}\right)} > 0$. We integrate the last relation

$$
\begin{aligned}
& F_1(t)\left(\widetilde{\lambda}(t)\right)^{-2} \\
&= F_1(1) + \int_1^t \left(\widetilde{\lambda}(s)\right)^{-2} \left\{ \frac{d}{dt}F_1(t)\Big|_{t=1} + 2\Lambda_1(k)F_1(1) + \int_1^s \int_{\mathbb{R}^n} \tau^{1-p}|u(x,\tau)|^p v(x,\tau)\, dx\, d\tau \right\} ds\,.
\end{aligned}
$$

Finally

$$
\begin{aligned}
F_1(t) = {} & \left(\widetilde{\lambda}(t)\right)^2 \Big[F_1(1) \\
& + \int_1^t \left(\widetilde{\lambda}(s)\right)^{-2} \left\{ \frac{d}{dt}F_1(t)\Big|_{t=1} + 2\Lambda_1(k)F_1(1) + \int_1^s \int_{\mathbb{R}^n} \tau^{1-p}|u(x,\tau)|^p v(x,\tau)\, dx\, d\tau \right\} ds \Big].
\end{aligned}
$$

Consider two first terms of the integrand

$$
\begin{aligned}
\frac{d}{dt}F_1(t)\Big|_{t=1} + 2\Lambda_1 F_1(1) &= \int_{\mathbb{R}^n} u_0(x) v_t(x,1) dx + \int_{\mathbb{R}^n} u_1(x) v(x,1) dx + 2\Lambda_1 \int_{\mathbb{R}^n} u_0(x) v(x,1) dx \\
&= \Lambda_1(k) \int_{\mathbb{R}^n} u_0(x)\varphi(x)\, dx + \int_{\mathbb{R}^n} u_1(x)\varphi(x)\, dx\,.
\end{aligned}
$$

Then

$$
\begin{aligned}
& \left(\widetilde{\lambda}(t)\right)^2 \left[F_1(1) + \int_1^t \left(\widetilde{\lambda}(s)\right)^{-2} \left\{ \frac{d}{dt}F_1(t)\Big|_{t=1} + 2\Lambda_1(k)F_1(1) \right\} ds \right] \qquad (6.22) \\
&= \left(\widetilde{\lambda}(t)\right)^2 F_1(1) + \left[\Lambda_1(k) \int_{\mathbb{R}^n} u_0(x)\varphi(x) dx + \int_{\mathbb{R}^n} u_1(x)\varphi(x) dx \right] \left(\widetilde{\lambda}(t)\right)^2 \int_1^t \left(\widetilde{\lambda}(s)\right)^{-2} ds\,.
\end{aligned}
$$

Lemma 6.10 ([34]) *There is T_1 such that*

$$
\widetilde{\lambda}^2(t) \int_T^t \widetilde{\lambda}^{-2}(s)\, ds \geq \frac{1}{32} t^k \quad \textit{for all} \quad t \geq T_1\,.
$$

Proof For all $T > 1$ we have

$$
\widetilde{\lambda}^2(t) \int_1^t \widetilde{\lambda}^{-2}(s)\, ds = \widetilde{\lambda}^2(t) \int_1^T \widetilde{\lambda}^{-2}(s)\, ds + \widetilde{\lambda}^2(t) \int_T^t \widetilde{\lambda}^{-2}(s)\, ds \quad \text{for all} \quad t \geq T\,. \qquad (6.23)
$$

There is for the large t the following asymptotic

$$
\sqrt{t} K_{\frac{1}{2-2k}}(\phi(t)) = \sqrt{\frac{\pi}{2}} \sqrt{1-k}\, e^{-\phi(t)} t^{k/2} \left(1 + o(1)\right)\,.
$$

Consider the second integral; for the sufficiently large T we have

$$
\begin{aligned}
&\widetilde{\lambda}^2(t)\int_T^t \widetilde{\lambda}^{-2}(s)\,ds \\
&\geq \frac{1}{2}e^{-2\frac{t^{1-k}}{1-k}}t^{2k}\int_T^t e^{2\frac{s^{1-k}}{1-k}}s^{-2k}\,ds \\
&= \frac{1}{2}e^{-2\frac{t^{1-k}}{1-k}}t^{2k}\frac{1}{4}\Bigg(2e^{\frac{2t^{1-k}}{1-k}}t^{-k}+ke^{\frac{2t^{1-k}}{1-k}}t^{-1}+2^{\frac{1}{1-k}}k\left(\frac{1}{k-1}\right)^{\frac{k-2}{k-1}}\Gamma\left(\frac{1}{k-1},\frac{2t^{1-k}}{k-1}\right) \\
&\quad -2e^{-\frac{2T^{1-k}}{k-1}}T^{-k}-\frac{ke^{-\frac{2T^{1-k}}{k-1}}}{T}-2^{\frac{1}{1-k}}k\left(\frac{1}{k-1}\right)^{\frac{k-2}{k-1}}\Gamma\left(\frac{1}{k-1},\frac{2T^{1-k}}{k-1}\right)\Bigg) \quad \text{for all} \quad t\geq T,
\end{aligned}
$$

where $\Gamma(a,z)=\int_z^\infty t^{a-1}e^{-t}\,dt$ is the incomplete gamma function. (See, e.g., [9, Sect. 6.9.2].) On the other hand, since $k=1-\varepsilon$, $\varepsilon>0$, we obtain for the incomplete gamma function the following asymptotic formula (see [9, Sect. 6.13.1])

$$
\begin{aligned}
\left(\frac{1}{k-1}\right)^{\frac{k-2}{k-1}}\Gamma\left(\frac{1}{k-1},\frac{2t^{1-k}}{k-1}\right) &= 2^{\frac{3-2k}{k-1}}e^{-\frac{2t^{1-k}}{k-1}}t^{k-2}\left(2+O(t^{k-1})\right) \\
&\leq ce^{-\frac{2t^{1-k}}{k-1}}t^{-1-\varepsilon} \quad \text{for all} \quad t\geq T.
\end{aligned}
$$

Consequently, for the sufficiently large $T_1>T$ we obtain

$$
\begin{aligned}
&\widetilde{\lambda}^2(t)\int_T^t \widetilde{\lambda}^{-2}(s)\,ds \\
&\geq \frac{1}{2}e^{-2\frac{t^{1-k}}{1-k}}t^{2k}\frac{1}{4}\left(e^{\frac{2t^{1-k}}{1-k}}t^{-k}-2e^{\frac{2T^{1-k}}{1-k}}T^{-k}-\frac{ke^{-\frac{2T^{1-k}}{k-1}}}{T}-2^{\frac{1}{1-k}}k\left(\frac{1}{k-1}\right)^{\frac{k-2}{k-1}}\Gamma\left(\frac{1}{k-1},\frac{2T^{1-k}}{k-1}\right)\right) \\
&\geq \frac{1}{16}t^k \quad \text{for all} \quad t\geq T_1.
\end{aligned}
$$

The estimate for the first term of (6.23) is evident. Lemma is proved. □

On the other hand, according to (6.19) we have

$$
F''(t)\gtrsim \lambda^{-p}(t)\left(\int_{|x|\leq R+\phi(t)-\phi(1)}|\varphi(x)|^{p/(p-1)}\,dx\right)^{1-p}t^{1-p}\,|F_1(t)|^p \quad \text{for large} \quad t,
$$

and, consequently, (6.22) and Lemma 6.10 imply

$$
\begin{aligned}
F''(t) &\gtrsim c_R(R+\phi(t)-\phi(1))^{-p-\frac{n-1}{2}(p-2)}t^{1-p} \\
&\quad \times\left|\frac{1}{32}t^k\left\{\Lambda_1(k)\int_{\mathbb{R}^n}u_0(x)\varphi(x)dx+\int_{\mathbb{R}^n}u_1(x)\varphi(x)dx\right\}\right|^p \quad \text{for} \quad t\geq T.
\end{aligned}
$$

Here $T > 1$ is sufficiently large number. It follows

$$\begin{aligned}
F(t) &= F(1) + \int_1^t F'(t_1)dt_1 = F(1) + \int_1^T F'(t_1)dt_1 + \int_T^t F'(t_1)dt_1 = \\
&= F(1) + \int_1^T \left\{F'(1) + \int_T^{t_1} F''(t_2)dt_2\right\} dt_1 + F'(T)(t-T) + \int_T^t \int_T^{t_1} F''(t_2)dt_2\, dt_1 \\
&\gtrsim F(1) + F'(1)(T-1) + F'(T)(t-T) \\
&\quad + \int_T^t \int_T^{t_1} c_R(R+\phi(t_2)-\phi(1))^{-p-\frac{n-1}{2}(p-2)} t^{1-p} \\
&\quad \times \left|\frac{1}{32} t_2^k \left\{\Lambda_1(k)\int_{\mathbb{R}^n} u_0(x)\varphi(x)dx + \int_{\mathbb{R}^n} u_1(x)\varphi(x)dx\right\}\right|^p dt_2\, dt_1 \\
&\gtrsim F(1) + F'(1)(T-1) + (t-T)\left\{F'(1) + \int_1^T F''(t_1)\, dt_1\right\} \\
&\quad + \left|\frac{1}{32}\left\{\Lambda_1(k)\int_{\mathbb{R}^n} u_0(x)\varphi(x)dx + \int_{\mathbb{R}^n} u_1(x)\varphi(x)dx\right\}\right|^p \\
&\quad \times \int_T^t \int_T^{t_1} c_R(R+\phi(t_2)-\phi(1))^{-p-\frac{n-1}{2}(p-2)} t_2^{1-p} \left|t_2^k\right|^p dt_2\, dt_1 ,
\end{aligned}$$

where $F'(1) = \int_{\mathbb{R}^n} u_0(x)\varphi(x)dx + \int_{\mathbb{R}^n} u_1(x)\varphi(x)dx$. Thus,

$$\begin{aligned}
F(t) \gtrsim F(1) + F'(1)(t-1) + &\left|\left\{\Lambda_1(k)\int_{\mathbb{R}^n} u_0(x)\varphi(x)dx + \int_{\mathbb{R}^n} u_1(x)\varphi(x)dx\right\}\right|^p \\
&\times \int_T^t \int_T^{t_1} \phi(t_2)^{-p-\frac{n-1}{2}(p-2)} t_2^{1-p} t_2^{kp} dt_2\, dt_1 .
\end{aligned}$$

Set

$$r = (1-k)\left[-p - \frac{n-1}{2}(p-2)\right] + 1 - p + kp + 2\,,\, q = (p-1)(1+n(1-k))\,.$$

We need $r \geq 1$, that is,

$$p \leq 2\frac{n-1+2/(1-k)}{n+3}\,.$$

We check the condition $(p-1)r > q-2$ of the Kato's lemma (see, e.g., [85, Lemma 2.1]), that is,

$$\frac{1}{2}(k-1)(n+3)p^2 + \frac{1}{2}(-k(n+1)+n+5)p + 1 - k > 0\,.$$

Since $k < 1$ we conclude $1 < p < p_{cr}(n,k)$. Theorem is proved. □

In particular, for the matter dominated universe with $k = 2/3$ we obtain

$$1 < p < p_{cr} := \frac{n+13+\sqrt{n^2+34n+193}}{2n+6}, \quad \text{if } n=3 \text{ then } 1 < p < p_{cr} := \frac{1}{3}\left(4+\sqrt{19}\right),$$

$$\text{and } 1 < p \leq 2\frac{(n-1)(1-k)+2}{(1-k)(n+3)} \quad \text{if } n=3 \text{ then } 1 < p \leq \frac{8}{3},$$

while for the radiation dominated universe with $k = 1/2$ we obtain

$$
\begin{aligned}
1 < p < p_{cr} &:= \frac{n+9+\sqrt{n^2+26n+105}}{2n+6}, && \text{if } n=3 \text{ then } 1 < p < p_{cr} := \frac{3+2\sqrt{3}}{3} \\
\text{and } \ 1 < p &\le 2\frac{(n-1)(1-k)+2}{(1-k)(n+3)} && \text{if } n=3 \text{ then } 1 < p \le \frac{8}{3}, \\
1 < p < p_{cr} &:= \frac{n+9+\sqrt{n^2+26n+105}}{2n+6}, && \text{if } n=4 \text{ then } 1 < p < p_{cr} := 2 \\
\text{and } \ 1 < p &\le 2\frac{(n-1)(1-k)+2}{(1-k)(n+3)} && \text{if } n=4 \text{ then } 1 < p \le \frac{10}{7}.
\end{aligned}
$$

The first case $1 < p < 1 + \frac{2}{n(1-k)}$ of the theorem means

$$
\begin{aligned}
1 < p < 1 + \frac{2}{n(1-k)}, \quad & \text{if } k = \frac{2}{3}, \ n=3 \ \text{ then } \ 1 < p < 3, \\
1 < p < 1 + \frac{2}{n(1-k)}, \quad & \text{if } k = \frac{1}{2}, \ n=3 \ \text{ then } \ 1 < p < \frac{7}{3}, \\
1 < p < 1 + \frac{2}{n(1-k)}, \quad & \text{if } k = \frac{1}{2}, \ n=4 \ \text{ then } \ 1 < p < 2.
\end{aligned}
$$

Consider now the difference

$$
1 + \frac{2}{n(1-k)} - 2\frac{(n-1)(1-k)+2}{(1-k)(n+3)} = -\frac{n(-k(n-5)+n-3)-6}{(1-k)n(n+3)}.
$$

For $n = 3$ we have

$$
1 + \frac{2}{n(1-k)} - 2\frac{(n-1)(1-k)+2}{(1-k)(n+3)} = \frac{1}{3} > 0 \quad \text{for all} \quad k \in (0,1).
$$

It remains to check the sign of $n(-k(n-5)+n-3)-6 > 0$.

For the semilinear generalized Tricomi equation $\partial_t^2 u - t^m \Delta u = |u|^p$ with the increasing coefficient, that is with $m \in \mathbb{N}$, the critical exponent $p_{crit}(m,n)$ and conformal exponent $p_{conf}(m,n)$ are suggested in [39]. We also mention interesting articles on the nonlinear higher order degenerate hyperbolic equations [70], the low regularity solution problem for the semilinear mixed type equation [72], and the local existence and singularity structures of low regularity solution to the semilinear generalized Tricomi equation with discontinuous initial data [71].

The Cauchy problem for the damped linear wave equations with a time-dependent propagation speed and dissipations, $u_{tt} - a(t)^2 \Delta u + b(t) u_t = 0$, where $a \in L^1(0,\infty)$, is considered in [24]. The analysis of results of [24] hopefully can lead to the global existence in the problem for the wave equation in the de Sitter spacetime and shed a light on the interval $(\sqrt{n^2-1}/2, n/2)$.

References

1. Abbasi, B., Craig, W.: On the initial value problem for the wave equation in Friedmann-Robertson-Walker space-times. Proc. R. Soc. Lond. Ser. A Math. Phys. Eng. Sci. **470**(2169), 20140361 (2014)
2. Allen, B., Folacci, A.: Massless minimally coupled scalar field in de Sitter space. Phys. Rev. D **35**(12), 3771–3778 (1987)
3. Anker, J.-F., Pierfelice, V., Vallarino, M.: The wave equation on hyperbolic spaces. J. Diff. Equat. **252**(10), 5613–5661 (2012)
4. Barros-Neto, J., Gelfand, I.M.: Fundamental solutions for the Tricomi operator, Duke Math. J. **98**(3), 465–483 (1999); Fundamental solutions for the Tricomi operator. II. Duke Math. J. **111**(3), 561–584 (2002); Fundamental solutions of the Tricomi operator. III. Duke Math. J. **128**(1), 119–140 (2005)
5. Bachelot, A.: On the Klein-Gordon equation near a De Sitter Brane, J. Math. Pures Appl. http://dx.doi.org/10.1016/j.matpur.2015.09.004 (2015)
6. Bachelot, A.: Waves in the Witten bubble of nothing and the Hawking wormhole. arXiv:1601.03682v1 (2016)
7. Baskin, D.: A parametrix for the fundamental solution of the Klein-Gordon equation on asymptotically de Sitter spaces. J. Funct. Anal. **259**, 1673–1719 (2010)
8. Baskin, D.: Strichartz estimates on asymptotically de sitter spaces. Annales Henri Poincaré **14**(2), 221–252 (2013)
9. Bateman, H., Erdelyi, A.: Higher Transcendental Functions, vol. 1, 2, McGraw-Hill, New York, (1953)
10. Birrell, N.D., Davies, P.C.W.: Quantum fields in curved space. Cambridge University Press, Cambridge, New York (1984)
11. Bros, J., Epstein, H., Moschella, U.: Particle decays and stability on the de Sitter universe. Ann. Henri Poincaré **11**(4), 611–658 (2010)
12. Bers, A., Fox, R., Kuper, C.G., Lipson, S.G.: The impossibility of free tachyons. In: Kuper, C.G., Peres, A. (eds.) Relativity and Gravitation, pp. 41–46. Gordon and Breach Science Publishers, New York (1971)
13. Bers, L.: Mathematical aspects of subsonic and transonic gas dynamics. Surveys in Applied Mathematics, vol. 3. John Wiley & Sons, Inc., New York (1958). Chapman & Hall, Ltd., London
14. Brevik, I., Simonsen, B.: The scalar field equation in Schwarzschild-de Sitter space. Gen. Relativ. Gravit. **33**(10), 1839–1861 (2001)
15. Catania, D., Georgiev, V.: Blow-up for the semilinear wave equation in the Schwarzschild metric. Diff. Integr. Equat. **19**, 799–830 (2006)
16. Choquet-Bruhat, Y.: Global wave maps on Robertson-Walker spacetimes. Modern group analysis. Nonlinear Dyn. **22**(1), 39–47 (2000)
17. Choquet-Bruhat, Y.: General relativity and the Einstein equations. Oxford Mathematical Monographs. Oxford University Press, Oxford (2009)
18. Choquet-Bruhat, Y.: Global wave maps on curved space times. In: Mathematical and quantum aspects of relativity and cosmology (Pythagoreon, 1998). Lecture Notes in Physics, vol. 537. Springer, Berlin (2000)
19. Choquet-Bruhat, Y., Chrusciel, P.T., Martín-García, J.M.: The light-cone theorem. Class. Quantum Gravity **26**(13), 135011, 22 (2009)
20. Costa, J.L., Alho, A., Natário, J.: Spherical linear waves in de Sitter spacetime. J. Math. Phys. **53**(5), 052501, 9 (2012)
21. Cole, J.D., Cook, L.P.: Transonic aerodynamics. Elsevier Science Pub. Co., Amsterdam, New York, North-Holland, U.S.A. (1986)
22. Darboux, G.: Lecons sur les systèmes orthogonaux et les coordonnées curvilignes. Principes de géométrie analytique. Les Grands Classiques Gauthier-Villars. Cours de Géométrie de la Faculté des Sciences. Éditions Jacques Gabay, Sceaux (1993)

23. Dohse, M.: Classical Klein-Gordon solutions, symplectic structures, and isometry actions on AdS spacetimes. J. Geom. Phys. **70**, 130–156 (2013)
24. Ebert, M.R., Reissig, M.: Theory of damped wave models with integrable and decaying in time speed of propagation. J. Hyperbol. Diff. Equat. in press
25. Ellis, G., van Elst, H.: Cargese lectures 1998: cosmological models. NATO Adv. Study Inst. Ser. C. Math. Phys. Sci. **541** 1–116 (1999)
26. Epstein, H., Moschella, U.: de Sitter tachyons and related topics. Comm. Math. Phys. **336**(1), 381–430 (2015)
27. Frankl, F.: On the problems of Chaplygin for mixed sub- and supersonic flows. Bull. Acad. Sci. USSR. Ser. Math. **9**, 121–143 (1945)
28. Galstian, A.: $L_p - L_q$-decay estimates for the Klein-Gordon equation in the anti-de Sitter space-time. Rend. Istit. Mat. Univ. Trieste **42**(suppl.) 27–50 (2010)
29. Galstian, A., Kinoshita, T., Yagdjian, K.: A note on wave equation in Einstein and de Sitter space-time. J. Math. Phys. **51**(5), 052501 (2010)
30. Galstian, A., Kinoshita, T.: Representation of solutions for 2nd order one-dimensional model hyperbolic equations. J. Anal. Math in press
31. Galstian, A., Yagdjian, K.: Microlocal analysis for waves propagating in Einstein & de Sitter spacetime. Math. Phys. Anal. Geom. **17**(1–2), 223–246 (2014)
32. Galstian, A., Yagdjian, K.: Global solutions for semilinear Klein-Gordon equations in FLRW spacetimes. Nonlinear Anal. **113**, 339–356 (2015)
33. Galstian, A., Yagdjian, K.: Global in time existence of the self-interacting scalar field in de sitter spacetimes
34. Galstian, A., Yagdjian, K.: The Strauss exponent for the semilinear equation on the Einstein-de Sitter spacetime
35. Germain, P.: The Tricomi equation, its solutions and their applications in fluid dynamics. Tricomi's ideas and contemporary applied mathematics (Rome/Turin, 1997) 7-26, Atti Convegni Lincei, **147**, Accad. Naz. Lincei, Rome (1998)
36. Germain, P., Bader, R.: Sur le problème de Tricomi. Rend. Circ. Mat. Palermo **2**(2), 53–70 (1953)
37. Gron, O., Hervik, S.: Einstein's general theory of relativity: with modern applications in cosmology. Springer, New York (2007)
38. Hawking, S.W., Ellis, G.F.R.: The Large Scale Structure of Space-time. Cambridge Monographs on Mathematical Physics, no. 1. Cambridge University Press, New York, London (1973)
39. He, D., Witt, I., Yin, H.: On the global solution problem for semilinear generalized Tricomi equations, I. arXiv:1511.08722v1 (2015)
40. Helgason, S.: Wave equations on homogeneous spaces. In: Lie group representations, III. College Park, Md., 1982/1983. Lecture Notes in Mathematics, vol. 1077, pp. 254–287. Springer, Berlin (1984)
41. Helgason, S.: Radon transforms and wave equations. In: Integral geometry, Radon Transforms and Complex Analysis (Venice, 1996). Lecture Notes in Mathematics, vol. 1684, pp. 9–121. Springer, Berlin (1998)
42. Higuchi, A.: Forbidden mass range for spin-2 field theory in de Sitter spacetime. Nuclear Phys. B **282**(2), 397–436 (1987)
43. Hintz, P.: Global well-posedness of quasilinear wave equations on asymptotically de Sitter spaces. arXiv:1311.6859v2 (2014)
44. Hintz, P., Vasy, A.: Semilinear wave equations on asymptotically de Sitter, Kerr-de Sitter and Minkowski spacetimes. Anal. PDE **8**(8), 1807–1890 (2015)
45. Ivanovici, O., Lebeau, G., Planchon, F.: Dispersion for the wave equation inside strictly convex domains I: the Friedlander model case. Ann. Math. **180**(1), 323–380 (2014)
46. Jamal, S., Kara, A.H., Bokhari, A.H.: Symmetries, conservation laws, reductions, and exact solutions for the Klein-Gordon equation in de Sitter space-times. Can. J. Phys. **90**, 667–674 (2012)
47. Kim, J.U.: An L^p a priori estimate for the Tricomi equation in the upper half space. Trans. Am. Math. Soc. **351**(11), 4611–4628 (1999)

48. Kluwick, A.: Transonic nozzle flow in dense gases. J. Fluid Mech. **247**, 661–688 (1993)
49. Kong, D.-X., Wei, C.-H.: Lifespan of smooth solutions for timelike extremal surface equation in de Sitter spacetime. arXiv:1311.3459v1 (2013)
50. Lau, S.R., Price, R.H.: Multidomain spectral method for the helically reduced wave equation. J. Comput. Phys. **227**(2), 1126–1161 (2007)
51. Lax, P.D., Phillips, R.S.: Translation representations for the solution of the non-Euclidean wave equation. Commun. Pure Appl. Math. **32**(5), 617–667 (1979)
52. Lupo, D., Payne, K.R.: On the maximum principle for generalized solutions to the Tricomi problem. Commun. Contemp. Math. **2**(4), 535–557 (2000)
53. Lupo, D., Payne, K.R.: Spectral bounds for Tricomi problems and application to semilinear existence and existence with uniqueness results. J. Diff. Equat. **184**(1), 139–162 (2002)
54. Lupo, D., Payne, K.R.: Critical exponents for semilinear equations of mixed elliptic-hyperbolic and degenerate types. Commun. Pure Appl. Math. **56**(3), 403–424 (2003)
55. Metcalfe, J., Taylor, M.E.: Nonlinear waves on 3D hyperbolic space. Trans. Am. Math. Soc. **363**, 3489–3529 (2011)
56. Metcalfe, J., Tataru, D., Tohaneanu, M.: Price's Law on Nonstationary Spacetimes. Adv. Math. **230**(3), 995–1028 (2012)
57. Morawetz, C.: The mathematical approach to the sonic barrier. Bull. Am. Math. Soc. (N.S.) **6**(2), 127–145 (1982)
58. Morawetz, C.: Mixed equations and transonic flow. J. Hyperbol. Differ. Equat. **1**(1), 1–26 (2004)
59. Moschella, U.: The de Sitter and anti-de Sitter sightseeing tour. In: Einstein, 1905–2005. Progress in Mathematical Physics, vol. 47, pp. 120–133. Birkhäuser, Basel (2006)
60. Näf, J., Jetzer, P., Sereno, M.: On gravitational waves in spacetimes with a nonvanishing cosmological constant. Phys. Rev. D **79**, 024014 (2009)
61. Nakamura, M.: The Cauchy problem for semi-linear Klein-Gordon equations in de Sitter spacetime. J. Math. Anal. Appl. **410**(1), 445–454 (2014)
62. Nocilla, S.: Applications and developments of the Tricomi equation in the transonic aerodynamics, mixed type equations. Teubner-Texte Math. **90**, 216–241 (1986). Teubner, Leipzig
63. Ohanian, H., Ruffini, R.: Gravitation and Spacetime. Norton, New York (1994)
64. Parker, L.E., Toms, D.J.: Quantum field theory in curved spacetime. Quantized fields and gravity. In: Cambridge Monographs on Mathematical Physics. Cambridge University Press, Cambridge, (2009)
65. Payne, K.: Interior regularity of the Dirichlet problem for the Tricomi equation. J. Math. Anal. Appl. **199**(1), 271–292 (1996)
66. Protter, M.H., Weinberger, H.F.: Maximum Principles in Differential Equations. Springer–Verlag, New York (1984). Corrected reprint of the 1967 original
67. Rendall, A.: Asymptotics of solutions of the Einstein equations with positive cosmological constant. Ann. Henri Poincaré **5**(6), 1041–1064 (2004)
68. Ringström, H.: Future stability of the Einstein-non-linear scalar field system. Invent. Math. **173**(1), 123–208 (2008)
69. Rodnianski, I., Speck, J.: The nonlinear future stability of the FLRW family of solutions to the irrotational Euler-Einstein system with a positive cosmological constant. J. Eur. Math. Soc. (JEMS) **15**(6), 2369–2462 (2013)
70. Ruan, Z., Witt, I., Yin, H.: The existence and singularity structures of low regularity solutions to higher order degenerate hyperbolic equations. J. Differ. Equat. **256**(2), 407–460 (2014)
71. Ruan, Z., Witt, I., Yin, H.: On the existence and cusp singularity of solutions to semilinear generalized Tricomi equations with discontinuous initial data. Comm. Contemp. Math. **17**(3), 1450028 (2015)
72. Ruan, Z., Witt, I., Yin, H.: On the existence of low regularity solutions to semilinear generalized Tricomi equations in mixed type domains. J. Differ. Equat. **259**(12), 7406–7462 (2015)
73. Shatah, J., Struwe, M.: Geometric Wave Equations. Courant Lecture Notes in Mathematics, vol. 2. New York University Courant Institute of Mathematical Sciences, New York (1998)

74. Smirnov, M.M.: Equations of Mixed Type. Translations of Mathematical Monographs, vol. 51. American Mathematical Society, Providence, R.I. (1978)
75. Strauss, W.A.: Partial Differential Equations. An introduction, 2ND edn. John Wiley & Sons, Ltd., Chichester (2008)
76. Tarkenton, G.M., Cramer, M.S.: Transonic flows of of dense gases. *ASME*, 93-FE-9 (1993)
77. Tricomi, F.: Sulle equazioni lineari alle derivate parziali di secondo ordine, di tipo misto. Rend. Reale Accad. Lincei Cl. Sci. Fis. Mat. Natur. **5**(14), 134–247 (1923)
78. Tolman, R.C.: Relativity, Thermodynamics, and Cosmology. Clarendon Press, Oxford (1934)
79. Vasy, A.: The wave equation on asymptotically de Sitter-like spaces. Adv. Math. **223**(1), 49–97 (2010)
80. Vasy, A.: Microlocal analysis of asymptotically hyperbolic and Kerr-de Sitter spaces (with an appendix by Semyon Dyatlov). Invent. Math. **194**(2), 381–513 (2013)
81. Yordanov, B., Zhang, Q.S.: Finite-time blowup for wave equations with a potential. SIAM J. Math. Anal. **36**(5), 1426–1433 (2005)
82. Yagdjian, K., Galstian, A.: The Klein-Gordon equation in anti-de Sitter spacetime. Rend. Semin. Mat. Univ. Politec. Torino **67**(2), 271–292 (2009)
83. Yagdjian, K.: The Cauchy Problem for Hyperbolic Operators Multiple Characteristics. Micro-Local Approach. Akademie Verlag, Berlin (1997)
84. Yagdjian, K.: A note on the fundamental solution for the Tricomi-type equation in the hyperbolic domain. J. Differ. Equat. **206**, 227–252 (2004)
85. Yagdjian, K.: Global existence in the Cauchy problem for nonlinear wave equations with variable speed of propagation. In: New Trends in the Theory of Hyperbolic Equations. Birkhäuser, Basel (2005). Oper. Theory Adv. Appl., **159**, 301–385
86. Yagdjian, K.: Global existence for the n-dimensional semilinear Tricomi-type equations. Commun. Partial Diff. Equat. **31**, 907–944 (2006)
87. Yagdjian, K., Galstian, A.: Fundamental solutions for the Klein-Gordon equation in de Sitter spacetime. Commun. Math. Phys. **285**, 293–344 (2009)
88. Yagdjian, K.: The semilinear Klein-Gordon equation in de Sitter spacetime. Discrete Contin. Dyn. Syst. Ser. S **2**(3), 679–696 (2009)
89. Yagdjian, K.: Fundamental Solutions for hyperbolic operators with variable coeffcients. Rend. Istit. Mat. Univ. Trieste **42**(Suppl.) 221–243 (2010)
90. Yagdjian, K.: Global existence of the scalar field in de Sitter spacetime. J. Math. Anal. Appl. **396**(1), 323–344 (2012)
91. Yagdjian, K.: On the global solutions of the Higgs boson equation. Commun. Partial Differ. Equat. **37**(3), 447–478 (2012)
92. Yagdjian, K.: Semilinear hyperbolic equations in curved spacetime. In: Fourier Analysis, Pseudo-differential Operators, Time-Frequency Analysis and Partial Differential Equations. Trends in Mathematics, pp. 391–415. Birkhäuser Mathematics (2014)
93. Yagdjian, K.: Integral transform approach to solving Klein-Gordon equation with variable coefficients. Math. Nachr. **288**(17–18), 2129–2152 (2015)
94. Yagdjian, K.: Huygens' principle for the Klein-Gordon equation in the de Sitter spacetime. J. Math. Phys. **54**(9), 091503 (2013)
95. Yagdjian, K.: Integral transform approach to generalized Tricomi equations. J. Differ. Equat. **259**, 5927–5981 (2015)